Hydrometeorological Hazards

Forthcoming Titles in the Series

Flash Floods Early Warning Systems: Policy and Practice
by Daniel Sempere-Torres

Coastal Storms: From Forecasting to Prediction
by Paolo Ciavola and Giovanni Coco

Hydrometeorological Hazards

Interfacing Science and Policy

Edited by

Philippe Quevauviller
Vrije Universiteit Brussel (VUB)
Department of Hydrology and Hydrological Engineering
Brussels, Belgium

WILEY Blackwell

This edition first published 2015 © 2015 by John Wiley & Sons, Ltd

Registered office:
John Wiley & Sons, Ltd, The Atrium, Southern Gate, Chichester, West Sussex, PO19 8SQ, UK

Editorial offices:
9600 Garsington Road, Oxford, OX4 2DQ, UK
The Atrium, Southern Gate, Chichester, West Sussex, PO19 8SQ, UK
1606 Golden Aspen Drive, Suites 103 and 104, Ames, Iowa 50010, USA

For details of our global editorial offices, for customer services and for information about how to apply for permission to reuse the copyright material in this book please see our website at www.wiley.com/wiley-blackwell.

Library of Congress Cataloging-in-Publication Data applied for.

A catalogue record for this book is available from the British Library.

Wiley also publishes its books in a variety of electronic formats. Some content that appears in print may not be available in electronic books.

Cover image: Big waves © johnnorth / iStockphoto

Set in 10/12pt TimesTenLTStd by Laserwords Private Limited, Chennai, India

1 2015

Contents

Series Foreword

The increasing frequency and severity of hydrometeorological extreme events are reported in many studies and surveys, including the 5th IPCC Assessment Report. This report and other sources highlight the increasing probability that these events are partly driven by climate change, while other causes are linked to the increased exposure and vulnerability of societies in exposed areas (which are not only due to climate change but also to mismanagement of risks and "lost memories" about them). Efforts are on-going to enhance today's forecasting, prediction and early warning capabilities in order to improve the assessment of vulnerability and risks and develop adequate prevention, mitigation and preparedness measures.

The Book Series on "Hydrometeorological Extreme Events" has the ambition to gather available knowledge in this area, taking stock of research and policy developments at an international level. While individual publications exist on specific hazards, the proposed series is the first of its kind to propose an enlarged coverage of various extreme events that are generally studied by different (not necessarily interconnected) research teams.

The Series encompasses several volumes dealing with various aspects of hydrometeorological extreme events, primarily discussing science–policy interfacing issues, and developing specific discussions about floods, coastal storms (including storm surges), droughts, resilience and adaptation. While the books are looking at the crisis management cycle as a whole, the focus of the discussions is generally oriented towards the knowledge base of the different events, prevention and preparedness, early warning and improved prediction systems.

The involvement of internationally renowned scientists (from different horizons and disciplines) behind the knowledge base of hydrometeorological events makes this series unique in this respect. The overall series will provide a multidisciplinary description of various scientific and policy features concerning hydrometeorological extreme events, as written by authors from different countries, making it a truly international book series.

The book on 'Prevention of hydrometeorological extreme events – Interfacing sciences and policies' is the first book of this Series; it has been written by policy-makers and scientific experts in the field. It offers the reader an overview of EU and international policies, discussions on science–policy interfacing, and a snapshot of the knowledge base of various types of events which are developed in separate volumes of the Series.

Philippe Quevauviller
Series Editor

Preface

Recent hydrometeorological extreme events (e.g. the Xynthia storm in France, the Liguria flash floods) have highlighted the increased exposure and vulnerability of societies and the need to strengthen the knowledge base of related policies. In particular, research undertakings are constantly enhancing forecasting, prediction and early warning capabilities in order to improve the assessment of vulnerability and risks linked to extreme climatic events, as well as to develop adequate prevention, mitigation and preparedness measures. These concern events such as extreme winds, storm surges, coastal and estuarine floods, taking into consideration the effects of climate change, which are threats faced by many countries in the world. In this respect, international policies and research cooperation are in full development, leading to new knowledge, innovative, cost-effective, technological or non-technological solutions and ecosystem-based approaches, as well as new forms of organisational and institutional/governance.

The book *Prevention of Hydrometeorological Extreme Events – Interfacing Sciences and Policies* is the first volume of a series which will gather scientific and policy-related knowledge related to climate-related extreme events. Invited authors are internationally recognised experts in their respective fields, who have built up worldwide networks in the framework of EU-funded research programmes. The present volume and the following ones in the series will hence reflect the most recent science and policy advances in the field.

Philippe Quevauviller

The Series Editor – Philippe Quevauviller

Philippe Quevauviller began his research activities in 1983 at the University of Bordeaux I, France, studying lake geochemistry. Between 1984 and 1987 he was Associate Researcher at the Portuguese Environment State Secretary where he performed a multidisciplinary study (sedimentology, geomorphology and geochemistry) of the coastal environment of the Galé coastline and of the Sado Estuary, which was the topic of his PhD degree in Oceanography gained in 1987 (at the University of Bordeaux I). In 1988, he became Associate Researcher in the framework of a contract between the University of Bordeaux I and the Dutch Ministry for Public Works (Rijskwaterstaat), in which he investigated organotin contamination levels of Dutch coastal environments and waterways. From this research work, he gained another PhD in chemistry at the University of Bordeaux I in 1990. From 1989 to 2002, he worked at the European Commission (DG Research) in Brussels where he managed various Research and Technological Development (RTD) projects in the field of quality assurance, analytical method development and pre-normative research for environmental analyses in the framework of the Standards, Measurements and Testing Programme. In 1999, he obtained an HDR (Diplôme d'Habilitation à Diriger des Recherches) in chemistry at the University of Pau, France, from a study of the quality assurance of chemical species' determination in the environment.

In 2002, he left the research world to move to the policy sector at the EC Environment Directorate-General where he developed a new EU Directive on groundwater protection against pollution and chaired European science-policy expert groups on groundwater and chemical monitoring in support of the implementation of the EU Water Framework Directive. He moved back to the EC DG Research in 2008, where he acted as research Programme Officer and managed research projects on climate change impacts on the aquatic environment and on hydrometeorological hazards, while ensuring strong links with policy networks. In April 2013 he moved to another area of work, namely Security Research, at the EC DG Enterprise and Industry where he is research Programming and Policy Officer in the fields of Crisis Management and CBRN.

Besides his EC career, Philippe Quevauviller has remained active in academic and scientific developments. He is Associate Professor at the Free University of Brussels and promoter of Master theses in an international Master on Water Engineering (IUP-WARE programme), which is under this function that he is acting as Series Editor of the *Hydrometeorological Extreme Events* Series for Wiley. He also teaches integrated water management issues and their links to EU water science and policies to Master students of the EurAquae programme at the Polytech'Nice (France).

Philippe Quevauviller has published (as author and coauthor) more than 220 scientific and policy publications in the international literature, 54 book chapters, 80 reports and 6 books and has acted as an editor and co-editor for 26 special issues of scientific journals and 15 books. He also coordinated a book series for Wiley on *Water Quality Measurements* which resulted in 10 books published between 2000 and 2011.

List of Contributors

Clara Armaroli
Dipartimento di Fisica e Scienze della Terra, Università di Ferrara, Ferrara, Italy

Nina Becker
Department of Economics, Helmholtz Centre for Environmental Research-UFZ, Leipzig, Germany, and Overseas Development Institute, London, UK

Laurens M. Bouwer
Deltares, Delft, The Netherlands

Philip Bubeck
adelphi, Berlin, Germany

Paolo Ciavola
Dipartimento di Fisica e Scienze della Terra, Università di Ferrara, Ferrara, Italy

Hugh Deeming
Engineering & Environment, Northumbria University, Newcastle upon Tyne, UK

Oscar Ferreira
FCT, CIMA, Universidade do Algarve, Faro, Portugal

Maureen Fordham
Engineering & Environment, Northumbria University, Newcastle upon Tyne, UK

Emmanuel Garnier
Institut Universitaire de France, CNRS and University of La Rochelle, France

Luis Garrote
Department of Agricultural Economics and Social Sciences, Technical University of Madrid, Madrid, Spain

Elisabetta Genovese
Centre International de Recherche sur l'Environnement et le Développement (CIRED), Nogent-sur-Marne, France

Philippe Gourbesville
Université Nice Sophia Antipolis, Polytech Nice Sophia, I-CiTy Lab, Nice, France

Colin Green
Flood Hazard Research Centre, Middlesex University, Hendon, UK

Stephane Hallegatte
Centre International de Recherche sur l'Environnement et le Développement (CIRED), Nogent-sur-Marne, France; Ecole Nationale de Météorologie, Toulouse, France, and The World Bank, Sustainable Development Network, Washington, DC, USA

Richard Harding
Centre for Ecology and Hydrology, Wallingford, Oxon, UK

Mitchell Harley
Dipartimento di Fisica e Scienze della Terra, Università di Ferrara, Ferrara, Italy

Ana Iglesias
Department of Agricultural Economics and Social Sciences, Technical University of Madrid, Madrid, Spain

Demetrio Innocenti
University of Antwerp, Institute of Development Policy and Management (IOB), Antwerp, Belgium, and The United Nations Office for Disaster Risk Reduction (UNISDR), Brussels, Belgium

Alison Kay
Centre for Ecology and Hydrology, Wallingford, Oxon, UK

Heidi Kreibich
Helmholtz Centre Potsdam – GFZ, German Research Centre for Geosciences, Potsdam, Germany

Zbigniew W. Kundzewicz
Institute for Agricultural and Forest Environment, Polish Academy of Sciences, Poznan, Poland, and Potsdam Institute for Climate Impact Research, Potsdam, Germany

Quentin Lequeux
Dipartimento di Fisica e Scienze della Terra, Università di Ferrara, Ferrara, Italy

Ivana Logar
Swiss Federal Institute of Aquatic Science and Technology (EAWAG), Dübendorf, Switzerland

Vasileios Markantonis
Faculty of Economics and Business Administration, Chemnitz University of Technology, Chemnitz, Germany

Volker Meyer
Department of Economics, Helmholtz Centre for Environmental Research, Leipzig, Germany

Elissaios Papyrakis
Institute for Environmental Studies, VU University Amsterdam, Amsterdam, The Netherlands; School of International Development, University of East Anglia, Norwich, UK, and International Institute of Social Studies, Erasmus University Rotterdam, The Netherlands

Clemens Pfurtscheller
Institute of Interdisciplinary Mountain Research, Austrian Academy of Sciences, Innsbruck, Austria

Jennifer Poussin
Institute for Environmental Studies, VU University Amsterdam, Amsterdam, The Netherlands

Valentin Przyluski
Centre International de Recherche sur l'Environnement et le Développement (CIRED), Nogent-sur-Marne, France

Philippe Quevauviller
Department of Hydrology and Hydrological Engineering, Vrije Universiteit Brussels, Brussels, Belgium

Nick Reynard
Centre for Ecology and Hydrology, Wallingford, Oxon, UK

Reimund Schwarze
Department of Economics, Helmholtz Centre for Environmental Research, Leipzig, Germany

Åsa Gerger Swartling
Stockholm Environment Institute, Stockholm, Sweden, and Stockholm Resilience Centre, Stockholm University, Stockholm, Sweden.

Jeroen C.J.M. van den Bergh
ICREA, Barcelona, Spain; Institute of Environmental Science and Technology, Universitat Autònoma de Barcelona, Barcelona, Spain; Faculty of Economics and Business Administration, VU University Amsterdam, The Netherlands, and Institute for Environmental Studies, VU University Amsterdam, Amsterdam, The Netherlands

Annegret H. Thieken
Institute of Earth and Environmental Science, University of Potsdam, Potsdam, Germany

Ap Van Dongeren
Deltares, Delft, The Netherlands

Henny A.J. Van Lanen
Hydrology and Quantitative Water Management Group, Wageningen University, Wageningen, The Netherlands

Jaap Van Thiel de Vries
Faculty of Civil Engineering and Geosciences, University of Delft, Delft, The Netherlands

Christophe Viavattene
Flood Hazard Research Centre, Middlesex University, Hendon, London, UK

Barbara Zanuttigh
Department of Civil, Chemical, Environmental and Materials Engineering, University of Bologna, Bologna, Italy

Part One

Setting the Scene

1.1

Strengthened Resilience from Historic Experience. European Societies Confronted with Hydrometeors in the Sixteenth to Twentieth Centuries

Emmanuel Garnier

Institut Universitaire de France, CNRS and University of La Rochelle, France

1.1.1 Introduction

In his seminal book published in 1992 on the evolution of society from a society of disasters to a society of risk, the sociologist Ulrich Beck clearly distinguishes between a "pre-modern" society qualified as "traditional", devoid of industries, and a "modern" society (Beck, 1992). In the first case, risk is non-existent, supplanted as it is by a social conviction: threats of all kinds which result from disasters are both natural and totally unpredictable. Against this traditional collective fatalism, he contrasts industrialized society which would redefine the relations which it maintains with its natural environment according to a relationship of domination (Man) and dominated (Nature). For Beck, by inventing the concept of risk, industrialization finally allowed its definition and quantification thanks to an improvement in instrumentation and to scientific progress. However, the historical reality observed in archives about hydrometeors is particularly enlightening. It indicates that the germs of a "risk"-based mentality can be observed very early on, in a time when societies and states remained nevertheless fundamentally agrarian and traditional. This historical work consequently aims to study

Hydrometeorological Hazards: Interfacing Science and Policy, First Edition. Edited by Philippe Quevauviller.

the "trajectories of vulnerability" of territories and European societies confronted with two types of hydrometeors: droughts and storm surges.

1.1.2 Five centuries of droughts

The results presented in this section fall within the framework of the EU project FP 7 'Fostering European Drought Research and Science-Policy Interfacing' (project number 282769). This project aims to reduce Europe's future vulnerability to and risk of drought by innovative in-depth studies that combine drought investigations in case study areas in water-stressed regions with drought analysis at the pan-European scale. In this perspective, it grants in particular an important role to the historical approach in helping us to understand better the frequency and severity of the droughts during the last 500 years as well as the reactions of the old societies.

Droughts are a factor of historic durability and, because of their impacts on societies, they left multiple indicators in the archives of the last 500 years. For the record, it is necessary to remind ourselves that the general term of 'drought' covers different notions. The most frequent meaning of the word is a rainfall deficit and an extreme climate event.

1.1.2.1 Historic material and methods of evaluation

Because of the unpredictable character and the absence of civil services specially dedicated to the study of these extreme events before the middle of the nineteenth century, historians have to make maximum use of the entire corpus of sources. The information we need is often hidden at random in the margins of some documentation and we cannot afford to neglect any type of archive if we want to hope to reconstruct long and relatively reliable chronologies (Garnier, 2010a).

Diaries drafted by private persons (priests, middle-class persons, aristocrats) and municipal chronicles are particularly useful. Besides the private, economic and political events, they are often very sensitive to the extreme events which engender a disaster, etymologically (*kata* and *strophe*) an upset. Certain authors provide an integrated approach to the drought by combining visual observations (heights of water on the hydrological scales on bridges), the phenology (state of the vegetation, fires), prices in the markets and even its social expression (scarcity, religious processions, riots).

The catholic church is a faithful ally of the historian studying droughts. *Ex voto*, small naive paintings hung in churches in France, Spain, Portugal and Italy, are iconographic testimonies which very frequently concern a meteorological abnormality. Often up to the eighteenth century, the extreme event (drought, storm, flood) was considered a demonstration of God's wrath. That is why the ancient societies asked the Church for an intervention.

Thanks to the religious processions, the historian has a relatively homogeneous series of the archival and historic plans because they emanate from the same lay or religious institution which registers them over long periods. These religious ceremonies allow the reconstruction of historic series included generally between 1500 and 1800, sometimes even beyond in the Spanish case (Barriendos and Martin-Vide, 1995; Barriendos, 2005). The Roman Catholic Church or the municipal authorities ordered these qualified ceremonies of rogations (*rogativas*) in Spain, or processions in Portugal and in France to avoid endangering the established order or the socioeconomic balance. In the case of drought, processions were organised *pro pluvia*, literally 'for the rain'.

Then, where wars spared cities, the historian has very precious municipal archives. They contain the registers of the municipal deliberations and the accounts. These documents begin frequently from the end of the fifteenth century. Deliberations and municipal accounts constitute an inexhaustible deposit of climatic data. The meteorological information is omnipresent in these registers and arises from an understandable desire to anticipate the risks of breaks in supplies, of diseases and riots. So any sustained drought sparks off a discussion within the government of the city. That is why the state or municipal authorities, from the fifteenth century, intervened by using diverse tools, such as processions, price controls, requisitions of wheat and, in the case of Valencia, imports of wheat.

Unlike temperature data, which appear around the 1700s, we have to wait until the beginning of the nineteenthth century to get instrumental data on rainfall or water flows. They result from the creation of scientific societies, such as the Royal Society of London, the Royal Academy of the sciences of Paris or the *Societas meteorologica palatina* of Mannheim in Germany. They may also have been produced by engineers especially devoted to the surveillance of rivers.

To address the lack of reliable instrumental data before 1800, the contents of archives offer two methodological solutions to estimate these natural events for which we have only textual descriptions. The first solution is to use all the chronological mentions of a drought appearing in archives. Concretely, it is a question of noting, for example, the first mention of a religious procession *pro pluvia*, then the municipal acts which evoke the drying up of the public fountains, the ban on drawing water from certain places, the lay-off of wheatmills and, in the most extreme cases, the problems of supply of wheat and wood via the waterway. Naturally, this list is not exhaustive. However, the location of these indicators in archives allows us to propose the duration in days for the very great majority of the droughts of the past. This choice is acceptable because today the World Meteorological Organization characterizes the drought according to the number of days without rain. Nevertheless, professional ethics require us to specify that the duration defined by the historian does not correspond to a total absence of rain but rather to its chronological perception by societies. Therefore the drought of the historian indicates instead a very dry and long episode with sufficiently important economic and human impacts for it to appear in the archives of the time.

Another methodological choice which can complete the evaluation by duration consists of creating an indexed scale of severity directly built according to the descriptive contents of the drought. Naturally, this results from a systematic inventory of the impacts engendered by the extreme event on societies. The list below shows the types of damage most commonly described in the archives on droughts:

- Rogation *pro pluvia*
- Early grape harvest dates
- Plants dry and die
- No harvest of hay
- Low water
- 'No rain' during the period

- Drying up of springs
- The ground is cracked
- We can ford a river
- Lay-off of wheatmills
- Forest fires
- The cattle die of thirst
- Riots around the springs
- Shortage or/and famine

Thus, the historian can observe the chronology of an event which is well recorded in archives. Figure 1.1.1 describes a particularly severe drought which affected the region of Valencia in 1725 and 1726.

From this inventory, a scale of severity between index −1 and index 5 can be realized, as shown in Table 1.1.1.

At index 1, the absence of precipitation (atmospheric drought) starts to be felt. If this continues, agriculture is affected and a fall in the levels of water is observed in the records (index 2). At indexes 3 and 4, the question of the resources becomes important. The situation deteriorates as the absence of precipitation affects societies with a high price of farm produce, a lay-off of wheatmills and a degradation of the ecosystems (index 4). The paroxysm of the social crisis is reached with index 5 when the drought becomes hydrological with a very clear deterioration in living conditions and an increase in social tensions over access to water.

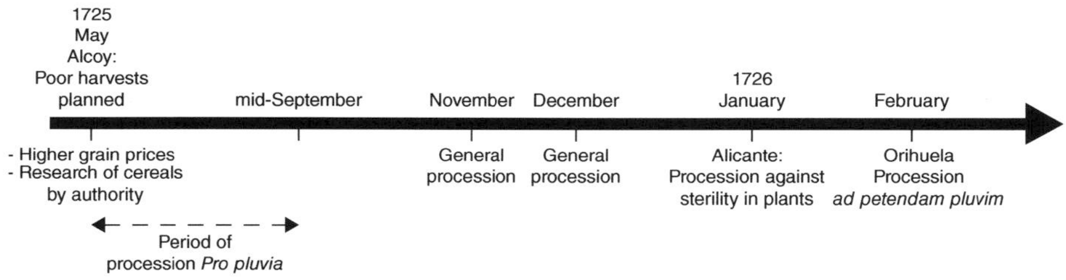

Figure 1.1.1 Estimate of the duration and severity of the Jucar drought of 1725–1726.

Table 1.1.1 Index of severity of the droughts (sixteenth to nineteenth centuries).

Index	Description
5	exceptional drought: no possible supply, shortage, sanitary problems, very high prices of wheat, forest fires
4	severe low-water mark: navigation impossible, lay-off of wheatmills, search for new springs, forest fires, death of cattle
3	general low-water (difficulties for navigation) and water reserves
2	local low-water in rivers, first effects on vegetation
1	absence of rainfall: rogations, evidences in texts
−1	insufficient qualitative and quantitative information but the event is kept in the chronological reconstruction

Table 1.1.2 Extract from historical database of the Jucar Basin.

Location	River	Year	Date	Comments	Relative duration	Severity index
Valencia	Jucar Basin	1412		*Que com per esguard de la seccada e sterilitat el any present, que per nostras paccats era estada en aquestra ciutat*	1 year	5
Orihuela	Segura	1726	February	*Rogativas ad petendam pluviam*	30 days	2
Valencia	Jucar Basin	1815–1816		*Las sequais extraordinarias de 1815 y 1816, las que dieron margen a abusos tan considerables y a tanta indigencia que à cado paso se veia uno asaltado, robado, espueto a mil insultos y usurpaciones*	2 years	4

These various types of drought are not independent of each other. The absence of rainfall results from an atmospheric drought. The ground dries out more quickly because of the wind and the heat. With varying rapidity, the water resources decrease, the drought becomes phreatic then hydrologic.

Finally, the method then consists of building a database, an extract of which is presented in Table 1.1.2 for the Jucar Basin. The database is divided into several headings. From left to right we have, successively, the location, the river affected by the drought, the date (month, and day where they are indicated by archives), an extract of the comments in the archives and, finally, the relative duration, deduced from the mentions in archives, and an index of severity, of which we shall speak again later.

Figure 1.1.2 shows the severity of droughts in the Jucar Basin from 1500 to 1900.

1.1.2.2 A comparative approach to historic European droughts

The French droughts of the Seine are compared to the Rhine valley and British areas because the Seine basin enjoys a moderate, climate intermediate between the western and semi-continental oceanic climates (Garnier, 2010b). In the case of the British droughts, multiple archives and printed sources were exploited in Cambridge and in London, in particular the remarkable diary of Samuel Pepys, secretary of the Admiralty under the reign of Charles II and James II.[1] The Rhine valley area (South Rhine, Bade-Wurtemberg, Switzerland and Alsace) is particularly well documented by the

[1] Cambridgeshire Archives, Memorials and Petitions presented to the Bedford Level Corporation. Cooper CH., Annals of Cambridge, Warwick, Cambridge, 1842–1853. Chronicle of London, Taylor, London, 1827. *Memoirs of Samuel Pepys*, Esq. F.R.S., Secretary to the Admiralty in the reigns of Charles II and James II, comprising his Diary from 1659 to 1669, deciphered by the Rev. John Smith, A.B., of St. John's College, Cambridge, from the original shorthand MS in the Pepysian Library and a Selection from his Private Correspondence. edited by Richard, Lord Braybrooke, London, Henry Colburn, 1825.

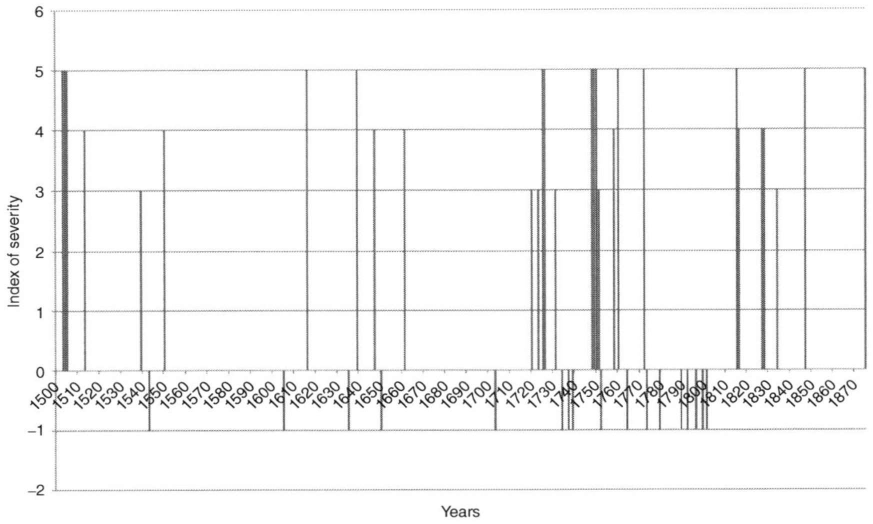

Figure 1.1.2 The severity of droughts in the Jucar Basin (Spain) 1500–1900. The columns indicate the severity of the drought according to the severity index scale.

religious chronicles, the meteorological newspapers and the municipal archives.[2] This rich corpus thus enables a reliable chronological reconstruction.

For 69 droughts of the Seine there were, respectively, 22 and 18 similar events in England and in the Rhine valley (Figure 1.1.3). The best correlation with the Seine basin results was across the Channel. The English droughts were identical to those in the Ile-de-France in 32% of cases for the period 1500–2009 versus 26% in the Rhine Valley. In the case of England, the highest correlations with the French droughts were in the sixteenthth (41%) and twentieth (40%) centuries while the best correlations with the Rhine valley area were in the seventeenth (46%) and nineteenth (31%) centuries.

The chronologies of the Rhine and England confirm rather broadly the big trends observed for the valley of the Seine. Three series are characterized by a first phase of droughts between 1500 and 1800 followed by a long, less dry cycle between 1800, then a clear resumption by 1960, in duration as in frequency.

Rather than trying a tedious and systematic comparison between the droughts of the sixteenth and twentieth centuries, the most extreme common droughts are studied in greater depth. The first is the disaster of 1556, very well described by the priests, the

[2] Hegel, Die Chronik der Stadt Straßburg, Leipzig, Verlag Hirzel, 1869, 498pp. Dietler, frère Séraphin, *Chronique des Dominicains de Guebwiller 1124–1723*, Société d'Histoire et du Musée du Florival sous la direction de Philippe Legin, Guebwiller, 1994, 359p. Mercklen, F.J., *Annales oder Jahrs-Gesachichten der Baarfüseren oder Minderen Brüdern S. Franc. Ord. Insgeneim Conventualen gennant, zu Than durch Malachias Tschamser*, Colmar, 1864, 2 volumes. Dostal, P., *Klimarekonstruktion der Regio TriRhena mit Hilfe von direkten und indirekten Daten vor der Instrumentenbeobachtung*, Berichte des Meteorologischen Institutes der Universität Freiburg, Freiburg, 2005, 165p. Glaser R., Klimarekonstruktion für Mainfranken, Bauland und Odenwald anhand direkter und indirekter Witterungsdaten, *Paläoklimaforschung*, 5, Stuttgart, New York, 1991, 138p. Muller C., *Chronique de la viticulture alsacienne au XVIIe siècle*, Riquewihr, J.D. Reber, 1997–2002, 5 volumes. Pfister C., Weingartner R., Luterbacher J., Hydrological winter droughts over the last 450 years in the Upper Rhine basin: a methodological approach, *Hydrological Sciences-Journal-des Sciences Hydrologiques*, n° 51–5, 2006, p. 966–985.

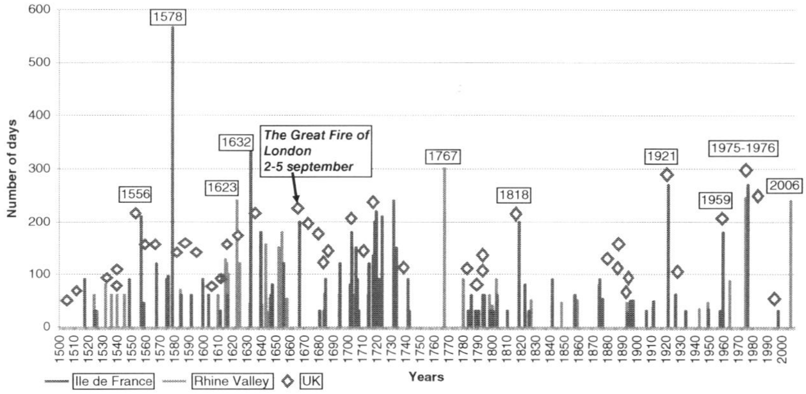

Figure 1.1.3 Chronology and duration (in days) of droughts in Ile-de-France, in England and in the Valley of the upper Rhine, 1500–2009.

bourgeois and the Parisian aldermen. Sources underline the gravity of the event when they speak of a 'very big hot season' and religious processions were organised almost everywhere in France (Garnier, 2010b). Even worse, they assert that not a single droplet fell between Good Friday and All Saints' Day. On the other side of the Channel, the situation is particularly difficult because the agricultural effects of the drought caused a rise in the price of wheat which rose sixfold. The Aldermen of London also evoke the drying up of fountains, wells and springs in their reports.

The most extreme event, however, occurred in 1666, bringing into play France and England, both victims of an exceptional drought whose consequences in London turned out to be catastrophic. The beginnings of the meteorological crisis are translated into big processions from May 17th. As usual, the Parisians and their aldermen turn to Saint Geneviève (protector of Paris) to obtain the rain.[3] In England, the progress of the crisis is much better known.[4] The drought began in November 1665 and peaked in August of the next year. Towards 15 August, Londoners were dumbfounded to discover that the Thames was reduced to a brook which was easily passable on foot. At the beginning of September, at the height of the heatwave and the drought, Samuel Pepys delivered a very prophetic meteorological vision when he compared the timber-framed houses of the capital to matches ready to ignite after several months without rain. On 2nd September, when a hot easterly wind was blowing, a fire broke out at a bakery and was immediately fanned by the wind. Illustrated in an anonymous picture in the London Museum, the great fire ravaged completely the medieval city inside the old Roman outer wall and threatened the aristocratic district of Westminster, which was finally spared. The blaze stopped on 5th September, but it was necessary to wait until 19th September and the first important rain before the embers were definitively extinguished. If the

[3] Archives nationales, H 1820/1.

[4] *Memoirs of Samuel Pepys*, Esq. F.R.S., Secretary to the Admiralty in the reigns of Charles II. and James II., comprising his Diary from 1659 to 1669, deciphered by the Rev. John Smith, A.B., of St. John's College, Cambridge, from the original shorthand MS in the Pepysian Library, and a Selection from his Private Correspondence, edited by Richard, Lord Braybrooke, London, Henry Colburn, 1825. Porter S, The Great Fire of London, John Wiley & Sons, Hoboken, 2009, 192pp.

disaster caused few direct deaths (eight), on the other hand, the material loss was very heavy. More than 13 000 houses, 87 churches, including St Paul's cathedral and the majority of the public buildings were consumed. More gravely still, it had been necessary to evacuate 80 000 inhabitants during the fire and it was then necessary to rehouse them. Having said that, the tragedy was also an unexpected opportunity for the King to eliminate a 'dangerous' population in social terms and to launch one of the largest town planning schemes of the modern era (sixteenth to eighteenth centuries).

Then, a continuous string of droughts also struck the valleys of the Seine, the Rhine and the Thames around the 1680s (1681, 1683, 1684, on 1685) followed by the 1700s (1701, 1705, 1709, on 1714) as so many precursory indicators of the European drought and heatwave of 1719. Considered as one of the hottest summers of the century in the United Kingdom, it raged mainly in England and in Wales with a rainfall deficit often mentioned in the English archives. A similar pattern can be found in the Germanic countries. The Dominican of Guebwiller (Vosges, Alsace) notices, after a still very cold and snow-covered April, a gradual rise in temperatures in the course of May. One month later, they considered them to be 'very hot'. In August, the situation was such that the magistrates of the Sovereign Council of Colmar (Alsace) sent investigators to find water reserves in the mountain from Vosge. By that time, the wheatmills of the city were closed for more than a fortnight threatening the city with scarcity[5] (Garnier, 2009). The drought of 1818, common to the basins of the Seine (200 days) and the Thames (90 days), coincides with an extremely dry phase, 1815–1822, in the Mediterranean countries. In the British Isles they spoke about 'the longest, the hottest and the driest summer in living memory'. In France, the farmers noticed that the ground was dry in mid-May and that most of the ponds almost disappeared in August. The Parisian observers indicated that the drought persisted until Christmas, a duration of the order of 200 days.

After 1818 a long phase of remission began and it is necessary to wait until 1893 for another event of European scale. During this year, respectively 28 and 45 days of drought were observed in Ile-de-France, in the Rhine Valley and especially in Great Britain. So we find only 21.8 mm of rain in Oxford between February and May (Wallace, 1997). The situation was also aggravated by a heatwave with disastrous health consequences in Europe. Using the climatological definition of spring (March, April and May), that year's spring was the warmest in the Central England Temperature Records (Manley, 1974). The twentieth century saw the return of severe droughts. The century began with a very severe drought in 1921 which affected a large part of the European continent and which even provoked a famine in Eastern Europe, aggravated in Russia by the political conditions connected to the civil war. From England to Italy, the precipitation was 40% lower than normal. Across the Channel the year 1921 was the driest since 1758 and it was characterized by a particularly long dry phase between April and August (Figure 1.1.4). In the major part of Great Britain and in Ireland, the total precipitation in June was less than 2.5 cm and in certain regions there was no rain at all. During the driest four months (April to July), the British precipitation was only 63% of the usual average.

In France, 1921 was the hottest and the driest year since the beginning of meteorological observations at Montsouris Park near Paris (1873). In total, only 270 mm of rain fell (against 625 mm on average). The drought was also accompanied by an exceptional period of sunshine with 2314 hours of sun (average: 1650 hours). The drought began in

[5] Archives municipales de Munster, CC 65.

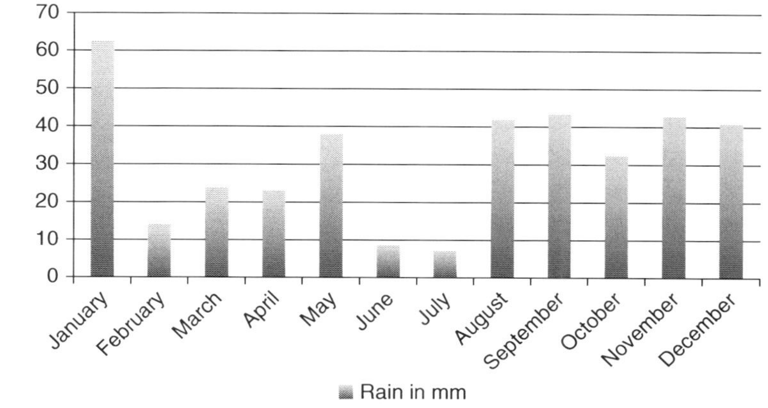

Figure 1.1.4 The precipitation for 1921 in Oxford. The columns indicate the monthly precipitation in millimetres. *Source*: Data from Radcliffe Observatory Oxford.

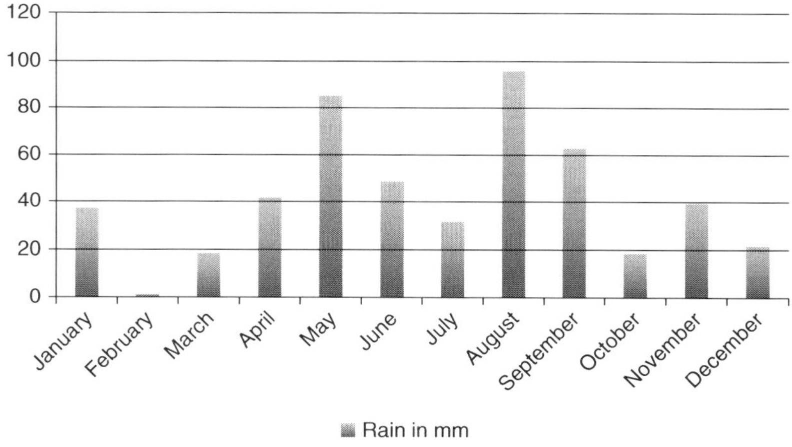

Figure 1.1.5 The precipitation of 1921 in Basel-Binningen (Switzerland). The columns indicate the monthly precipitation in millimetres. *Source*: Data from http://www.infoclimat.fr/climatologie/annee/1921/bale-binningen/valeurs/06601.html, accessed 20th May 2014.

autumn, 1920 and got worse in February. As in Great Britain, March and April were exceptionally dry and brightened up so that, in June, the drought became dramatic. In Paris, it was the driest June since 1810 and between 22nd May 22 and 11th July 11, only 2 mm of rain fell.

In the region of the upper Rhine, the Rhine reached its lowest level recorded in Cologne since records began (1816). On 11th May it reached the extraordinarily low level of only 1.22 metres.[6] As in France, the driest month in Basel was February and the following months until winter remained very dry, with the exception of May, August and September, which were relatively wetter (Figure 1.1.5).

[6] Kölner Stadt-Anzeiger Newspaper (May 12th, 1921)

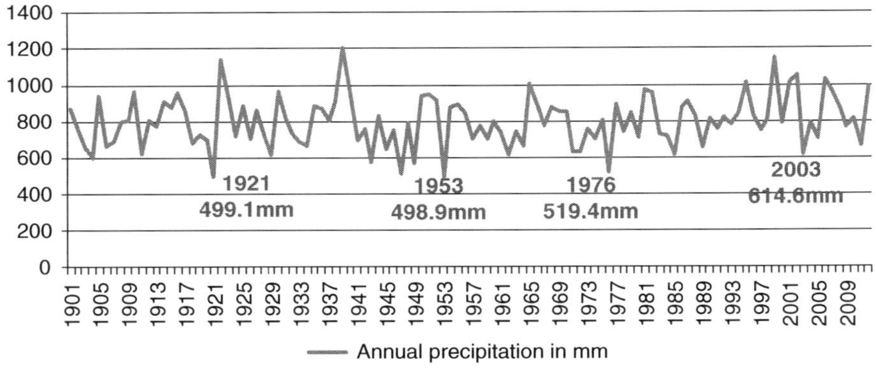

Figure 1.1.6 Annual precipitation at Basel-Binningen between 1901 and 2012. The curve indicates the annual precipitation in mm. *Source*: Data from http://www.infoclimat.fr/climatologie/annee/1921/bale-binningen/valeurs/06601.html.

Considered in a wider chronological context, the drought of 1921 appears to be one of the most extreme events of the last 100 years. Figure 1.1.6 presents the curve of the annual precipitation recorded at the station of Basel-Binningen (Switzerland) between 1902 and 2012. It reveals that only the drought of 1953 (498.9 mm) was more severe than that of 1921 (499.1 m) while those, more contemporary, of 1976, 2003 and 2011 showed higher precipitation rates.

The long term historical reconstruction of droughts is very useful for the climatologist and the hydrologist. Indeed, history can be a means of giving more reliability to the models established using data from 50 years ago or more. While the various scenarios of the IPCC (SREX report) predict an increase in extreme events, it is necessary to place them in a historical context.

1.1.3 The European coast confronted with surges: A first ever?

After the disaster in February, 2010 caused by the storm Xynthia on the French Atlantic coast, the media and the majority of the policy makers rather systematically used the terms 'new', 'exceptional', even 'unpredictable' to qualify it.[7] Nevertheless, only a few years before this disaster, extreme events (Katrina in 2005 in the United States, the tsunami of 2004 in the Indian Ocean) had already underlined the devastating effects provoked by coastal hazards. With 185 000 km of coast, Europe is particularly vulnerable to this type of risk because, within this coastline, there is a concentration of high population densities, important urbanization and major centres of the world economy. Economically, it is impossible to design, to finance and to build any engineering capable of definitively protecting these strategic territories. Faced with this reality, historical experience can contribute to assisting the disaster and emergency services by giving a perspective on the severity and the frequency of the surges which struck

[7] Le Monde Newspaper (26th February 2010).

Table 1.1.3 Preliminary data on surges and tsunamis on the French coasts, 1500–2010.

Coast	Number of events	State of research
Poitou-Charentes-Pays de Loire	32	very complete
Normandie	8	to be completed
Méditerranée	25	to be completed

the littoral societies in the past and on the measures which they were able to develop in certain cases to reduce the impact of these surges. The results which follow are embedded in research conducted within the framework of the project FP 7 RISC-KIT (Resilience-Increasing Strategies for Coasts-toolKIT).

1.1.3.1 The French coasts

Since 2010 a first phase of research in archives produced an initial inventory of surges, which was nevertheless uneven, depending on the coast. The Atlantic coast affected by Xynthia in 2010 is today perfectly known thanks to in-depth research (Garnier, 2010c; Garnier and Surville, 2011).

The historic results (Table 1.1.3) contradict the idea that the storm of February 2010 was a totally new hazard in France for centuries. With 63 events distributed on the Mediterranean, Norman and Atlantic coasts, surges are unmistakably phenomena of historical durability even if, effectively, they were not necessarily as severe as Xynthia.

Figure 1.1.7 shows all the risks of surges counted in archives for the littoral portion between the estuary of the Gironde and that of the Loire during 1500–2010. It distinguishes between a surge (hydrometeor) and a tsunami (of seismic origin), a phenomenon which regularly affected these coasts, in particular La Rochelle on 6th September 1785 (Garnier and Surville, 2010, 2011). Two more extreme sequences appear clearly. The first occurs during the eighteenth century (7 surges and 2 tsunamis) and the second during the period 1850–1950 (9 surges and 1 tsunami). The second half of the twentieth century is characterized by a disappearance of the maritime hazard. The Mediterranean and Norman examples confirm widely the Atlantic example where surges have occurred regularly since 1500. We observe a relative similarity between them with an outbreak of extremes between 1800 and 1950 then a disappearance of the hazard from the 1950s. This remission during the second half of the twentieth century doubtless explains the misleading feeling of 'novelty' in February, 2010.

1.1.3.2 European comparison

A comparison on the European scale is now imperative to understand better the geographical and chronological variability of the risk from the sea. The most reliable French chronology of the French Atlantic Ocean is thus compared with that of the North Sea (UK, Germany, Netherlands, Denmark) (Lamb, 1991), Figure 1.1.8.

There is surprising similarity between both maritime spaces, with 29 events detected on the French coasts against 27 in the North Sea. The secular distribution shows clearly

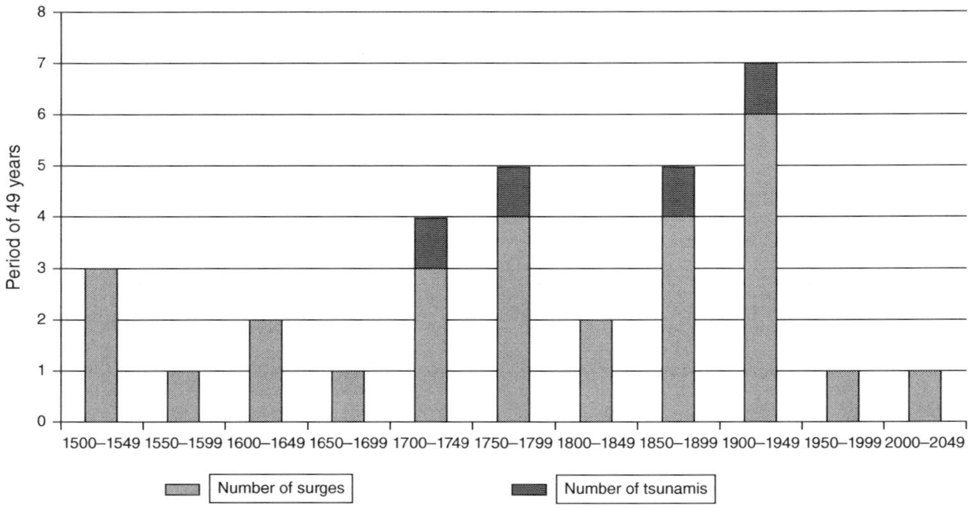

Figure 1.1.7 Surges on the Poitou-Charente coast 1500–2010. The columns indicate the number of events for a period of 49 years. Surges are indicated by the lighter shaded columns and tsunamis by the darker ones.

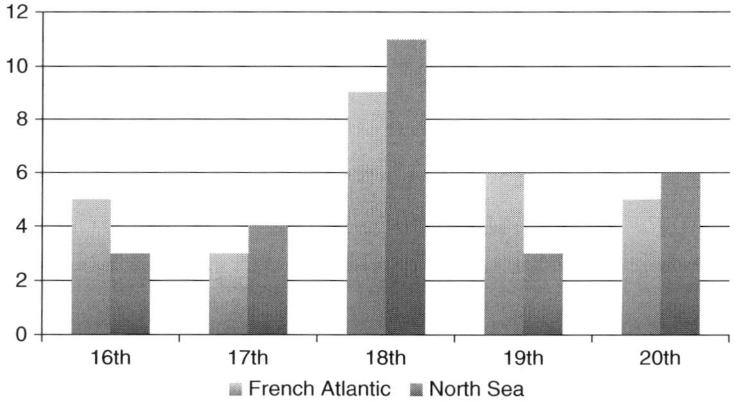

Figure 1.1.8 Secular distribution of surges in France (Atlantic Coast) and on the North Sea Coasts (UK, Germany, Netherlands, Denmark) 1500–2000. The left-hand columns indicate the number of events for France and the right-hand ones those for the North Sea.

that, in both cases, the eighteenth century was most affected by hazards with 20 surges (approximately a third of the total), followed in decreasing order by the twentieth and nineteenth centuries. The observation of this chronology, distributed this time by periods of 50 years, underlines strong disparities between both series, more particularly during the contemporary sequence. Their trends are totally contradictory in the twentiethth century. Years 1900–1950 are struck by four surges of high severity on the French coasts while the coasts of the North Sea undergo only two floods. On the other hand, the scenario is totally different for the years 1950–2000. Unlike France, Northern Europe faced the storms of 1953 (Netherlands, Great Britain) and of 1962 (the North of Germany).

1.1.3.3 Memory of the big floods in Northern Europe

The surges occurring between 1500 and 2010 are shown in Figure 1.1.9. Among the group of tragic surges in Europe over this period, we underline here the particularly disastrous impact which certain events had on the littoral societies of northern Europe. The long history of littoral events was subject to and is always subject to the workings of human memory. The conservation and transmission of the memory of the events (via schools, monuments, marks, and anniversary ceremonies) is the opposite of what happened in the French case. In France, having at first denied the risk of surge, today we try hard to erase the tracks and memory of the Xynthia disaster. The wish of local and national authorities is to avoid any feeling of fear or panic among the population.

The 'Mandränke' of October, 1634 The storm wave of the night of 11–12 October 1634 known by the names Mandränke (tide) and Buchardiflut (storm Buchardi) was one of the most devastating storms of the last five centuries. It occurred at the worst time for the region. After an epidemic of plague engendered by the Thirty Years War (1618–1648), for a decade it suffered from the assaults of icebergs (Little Ice Age period) which partially destroyed the network of coastal dykes. Buchardi broke out on the coast of Schleswig-Holstein and Denmark. It gobbled up the island of Strand and caused the death of 8000 to 15 000 persons and about 50 000 head of cattle. The tremendous power of the waters ravaged a coastal area of about 80 km between the island of Nordstrand and the Danish border.

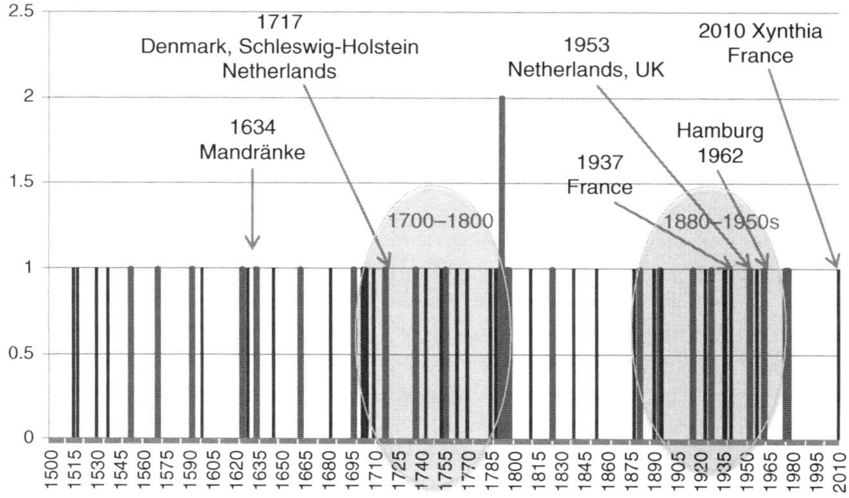

Figure 1.1.9 The European surges between 1500 and 2010. The red columns indicate the events of the North Sea and the blue the French events (Atlantic Coast).

In 1630 the government had commissioned the Dutch engineer Jan Adriaansz Leeghwater to restore and strengthen the coastal defence system. These four years of expensive work could not resist the violence of the sea which destroyed the defences in only a few hours. At Eiderstedt, in the southern part of the island of Nordstrand, about 2000 inhabitants died in spite of the protection provided by dykes 6 m in height. Inside the church of Klixbüll, the survivors of the disaster transmitted the memory of the flood by engraving a mark on the wall of the building situated at 4.30 m above ground level.

To help reduce contemporary vulnerability, it is carefully protected and maintained to transmit to new generations the memory of the risk. Finally, the trauma was such that many survivors preferred to leave the region, emigrating on a massive scale, in particular to the Netherlands. The flood of 1634 drew a new littoral landscape with the disappearance of a quarter of the region and the appearance on the west coast of the islands of Nordstrand, Pellworm, Halligen Südfall and Nordstrandischmoor. Fields invaded by the salt water remained unproductive for several years and the paralysis of the regional economy was total. Estimates and invoices preserved in archives reveal that these areas were restored with great difficulty after the disaster. In 1640, the inhabitants were still reconstructing houses and churches damaged by the storm Buchardi six years earlier (Gottschalk, 1977).

The north sea surge of 1953 This extreme event is registered in contemporary European memory as the most severe flood of oceanic origin during the twentieth century. The event occurred on the night of 1st February 1953, when a storm from the Northwest affected simultaneously the Netherlands, the Eastern coast of England, Belgium and Germany. It caused the death of 2100 persons of which 1800 were in the Netherlands. The conjunction of a strong tidal range and a violent wind (as during Xynthia) contributed to pushing waters which reached a height of 4.5 m over the average level of the sea. In Holland, dykes were broken in numerous places along the west coast and the sea invaded vast areas. The closest houses were totally destroyed by waves and their owners were taken completely by surprise while they were sleeping. This reality explains the heavy human cost, to which was added about 72 000 evacuees, and a badly damaged road and communication network. The omnipresent sea forced the rescue services and survivors to move around by boat.

For the English East coast, the UK Environment Agency recorded that 300 individuals died, 24 000 houses were destroyed and 40 000 people were evacuated. In Belgium, there were numerous submerged or worn-out dykes, provoking the flooding of the regions of Ostend and Antwerp and the death of 40 people. Shortly after the disaster, the British and Dutch authorities would learn the lessons from their lack of preparation. The former set up the Thames Barrier to protect London and the latter developed the famous Deltawerken, or 'Delta plan', consisting of closing the mouths of the main local rivers with dykes, with the exception of the canals of Nieuwe Waterweg and Western Schelde, because of their economic importance for the ports of Rotterdam and Antwerp. For the benefit of the populations exposed to the risk, the authorities favour the creation and the teaching of a culture of risk awareness in diverse and original forms.[8]

[8] http://www.deltawerken.com/, accessed 23rd May 2014. https://www.gov.uk/browse/environment-countryside/flooding-extreme-weather, accessed 23rd May 2014.

1.1.4 A memory of risk or a culture of survival?

The recent report IPCC-SREX (IPCC, 2012) underlines that the major part of the data "history" relating to extreme events limits itself to a recent period, generally since 1950. As a consequence, the hypothesis of an increase in these events is far from being clearly established (IPCC, 2012). Historical experience thus gives meaning to the often neglected notion of variability, geographical or historical. While the French people perceived Xynthia as new, the British, the Dutch people and the Germans knew that this littoral risk had not gone away because they had faced it in 1953 and in 1962.

Contrary to all expectations, risk in its social dimension was not formerly designated as fate but rather as a state of waiting resulting in the anticipation of a crisis capable of upsetting the community. It implies a perception of the danger, a translation of the memory of a past disaster from which emerges the collective consciousness of a potential threat. Consequently, the community takes care to dig up and to maintain their memory in diverse forms, to adapt its environment as a consequence, and in certain cases to develop alert measures.

1.1.4.1 Living with droughts in the Cyclades: The Syros island

In numerous Mediterranean regions, water resources are exploited so intensively that there is risk of causing a major water shortage. In an uncertain climatic context, the situation could get worse if an outbreak of droughts occurs, as is predicted by numerous climatic models. Once again, the IPCC-SREX report underlines the absence of transverse knowledge on this subject (IPCC, 2012). There is a lack of hydrological data, of historical context and of knowledge regarding traditional practices, all of which could help consolidate or contradict the current climatic models. Besides the pressure exercised by climate change, the transformations in the use of land constitute an aggravating factor for the Mediterranean environment and water resources. These changes in the use of land proceed from anthropogenic interventions engendered by agriculture, urbanization or additionally by natural events, such as fires (Sofios, Arabatzis and Baltas, 2008).

Although covered mainly with brush, the island of Syros is popular with locals and sailors for the freshness and the lushness that distinguishes it from other islands of the Cyclades. Historical evidence all emphasizes the historical self-sufficiency of the island in water resources and the quality and flow of the fountain near Ano Syros (old city), later called by the orthodox the Saint Athanassios spring. In ancient times, this spring was already known to pilgrims who came to purify themselves before going to Delos (testimony of Homer). For the island as a whole, agriculture was made possible by the many springs distributed mainly in the southern and western parts, as evidenced by the Abbot Della Rocca in 1790.[9] According to him, without them it would have been impossible to irrigate vegetable gardens and fruit trees consisting of orange and lemon trees. An important turning point in both demographic and environmental terms came in 1821 with the creation *ex nihilo* of what would become the town of Hermoupolis along the deserted cove dominated by the perched medieval town of Ano Syros. This

[9] Della Rocca (Abbot of Syra), Paris, Imprimerie Bleuet, 1790.

Figure 1.1.10 Terraces of crops (*anavathmi*) in the north of Syros.

spontaneous agglomeration was rapidly populated with the influx of refugees driven out by the Ottomans during the Greek War of Independence (1821–1829). Very quickly, urban and city dwellers faced a major problem with water.

In the face of these challenges, namely the economic crisis, the demographic development and the access to water, the experience of history lends a clear legitimacy to future strategies of sustainable development. De facto, archives as well as the testimonies collected from old people highlight the benefits of traditional management of water and droughts compared with contemporary practices.

In an article published in March 1960 in the local newspaper 'Courage' about the water problem, the past role of *anavathmi*, the terraces of cultures used for centuries in the Cyclades (Figure 1.1.10) has been evoked. Already endangered because of the rural exodus, the newspaper praises their virtues by explaining that they prevent the fast flow of rainwater towards the sea thanks to the culture of vegetables, some barley and to the arboriculture (olive trees and fig trees mainly). Because of its mountainous character, particularly in the northern part, and of its semi-arid climate, the construction of terraces was imperative from antiquity to make best use of the low precipitation, maintain the soil and develop farming (Koutsoyannis *et al.*, 2008). The terraces still visible today are maintained by a retaining wall, the construction of which requires the reorganization of the preexisting ground to the full height of the wall. Then, the farmer sets up the base of the wall and a heap of rubblestones to direct the drainage towards the interior of the wall. In this way, sediments carried by rainfall accumulate behind the walls of terraces laid out in a cascade formation (Harfouche, 2005).

Archives and chronicles of the eighteenth century show the existence of an agricultural system of original rotation of crops. It includes a part under the plough (*engheria*) and another one lying fallow (*pastra*). These lands, sowed alternately, were cultivated by means of the swing plough which enabled ploughing in a crossed pattern to retain the humidity and the rainwater, the major part of which falls in October and November.[10] The land which was fallow for one year was not wasted because vegetables were sown there.

More important still at this time of discussion on the adaptation of societies confronted with the water shortage in Africa and in the Mediterranean regions, the oral testimonies revealed the existence of plants called *anhydra*, in other words, plants which were not watered. Produced for centuries in the Cyclades, these plants included melons, tomatoes and French beans which were cultivated on the cooler north-facing hillsides. To compensate for the absence of irrigation, farmers prepared the ground by digging so that it was not too hard and so that the night humidity could penetrate. Endangered today because they are now only cultivated by elderly people, who still preserve seeds, these plants produce on average 4–5 kg of vegetables and fruit a day on a cultivated surface of less than 5 acres.

As a paradoxical effect of the economic crisis, terraces which had been largely abandoned over these last few decades and reconquered by bushes and shrubs, have now benefitted from the return of the native island inhabitants who had previously migrated to Athens. Unemployed and impoverished retired people have returned to the family farms to run them again after a long period of abandonment. Deprived of financial means, they try to reuse the traditional hydraulic techniques but the task turns out to be difficult because the knowledge has been lost, sometimes irretrievably.

Most of our troubles are still our work (Jean-Jacques Rousseau, 1755)

This quotation of the philosopher Rousseau on the earthquake of Lisbon on All Saints' Day 1755 remains inescapable in the contemporary reflection on vulnerability. Not only were his words written within the context of a marine flood, a tsunami caused by three successive shocks but, better still, they gave rise to one of the first public denunciations of the vulnerability of a littoral city.[11] The speech of the philosopher resulted in an extremely radical proposal because he proposed a limit on the size of conurbations on coasts and called upon a principle the paternity of which is often claimed by our modern societies: the precautionary principle.

1.1.4.2 The Atlantic flood of 1937 as a revelation of coastal weaknesses

On its front page of 15 March 1937, the newspaper Ouest-France reported a "real tidal wave hits the Atlantic Coast" on the night of 13 to 14 March. Indeed, the phenomenon impacted on a maritime area which stretched from the Basque Country to the south of Brittany (Figure 1.1.11). The information revealed in the press of the time indicates

[10] Pitton de Tournefort J., Paris, Imprimerie Royale, 1717; Ambelas T.D., Hermoupolis, Renieri Printezi Press, 1874.
[11] Rousseau, *Lettre sur la Providence*.

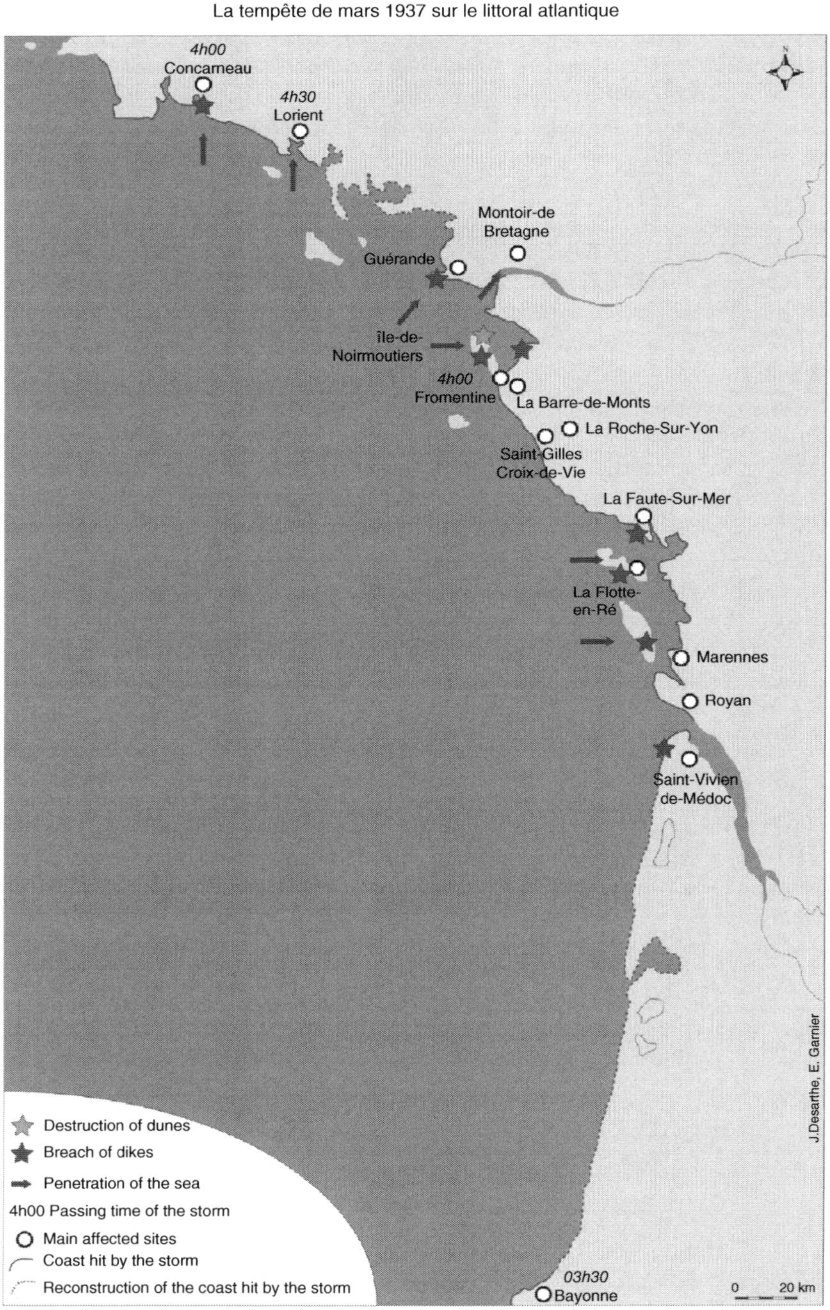

Figure 1.1.11 Geographical impact of the storm of March 1937 on the French coast.

that the direction of the the storm was, as in February 2010, on a south-southwest axis. This deep depression of 2010 went up the Bay of Biscay and hit France at night. The route of the storm of 1937 is also known. The storm passed by the island of Madeira then hit the Galician, Basque and finally the French coast at night. As the map shows very well, the intrusion of the sea occurs mainly in Southern Brittany, in Vendée in the sectors of Noirmoutier, the Anse de l'Aiguillon and finally in Charente-Maritime. For each of these regions, the destruction of the sea defences in the form of dykes allowed the invasion of the ocean (Garnier, Henry and Desarthe, 2012).

In February 2010, 29 persons died in La Faute-sur-Mer, drowned in a recent housing development. In 1937, the dyke of La Faute-sur-Mer to the Aiguillon-sur-Mer was also destroyed and military reinforcements were needed to repair it as a matter of urgency with the help of the inhabitants. At that time, there were no deaths because of a littoral development pattern quite different from that of today. As the extract of the map of 1889 (Figure 1.1.12) shows, the city of La Faute-sur-Mer which we know today did not

Figure 1.1.12 Urbanization and vulnerability in the municipalities of La Faute-sur-Mer and the Aiguillon-sur-Mer (1889–2010). The upper map is dated 1889 and the circle indicates the approximate location of the current city of La Faute-sur-Mer. At that time, it was uninhabited and consisted of protective dunes. The lower image is dated 2010 and shows the phenomenon of urbanization and of artificialisation of the coast.

exist and the map shows only some fishermen's huts on the future site of the current seaside town. The original town of Aiguillon-sur-Mer had developed behind the natural protection of two cordons of dunes.

More than 100 years later, the map of 2010 underlines the aberrations regarding town planning. In La Faute-sur-Mer, 3000 houses were built by draining swampy zones situated below sea level and by destroying the protective dunes. The latter were replaced by artificial dykes which were supposed to be impossible to submerge. One of them was protecting a new housing development but it was unfortunately destroyed by Xynthia in February 2010, causing the death of 29 persons. By concealing the risks, the planners and the elected representatives thus made these sectors less resilient and aggravated the exposure to littoral risk.

1.1.4.3 The lessons of history regarding impact strength

The storm of March 1937 caused severe structural damage on the Atlantic coast. Infrastructure (bridges, roads, railroad) was drastically damaged. Fields were unusable for several months because of the salt deposited by the sea. On the other hand, archives mention no deaths in spite of a vast and very violent flood, the demographic effects of which should have been even worse because it struck at night. This surprising fact was due to a very flexible and effective alert-prevention system which was based on local reactions and which meant that the first few hours of the disaster were dealt with without the intervention of public help. The communities of the time, as during the storm Lothar in December 1999 or Klaus in January 2009, were cut off from the rest of the world because winds had destroyed the electricity network as well as the telephone network. Having preserved the memory of the littoral risk under diverse forms such as *ex voto* in churches (Figure 1.1.13) or the marks of floods (Figures 1.1.14 and 1.1.15), the European littoral societies could look after themselves during the first hours of the crisis, often the most dangerous for the inhabitants.

Figure 1.1.13 *ex voto* reminding of a flood (surge) in 1799. La Rochelle (Charente-Maritime, France).

Figure 1.1.14 Mark of historic surges. Brockdorf (Schleswig-Holstein, Germany). *Source*: Mark Hengerer, 1988.

Figure 1.1.15 Flood plaque, Goldcliff parish church, Newport (Wales, UK).

In particular, they used the network of bell towers of churches (storm of 1937 in France) to alert the populations of the islands and the mainland, almost in real time. So, the populations exposed to the risk took refuge within safety zones which the archives describe as 'unflooded in living memory'. Others took refuge on the first floor of their house, a precautionary principle often forgotten today. In Vendée and in Charente-Maritime for example, the majority of houses built over the past 30 years are ground level houses devoid of openings in the roof and equipped with electric shutters and doors.

1.1.5 Conclusion

In the face of this historical experience, what can we say about risk management today? De facto, the groups, observers and committees dedicated to risk prevention have multiplied over the past few years, under pressure of the European authorities in particular. They consist mainly of representatives of states and engineers. Their scientific component is generally constituted by specialists from the hard sciences (hydrologists, climatologists, meteorologists). In these conditions, their recommended remedies are based on models which are poorly understood by the citizens and generally involve a recalibration of the coastal defences (dams, reservoirs, dykes) which are expensive and unaffordable in times of economic crisis.

The various historical examples evoked in this chapter show how much the need for memory can constitute one of the answers for societies confronted with an increasing vulnerability, and with a greater severity of hydrometeors. Considering this, the historian answers that the destructive power of these extreme events can be probably limited by envisaging a culture of "conservation" based on the capacity of a community to acquire and to develop, through history, the memory of its own inheritance in matters of survival.

While anticipating the climate of the next 30 years can only be guesswork, the construction of a European memory of historical extremes and the adaptation of agrosystems, urban and littoral territories, would certainly be very helpful. It would favour the emergence of a new era, less expensive in lives and money: an era of more resilient societies. Aware of the risk, these societies would thus be better prepared to deal with the shock of extreme events, which are certain to occur again.

References

Barriendos, M. and Martin-Vide, J. (1995) The use of rogation ceremony records in climatic reconstruction: a case study from Catalonia (Spain), *Climatic Change*, **30**, 1995, 201–221.

Barriendos, M. (2005) Climate and culture in Spain. Religious responses to extreme climatic events in the Hispanic Kingdoms (16th–19th centuries) in *Kulturelle Konsequenzen der 'Kleinen Eiszeit'* (eds W. Behringer, H. Lehmann and C. Pfister), Vandenhoeck & Ruprecht, Göttingen, 379–414.

Beck U. (1992) *Risk Society: Towards a New Modernity*, Sage, London.

Garnier E. (2009) *Le renversement des saisons. Climats et sociétés en France 1500–1850*, Habilitation à Diriger des Recherches, Université de Besançon, 356 pp.

Garnier E., ed. (2010a) *Climat et Histoire, XVIe–XIXe siècles*, numéro thématique 57-3, *Revue d'Histoire Moderne et Contemporaine*, Belin, Paris, 159 pp.

Garnier E. (2010b) Les dérangements du temps, 500 ans de chaud et de froid en Europe, *La Houille Blanche*, **4**, 26–42.

Garnier E., ed. (2010c) *Historical Report about the Xynthia Storm for the French Parliament* (4 July 2010).

Garnier E. and Surville F., eds (2010) *Climat et Révolutions Autour du ournal du Négociant Jacob Lambertz (1733–1813)*, Saintes, Le Croît vif Saintes, pp. 132–134.

Garnier E. and Surville F. (2011) *La Tempête Xynthia Face à l'Histoire. Submersions et Tsunamis sur les Littoraux Français du Moyen Age à Nos Jours*, Le Croît vif,, Saintes, 174 pp.

Garnier E., Henry N. and Desarthe J. (2012) Visions croisées de l'historien et du courtier en réassurance sur le submersions. Recrudescence de l'aléa ou vulnérabilisation croissante ?, in *Gestion des Risques Naturels. Leçons de la Tempête Xynthia*, (eds S. Hallegate and V. Przyluski) Editions Quae, Paris, pp. 107–130.

Gottschalk M.K.E. (1977) *Stormvloeden en Rivieroverstrommingen in Nederland*, vol. 3 (1600–1700), LAMB, Assen and Amsterdam, p. 47.

Harfouche R. (2005) Retenir et cultiver le sol sur la longue durée: les terrasses de culture et la place du bétail dans la montagne méditerranéenne. *Anthropozoologica,* **40**, 45–80.

IPPC (2012), *Managing the Risks of Extreme Events and Disasters to Advance Climate Change Adaptation*, Special Report of the Intergovernmental Panel on Climate Change, Cambridge University Press, 582 pp.

Koutsoyannis D., Zarkadouas N., Angelakis A.N. and Tchobanoglous G. (2008) Urban water management in ancient Greece: legacies and lessons. *Journal of Water Resources Planning and Management*, **134**, 45–54.

Lamb H. (1991) Historical Storms of the North Sea, British Isles and Northwest Europe, *Cambridge University Press*, 204 pp.

Manley G. (1974) Central England temperatures: monthly means 1659 to 1973, *Quarterly Journal of the Royal Meteorological Society*, **100**, 389–405.

Sofios S., Arabatzis G. and Baltas E. (2008) Policy for managment of water resources in Greece. *Environmentalist*, **28**, 185–194.

Wallace J.G. (1997) *Meteorological Observations at the Radcliffe Observatory Oxford: 1815–1995*, School of Geography, Oxford, Research Paper 53.

1.2
Current Understanding of Climate Change Impacts on Extreme Events

Richard Harding, Nick Reynard and Alison Kay
Centre for Ecology and Hydrology, Wallingford, Oxon, UK

1.2.1 Introduction

Water related issues are the most prominent amongst the commonly identified impacts of climate change (see Figure 1.2.1). Further, water related extremes – storms, floods and droughts are the likely most immediate manifestations of a changing climate. Floods and droughts cause immense (and often quantifiable) losses and disruption and generate considerable public attention.

The amount of damages reported for floods for the last two decades (625 billion 2012 US$) rank them in third position after storms (913 billion 2012 US$) and earthquakes and tsunamis (770 billion 2012 US $). The frequency of reported severe flooding has increased dramatically between the years 1999–2000 and 2005–2006[1]. Climate change may be partly to blame for this increase but socio-economic effects are equally likely to be important. Flood exposure, particularly in coastal cities, is likely to increase in the future due to growing population and assets, climate change and rising sea levels (Hallegatte *et al.*, 2013).

Following a flood event the question inevitably asked is 'is this climate change?'. This is an important and relevant question but needs to be more precisely phrased. Extremes have always happened so the more scientific question is 'is the likelihood of this extreme event increased as a result of increasing greenhouse gases?'.

[1] Based on EM-DAT: the OFDA/CRED International Disaster Database (www.em-dat.net) Université Catholique de Louvain, Brussels, Belgium).

Hydrometeorological Hazards: Interfacing Science and Policy, First Edition. Edited by Philippe Quevauviller.

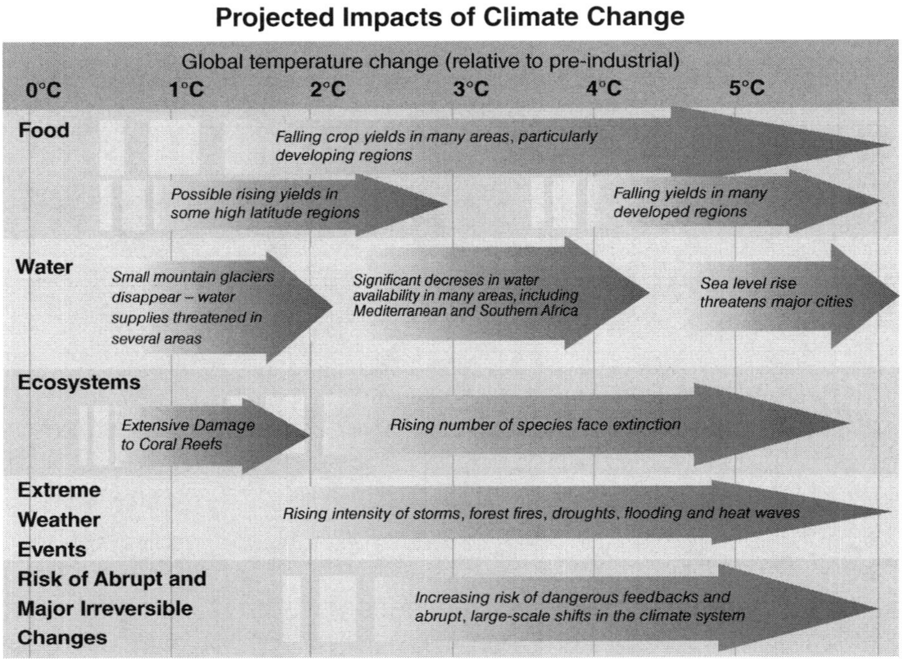

Figure 1.2.1 Major impacts of climate change. *Source*: Redrawn from IPCC 2007. Climate Change 2007: Impacts, Adaptation and Vulnerability. Contribution of Working Group II to the Fourth Assessment Report of the Intergovernmental Panel on Climate Change, M.L. Parry, O.F. Canziani, J.P. Palutikof, P.J. van der Linden and C.E. Hanson (Eds), Cambridge University Press, pp. 7–22.

1.2.2 Global water balance, past and future

1.2.2.1 Humidity and evaporation

Increasing greenhouse gases (GHGs) are very likely to influence the hydrological cycle. The most immediate impact of increasing GHGs is increasing temperature – the observed 0.85 °C increase in air temperature in the twentieth century is almost certainly a result of increasing CO_2 levels. It is likely there will be at least a further 1.5 to 2 °C increase in the current century (IPCC, 2013), although the exact amount will depend on future emissions which, of course, are determined by how humanity responds to this challenge. The amount of water vapour the air can hold at saturation increases exponentially with temperature (7% per °C). This is very likely to lead to increasing evaporation from ocean regions and to increase evaporative demand over land. The global near-surface air specific humidity has increased since the 1970s. However, during recent years the near-surface moistening over land has abated. As a result, fairly widespread decreases in relative humidity near the surface are observed over the land (see e.g. Willet *et al.*, 2008). Increasing temperatures and decreasing relative humidity would be expected to increase evaporation, however, this trend is compounded by changes in wind speed over the ocean and radiation and soil water over the land.

The overall consensus is that evaporation has increased and this is contributing to the observed increased atmospheric humidity (Kay, *et al.*, 2013).

1.2.2.2 Precipitation

Increasing evaporation and atmospheric humidity should lead to increasing precipitation – over the long term evaporation and precipitation must balance. The signal of increasing rainfall has, however, not yet been observed in the global mean rainfall data sets, most of which show only a small, but statistically insignificant, increase in the last two decades. However, regional changes have been observed. The high and mid-latitudes in the northern hemisphere show a steady increase from 1900 to 2010, amounting to about 30 mm. In the last two decades the precipitation in the tropics has also increased overall. Generally, studies indicate that precipitation over wet areas has increased but decreased over drier areas (Zhang *et al.*, 2007; Allen *et al.*, 2010). In the southern hemisphere there is considerable decadal variability, often associated with ENSO (El Niño southern oscillation) effects, and, hence, it is difficult to discern long-term trends.

In many regions of North America and Europe the balance between snowfall and rain has also changed, with precipitation falling as rain rather than snow, particularly on the southern margins of the seasonal snow regions (Kunkel *et al.*, 2009; Serquet *et al.*, 2011).

1.2.2.3 River run-off

Many if not most large rivers are strongly affected by human influences, such as dam construction, water extraction and irrigation, so trends in river flow must be interpreted with caution. Dai *et al.* (2009) found that only about one-third of the top 200 rivers (including the Congo, Mississippi, Yenisey, Paraná, Ganges, Columbia, Uruguay, and Niger) showed statistically significant trends during 1948–2004, with the rivers having downward trends (45) outnumbering those with upward trends (19). The changes in river flow are qualitatively consistent with regional changes in precipitation and temperature, Dai *et al.* (2009). Examples are the Yellow River in northern China since the 1960s (Piao *et al.*, 2010) where precipitation and river flows have decreased and many parts of the United States (Groisman *et al.*, 2004), and in the Yangtze River in southern China (Piao *et al.*, 2010) where both precipitation and river flows have increased.

At the higher latitudes, Peterson *et al.*, (2002) report increases in the annual discharge from the six largest Eurasian rivers flowing into the Arctic Ocean by 7% between 1936 and 1999. Rising minimum daily flows also have been detected in northern Eurasian rivers (Smith *et al.*, 2007) and increasing winter base flow and mean annual stream flow (resulting from possible permafrost thawing) have also been reported in Northwest Canada (St. Jacques and Sauchyn, 2009). However, for the Arctic drainage areas, upward trends in streamflow are not always accompanied by increasing precipitation, especially over Siberia, based on available data, although recent surface warming and associated downward trends in snow cover, soil ice content, and changes in evaporation over the northern high latitudes may have contributed to increased run-off in these regions (Adam and Lettenmaier, 2008).

Recently, Stahl *et al.* (2010) and Stahl and Tallaksen (2012) investigated streamflow trends based on a data set of near-natural streamflow records from more than 400 small catchments in 15 countries across Europe for 1962–2004. A regional coherent pattern of annual streamflow trends was revealed with decreasing streamflows in southern and eastern regions, and generally positive trends elsewhere.

1.2.3 Global extremes – Rainfall, floods and droughts

1.2.3.1 Precipitation

Increasing evaporation and atmospheric humidity will impact on the energy exchanges within the atmosphere. The potential for increasing latent heat release associated with cloud condensation is likely to increase storminess and enhance convective rainfall. This has in fact been observed in the northern hemisphere, where data are available, two thirds of areas show significant increases in daily precipitation intensity (defined typically by the 95th percentile of daily rainfall, Min *et al.*, 2011). Regionally there are significant upward trends in northern and mid-Europe, north and Central America and India. Within Europe not only have the absolute precipitation extremes in the winter increased (by between 4 and 10% per decade) but there has been an increase in the number of wet spells lasting longer than 2 to 3 days with high precipitation intensities (Zolina, 2009). Elsewhere, for large areas of high latitude regions, Africa and South America, there are not sufficient long-term records of daily rainfall available to identify trends. These studies refer to daily precipitation extremes. There have been few studies at sub-daily scales to make an equivalent assessment.

1.2.3.2 Storminess

There is much debate in the literature as to whether there has been an increase in storminess in mid-latitudes. It is likely there has been a poleward shift in the extra-tropical storm tracks but inconclusive assessments as to whether the intensity of storms has increased.

There is also uncertainty about whether the frequency and size of tropical storms and hurricanes is changing and has been affected by increasing GHGs. The humidity and energy argument above suggests there will be an enhancement and indeed the numbers of Atlantic tropical storms and hurricanes has increased in the last two decades, but a long term trend or direct causal link has yet to be established. One reason for this uncertainty is the difficulty of establishing a long time series extending before the advent of satellite monitoring. Another reason for uncertainty may lie in the inability of climate models to capture sub-grid processes, such as tropical storms, rather than the actual lack of a causal link.

Increasing evaporation, rainfall and storminess as a response of the atmosphere to increasing GHGs and temperatures are commonly referred to as the acceleration of the global water cycle. This is a well established thermodynamic effect. To predict regional distributions of rainfall in the future we need to understand the dynamics of the atmosphere and, in particular, how the tropical and extra-tropical depressions and anticyclonic systems will form and move. To simulate these systems we need the gridded meteorological models. These are based on the weather forecasting models but run over centuries, these climate (or increasingly earth system) models are vital but imperfect tools. The main limitation of these complex models is that to enable them to simulate long periods of time on current computers they need to operate with coarse grid scales. In the past these grids have been a few degrees of latitude and longitude (a few hundred kilometres) although as the supercomputers have become faster their grids are reducing to at least below 100 km. However, even this finer grid fails to capture the medium- or small-scale rain systems – the effect of which have to be parameterised. This means the regional rainfall patterns generated by these models tend to be imperfect and vary considerably between models.

1.2.3.3 Droughts

Droughts have profound impact not only on water supply but also on food production, biodiversity and human well being in general. It is generally accepted that, along with increasing rainfall intensity, climate change will also bring increasing occurrence of drought (see e.g. IPCC, 2007). There has been a suggestion that drought occurrence has already increased through the twentieth century (Dai, Trenberth and Qian, 2004) but the magnitude of this change has been shown to depend on the methodology used to estimate evaporative losses (Sheffield, Wood and Roderick, 2012), hence suggesting that prediction of future drought depends critically on its definition and method of calculation.

In Europe there has been a general trend towards drier summers and wetter winters (see e.g. Osborn and Hulme, 2002) accompanied by an increase in temperatures, which influences snowmelt. Thus in a study of flows from small, undisturbed catchments Stahl *et al.* (2010) found that in catchments which have low flows in the summer, the low flows decreased, but in catchments with winter minima, associated with winter snow accumulation, these minimum flows increased with the increasing frequency of winter melt events. The presence of groundwater in a catchment also has an influence, buffering the summer flows, and in these catchments there is less evidence of increasing hydrological drought.

Types of Drought

Drought can be defined as a shortage of water resulting from a prolonged period of abnormally low precipitation. Drought is a complex phenomenon which varies spatially in its extent, duration, frequency and severity. Droughts can be of many types, depending on the user perspective, for example:

- A meteorological drought
- A hydrological flow drought
- An agricultural drought, linked to the availability of water in the soil.

A drought is a departure from the mean – thus a very dry country (like Libya with 180 mm of rain a year) is not in permanent drought. In fact a wet country may be more vulnerable to temporary deficiencies in rainfall.

A *meteorological drought* can be quantified in terms of a cumulative rainfall deficit over a set period or some index of the length of periods without rain.

A *hydrological drought* is a deficiency of river flow (or in some cases ground water or reservoir level). A hydrological drought indicator can be identified by the number of days the measured (or modelled) river flow drops below the mean 90th percentile (or other threshold). For a region, the regional deficiency index (RDI) is the proportion of stations below the threshold on any one day. A water resource drought might be more complex as it balances the supply with demand.

An *agricultural drought* might be linked to losses of crop productivity but a more general index might be related to soil moisture deficit. Unfortunately soil moisture is not measured routinely everywhere and will depend on soil and crop type. The Palmer drought severity index (PDSI) is an attempt to generalise and attempts to quantify the cumulative deficit (relative to mean local condition) in soil moisture by incorporating estimates of evaporation and precipitation into a hydrological accounting system. Unfortunately it is dependent on how the evaporation is calculated.

1.2.3.4 Floods

The Working Group II of the IPCC concluded in the 4th Assessment Report that there was no general global trend in the incidence of floods (Kundzewicz *et al.*, 2007). The SREX report (IPCC, 2012) went further to suggest that there was low agreement and thus *low confidence* at the global scale regarding changes in the magnitude or frequency of floods or even the sign of changes. Trends in floods are, of course, strongly influenced by changes in river management. While the most evident flood trends appear to be in northern high latitudes, where observed warming trends have been largest, in some regions no evidence of a trend in extreme flooding has been found, e.g., over Russia, based on daily river discharge (Shiklomanov *et al.*, 2007). Other studies for Europe (Hannaford and Marsh, 2008; Renard *et al.*, 2008; Petrow and Merz, 2009; Stahl *et al.*, 2010) and Asia (Jiang *et al.*, 2008; Delgado *et al.*, 2010) show evidence for upward, downward or no trend in the magnitude and frequency of floods, so there is currently no clear and widespread evidence for observed changes in flooding except for the earlier spring flow in snow dominated regions (IPCC, 2012). In summary, there continues to be a lack of evidence and thus *low confidence* regarding the sign of trend in the magnitude and/or frequency of floods on a global scale.

In a further analysis of the European flow data set, increases of the annual 7-day maximum flow were found in north and west Europe and decreases in the south and east (Stahl *et al.*, 2011). Thus, while increasing trends of heavy rains have been documented, a ubiquitous increase in flood maxima is not evident in the data (Pinskwar *et al.*, 2012). Increasing rainfall intensities would be expected to lead to increasing flash floods but any trend may be masked by changes in land use and river regulation, exacerbated by the paucity of long records with high time resolution observations (see IPCC, 2012). There have undoubtedly been a large number of floods, and other extreme events, in the first years of the twenty-first century. As discussed above, the oft-asked question of whether an event is caused by climate change may be the wrong question. All weather events will be affected by climate change because the environment in which they occur is warmer and moister than it used to be (Trenberth and Fasullo, 2012).

In a possibly unique study Pall *et al.* (2011) and Kay *et al.* (2011a) looked at the floods that occurred in the UK in the autumn of 2000, the wettest autumn in England and Wales since records began. To assess the contribution of anthropogenic GHG emissions to the risk of floods, a several thousand member ensemble of atmospheric climate model runs with realistic industrial levels of atmospheric CO_2 was compared with a second ensemble for which anthropogenic emissions of GHGs were assumed not to have occurred. The simulated precipitation from these two ensembles was fed into a simple rainfall/run-off model with the conclusion that the greenhouse warming increased the flood risk by approximately a factor of two. The paper by Kay *et al.* (2011a) fed the rainfall ensemble from the climate model into a catchment-based hydrological model for eight catchments across the UK, in regions highly affected by flooding in Autumn 2000. Their results broadly supported the findings of Pall *et al.* (2011) that anthropogenic influences had increased the chance of the autumn 2000 event occurring in seven of the eight catchments. The eighth showed a decrease, thus overall the flood risk increased with increasing GHGs, but with significant uncertainty due to the variability of catchment characteristics and climate model rainfall distributions.

1.2.4 Future global predictions

Despite their limitations there are some common features of the climate model simulations for the twenty-first century which provide some indication of our future climate. To provide a consensus and some measure of the uncertainties on climate model simulations the World Climate Research Programme (www.wcrp-climate.org/) has sponsored a series of Coupled Model Intercomparison Projects, the most recent of which is CMIP5 (Taylor, Stouffer and Meehl, 2012; IPCC, 2013). More than 20 modelling groups performed CMIP5 simulations using more than 50 models. The simulations in CMIP5 use consistent future scenarios, namely four representative concentration pathways (RCPs) (Moss *et al.*, 2010), designed to cover a wide range of possible magnitudes of climate change in models. The major relevant findings of the CMIP5 studies are:

- As temperatures rise, specific humidity increases but relative humidity is projected to remain roughly constant. The larger projected warming over the land compared to the ocean will change atmospheric circulations and likely result in increased moisture transport from the oceans to the land.
- It is virtually certain that, in the long term, global precipitation will increase with increased global mean surface temperature. Changes in average precipitation in a much warmer world will not be uniform, with regions experiencing increases, or decreases or no significant change at all. The high latitudes are very likely to experience greater amounts of precipitation due to the additional water carrying capacity of the warmer troposphere as well as increased transport of water vapour from the tropics. Many mid-latitude arid and semi-arid regions are likely to experience less precipitation and many moist mid-latitude regions are likely to experience more precipitation. The largest precipitation changes over northern Eurasia and North America are projected to occur during the winter.

- A shift to more intense individual storms and fewer weak storms is *likely* as temperatures increase. Over most of the mid-latitude land-masses and over wet tropical regions, extreme precipitation events will *very likely* be more intense and more frequent in a warmer world.
- It is *likely* that the area encompassed by monsoon systems globally will increase over the twenty-first century. While monsoon winds are *likely* to weaken, monsoon precipitation is *likely* to intensify due to the increase in atmospheric moisture.
- Regional and global scale projections of soil moisture and drought remain relatively uncertain compared to other aspects of the hydrological cycle. Drying in the Mediterranean, south-western US and southern African regions is consistent with projected changes in Hadley circulation and increased surface temperatures, so surface drying in these regions is likely as global temperatures increase.
- Annual surface evaporation over the ocean is projected to increase as global temperatures rise and is projected to change over land following a similar regional pattern of precipitation and soil moisture. Evaporation will decrease over the south-western US/north-western Mexico, southern Africa and land bordering the Mediterranean. Evaporation will increase over land in the northern high latitudes, consistent with the increase in precipitation and an overall warming leading to increasing potential evaporation. There is *high confidence* in patterns of these changes but there is *low confidence* in the magnitude.
- Under the higher emission scenarios an increased risk of drought is *likely* (*medium confidence*) in presently dry regions. There is, however, some uncertainty, for example increasing CO_2 may decrease evaporation directly, increasing water use efficiency and hence reducing the severity of future hydrological droughts (Prudhomme *et al.*, 2013b).
- Decreases in run-off are *likely* in southern Europe, the Middle East, and south-western US. The CMIP5 models project consistent increases in high latitude run-off, but confidence in this projection is tempered by large biases in their simulation of present-day snow cover.

In summary, the latest projections from climate models indicate a world of changing rainfall patterns, more extreme rainfall and more droughts in regions already experiencing water scarcity. The responses of river flows, soil moisture and water resources will depend on the regional detail of both the regional climate change and catchment characteristics. There is still much uncertainty, particularly regionally, in components of the future water cycle and this suggests we need a probabilistic and risk-based analysis of future extreme events.

The issue of uncertainty remains highly pertinent, particularly for projections of changes in climate and the subsequent impact on hydrological extremes. There are three main sources of uncertainty; emissions, modelling and natural variability (Deser *et al.*, 2012). More specifically, for addressing the impacts of change there is a cascade of uncertainty from the range of potential GHG emissions, to the Global Climate Model (GCM) structure and parameterisation, to the methods chosen to spatial downscale (which might also include Regional Climate Model structure and parameterisation) to the uncertainty associated with the hydrological model structure and parameterisation (Chen *et al.*, 2011; Kay *et al.*, 2009; Prudhomme and Davies, 2009; Wilby and Harris, 2006), Figure 1.2.2. Of these, GCM uncertainty has been found to be the largest single

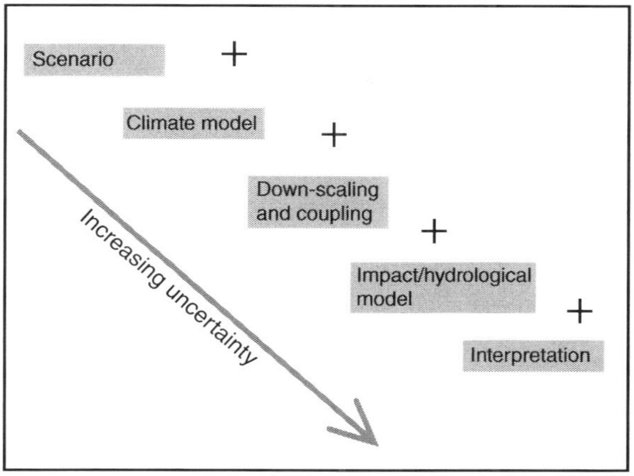

Figure 1.2.2 Sources of uncertainty in climate change impact assessments.

source, although other sources can also be significant, particularly that due to (current and future) natural variability, which can make the results from time-slice regional climate runs misleading (Kay and Jones, 2012a).

Larger climate projection ensembles have been produced (e.g. UKCP09: Murphy *et al.*, 2009) to enable risk-based approaches to decision-making (e.g. Christierson *et al.*, 2012), but these are still subject to significant uncertainty (e.g. Kay and Jones, 2012b). An alternative to the traditional top-down impact studies are bottom-up, vulnerability-based, approaches, which focus on sensitivity testing and adaptation options rather than climate change projections (van Pelt and Swart, 2011). A case study of such an approach to inform climate change policy development for flood management in Great Britain is described in Section 1.2.6.

1.2.5 Regional drought and water resources

1.2.5.1 Drought in the twentieth century

Drought can cause serious problems across much of Europe. Many droughts are localised and short, but others are widespread and cause environmental and social effects that cross national boundaries. The European Drought Catalogue (spanning 1961–2005) defines for 23 homogenous regions in Europe, the time series of regional streamflow deficits; see Figure 1.2.3 and Hannaford *et al.* (2011). This enables a characterization of major drought periods in terms of duration, seasonality and spatial coherence in the various regions. An example of the catalogue is given for two contrasting regions in Figure 1.2.4.

Trend detection studies have been performed to identify possible changes in historical streamflow series. These included an assessment of hydrological change (annual mean, monthly mean and low streamflow) in small basins at the sub-grid scale of climate models based on the newly assembled and updated streamflow data set for Europe (Stahl *et al.*, 2010). Figure 1.2.5 (top left) shows a regionally coherent picture of observed

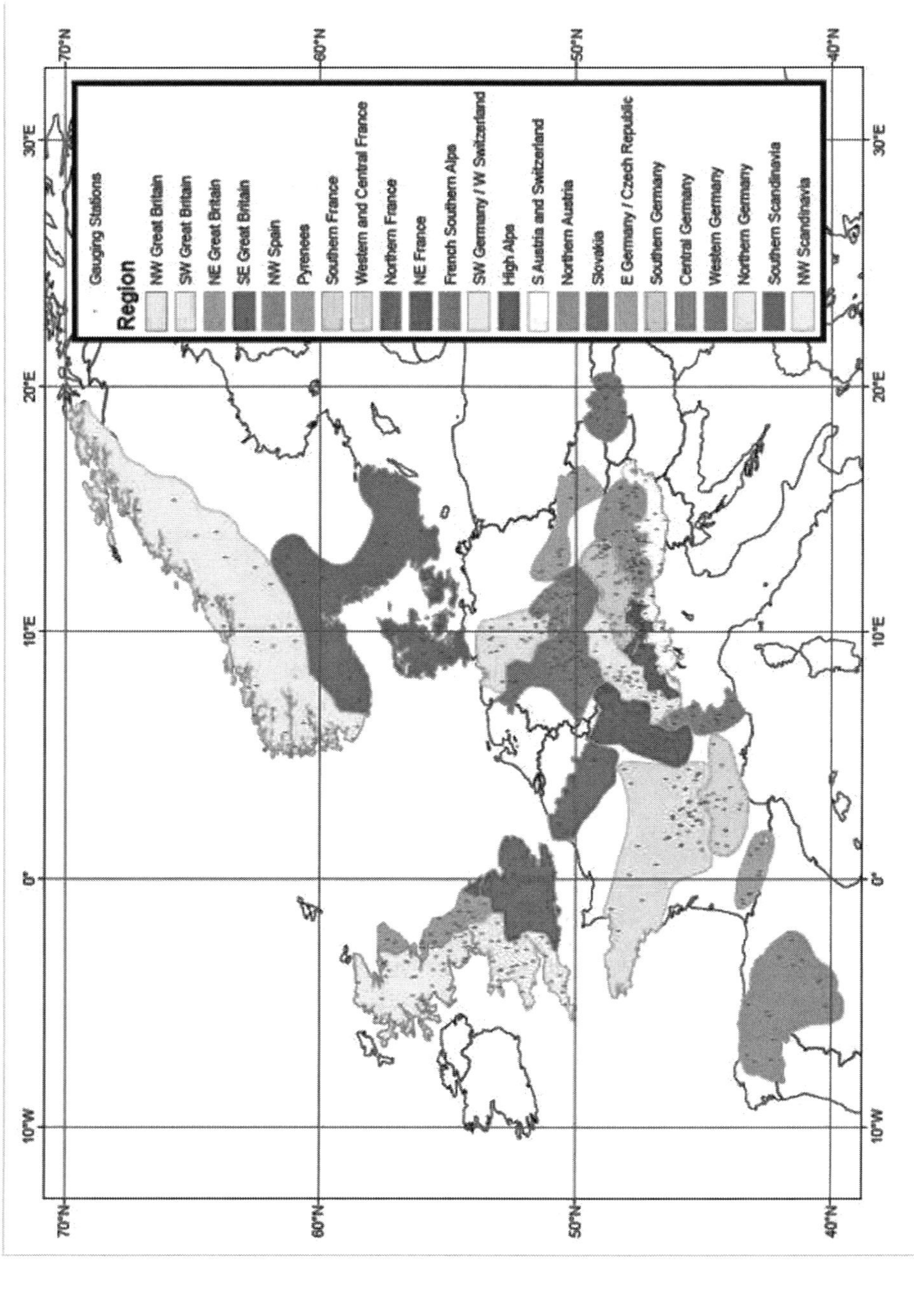

Figure 1.2.3 Regions used in the drought catalogue for Europe. *Source*: Parry *et al.*, 2011. See plate section for colour representation of this figure.

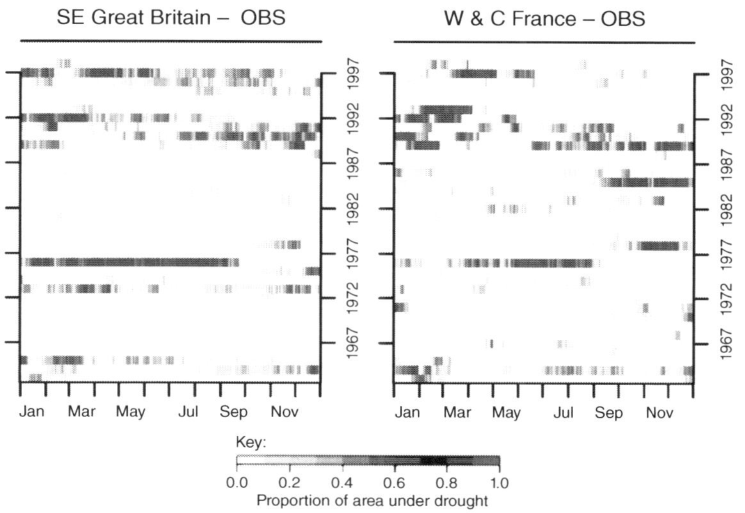

Figure 1.2.4 Drought catalogue derived from observed river flow gauges for two contrasting regions of Europe: Southeast Great Britain and West and Central France. The months of the year are shown on the *x*-axis, years on the *y*-axis. The shades of grey show the Regional Deficiency Index, a measure of the proportion of the region experiencing a flow deficiency. *Source*: Parry *et al.*, in press. See plate section for colour representation of this figure.

annual streamflow trends, with negative trends in southern and eastern regions, and generally positive trends elsewhere. In a follow-up study trend maps for annual and monthly run-off, and high and low flows across the whole of Europe (filling the white spaces on the map) are presented based on an ensemble of eight large-scale hydrological models. Modelled trends were validated against trends from 293 discharge records, showing that the ensemble mean provides the best representation of trends. Estimates of change are particularly reliable for annual run-off, winter run-off, and high flows. The new trend maps reveal valuable details of a pronounced gradient between positive (wetter) trends in the Northwest and negative (drier) trends in the Mediterranean and in the Southeast (Figure 1.2.5), and provide a considerable improvement over previously published maps of observed trends covering only parts of Europe (Stahl *et al.*, 2011; Stahl and Tallaksen 2012). The broad, continental-scale patterns of change are mostly congruent with the hydrological responses expected from future climatic changes, as projected by climate models.

1.2.5.2 Twenty-first century drought

To assess the impact of climate change on hydrological droughts a multi-model experiment was undertaken, including seven global impact models (GIMs) driven by climate data from five global climate models (GCMs) from CMIP5 under four different representative concentration pathways (RCPs). Drought severity was defined as the fraction of land under run-off deficit (run-off less than a drought threshold) and is a measure

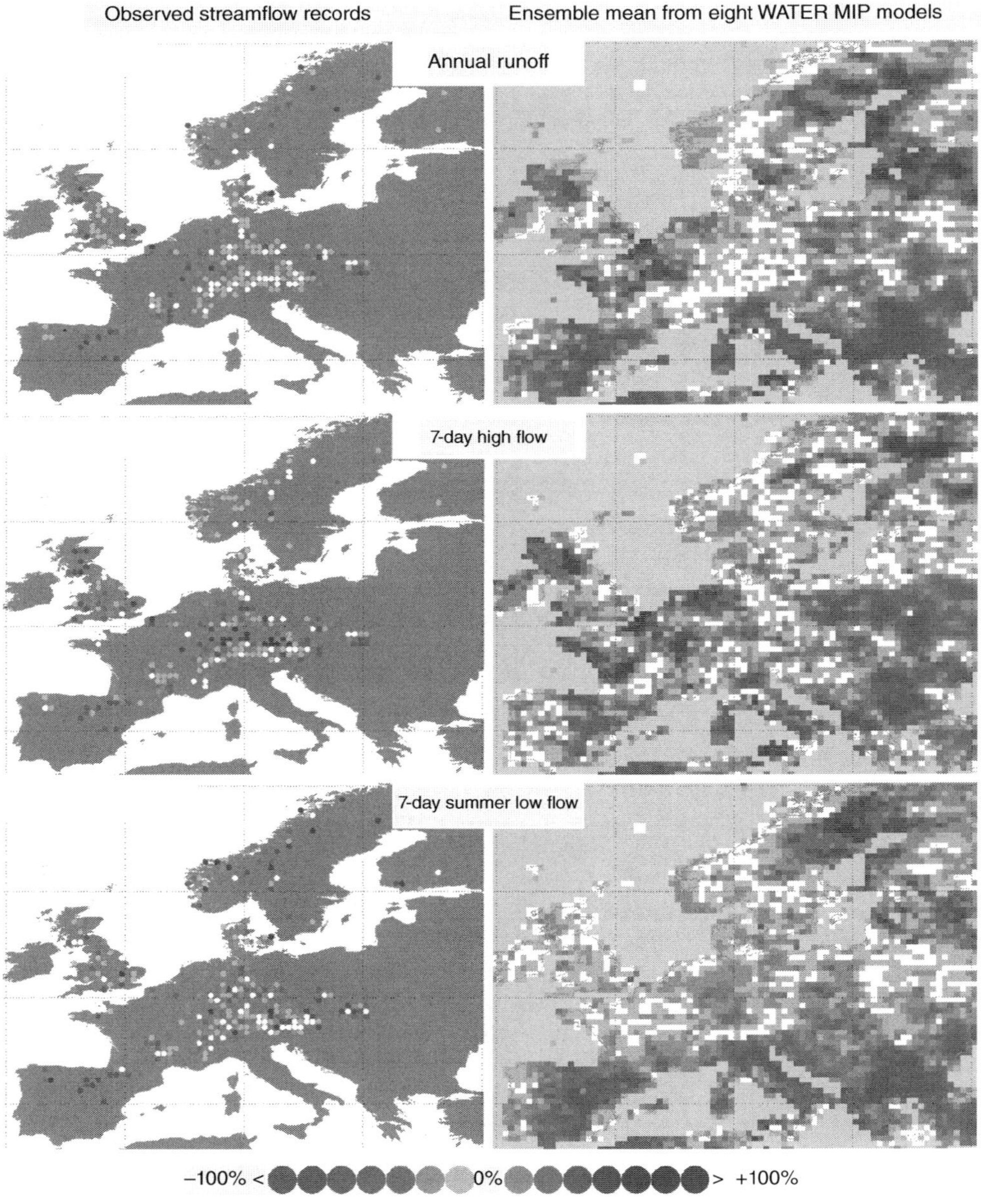

Figure 1.2.5 Trends in annual run-off, high and low flows in Europe: left: observed and right: as simulated by a multi-model ensemble of eight large-scale hydrological models. *Source*: Stahl *et al.*, 2011. © American Meteorological Society. Used with permission. See plate section for colour representation of this figure.

of the time-integrated effect of several interlinked processes and stores, including precipitation, evaporation and soil moisture storage. Results show a likely increase in the global severity of drought at the end of the twentyfirstst century, with systematically greater increases for the RCPs describing stronger radiative forcings. Under RCP8.5, droughts exceeding 40% of the non-arid parts of the land area are projected by nearly half of the simulations. This increase in drought severity has a strong signal-to-noise ratio (across all the model combinations) at the global scale, and Southern Europe, Middle East, South East United States, Chile and South West Australia are identified as possible hotspots for future water security issues, Figure 1.2.6. The uncertainty due to GIMs is greater than that from GCMs, particularly if including a GIM that accounts for the dynamic response of plants to CO_2 and climate, as this model simulates little or no increase in drought frequency. In particular when the CO_2 fertilization effect is included in the hydrological models the increase in drought frequency all but disappears. This study demonstrates that different representations of terrestrial water cycle processes in GIMs can be responsible for considerable uncertainty in the response of hydrological drought to climate change. When assessing the impact of climate change on hydrology it is hence critical to consider a diverse range of GIMs to better capture the uncertainty associated with the models (Prudhomme *et al.*, 2013b).

1.2.6 Case study: Science to support policy for flood management under climate change

For many years in the UK the guidance on climate change for flood managers was based on a sensitivity allowance approach, reflecting the high degree of uncertainty in future projections of change in high river flows. The advice for flood managers was to add a factor of 20% to peak flows to assess the sensitivity of proposed flood management options to climate change and if the schemes were still viable within this range then they may be assumed to be offering an appropriate level of protection up to 2050 (Defra, 2006). One of the main issues with this type of guidance, based on a top-down impact study, is that the entire analysis would be need to be repeated as new climate projections are produced. Moreover, the original guidance was to be applied to all catchments across England and Wales, regardless of location or individual catchment properties.

In 2006 the UK Department for Environment, Food and Rural Affairs (Defra) funded some research to develop a bottom-up, sensitivity-based approach for changes in flood peaks in Great Britain (Reynard *et al.*, 2009; Kay *et al.*, 2011b). The sensitivity domain involved changes in mean annual precipitation and seasonality of precipitation, as well as changes in temperature and potential evaporation (Prudhomme *et al.*, 2010). Hydrological modelling was performed, using the sensitivity domain, for 154 catchments across Britain, resulting in response surfaces for each catchment showing changes in flood peaks at four return periods (2-, 10-, 20- and 50-years) (Prudhomme *et al.*, 2013b). These response surfaces were then grouped by similarity, resulting in nine response types. Families of response types were then characterized by catchment properties (Prudhomme *et al.*, 2013c), enabling estimation of the response type of any catchment where the required catchment properties are available. This enables easy catchment impact estimation, even for multiple large ensembles of climate change projections, by overlaying projections on average response surfaces of the

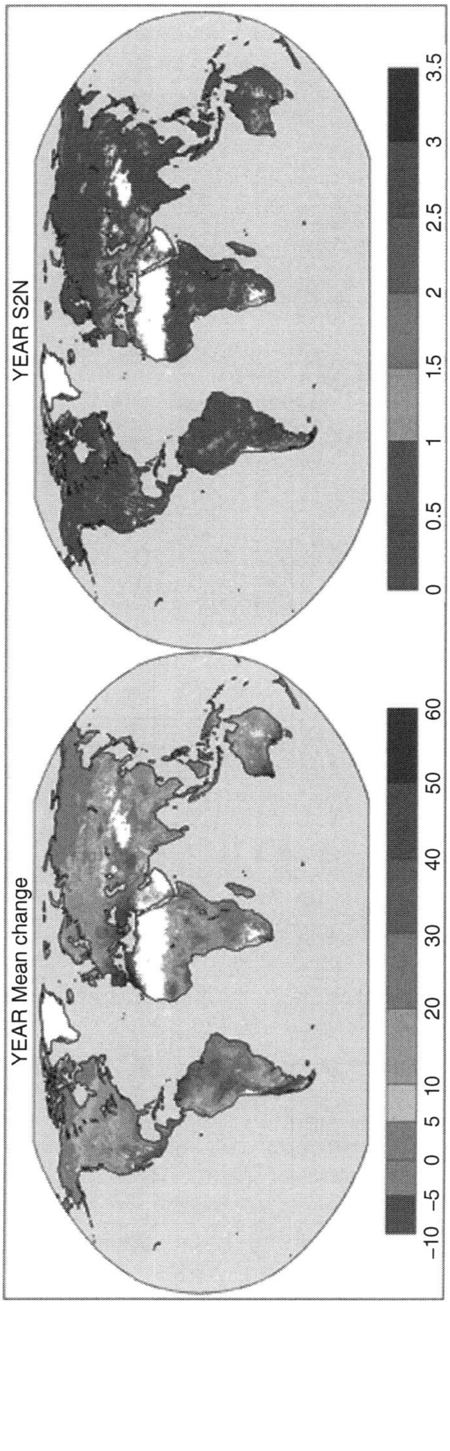

Figure 1.2.6 Percentage change in the occurrence of days under drought conditions for the period 2070–2099 relative to 1976–2005, based on a multimodel ensemble MME experiment under RCP8.5 from five global climate models and seven global impact models: (a) MME mean change and (b) associated signal-to-noise ratio (S2N, MME mean change divided by its inter-quartile range). *Source*: Prudhomme *et al.*, 2013b, Fig. 1. See plate section for colour representation of this figure.

Table 1.2.1 New allowances for changes in peak river flows in the Thames catchment (Environment Agency, 2011).

	Total potential change anticipated for the 2020s (%)	Total potential change anticipated for the 2050s (%)	Total potential change anticipated for the 2080s (%)
Upper end estimate	30	40	70
Change factor	10	15	25
Lower end estimate	−15	−10	−5
Thames H++	40	55	90

appropriate type. This technique was then applied using a set of more than 1000 gauged catchments across England and Wales and the UKCP09 probabilistic climate change projections (Murphy *et al.*, 2009), which consist of sets of 10 000 monthly changes in a number of climate variables for a 25 km grid or for river-basin or administrative regions across the UK. The UKCP09 projections for each river-basin region were overlaid on the average response surfaces for each response type, resulting in probabilistic impacts for each response type in each region (Kay *et al.*, 2014a,b). For each region, a regional probabilistic impact estimate was then produced, by weighting the response-type impacts with the number of catchments estimated as having each response type. These regional impacts ranges were used to support the introduction of regional allowances for changes in flood peaks under climate change (Environment Agency, 2011). The derived changes for the Thames catchment are shown in Table 1.2.1. For each region there are "anticipated changes" for three time periods (2020s, 2050s and 2080s), and for four "scenarios" spanning the estimated uncertainty in the methodology. The H++ scenarios were derived assuming upper end estimates of change for catchments within the region that show significantly greater increases to river flood flows than more standard catchments in the region.

1.2.7 Adaptation planning

It is now widely accepted that businesses must be resilient against both climate variability and possible future climate change. The Climate Change Act 2008 made the UK the first country in the world to have a legally-binding long-term framework for both mitigation and adaptation to climate change. It created a framework for building and reinforcing the UK's ability to adapt to climate change.

Adaptation actions should be robust, flexible and/or reversible. If possible they should be "no-regret" actions and offer some headroom or safety margin. They should provide "soft" rather than "hard" options (particularly relevant to issues around floods and droughts), which can offer the required flexibility and reversibility. Adaptation actions should also be aware of, and when appropriate take account of, adaptation being taken by others. Any action taken should be cost-effective and proportionate to the risk. All these characteristics require some underpinning science to understand the scale and the nature of the adaptive response as well as its consequences.

Within the recent exponential increase in the academic literature on adaptation, papers related to the water sector account for a relatively high proportion (~25%) (Wilby and Keenan, 2012; Figure 1). A recent special issue in the journal *Environmental Science and Policy* (Quevauviller, 2011) focused on adaptation to reduce European water-related risks under a changing climate, covering the water cycle and ecosystems, water security and droughts, and extreme floods. A particular issue for the water sector is the need for consideration at the river catchment/basin scale, as local adaptation measures can have effects up/downstream (Huntjens *et al.*, 2012), but this can be problematic as river catchment/basin boundaries often do not coincide with local authority or devolved administration boundaries in Britain, or even country boundaries elsewhere.

A study of observed adaptation measures in the UK (Tompkins *et al.*, 2010) found higher levels of adaptation in the water supply and flood management sectors, and they suggest this may be because these sectors have longer lead-times for capital-intensive investment, or strict regulation, or are more susceptible to current weather variability and extremes. They also found that adaptation measures seldom seemed to be implemented in response to climate change alone; other influences included population pressures and cost savings. Although they found that UK adaptation so far has been predominantly government-driven, they noted an important trickle-down effect.

There is much debate over whether or not climate models are the appropriate tools or provide the right information to inform adaptation (Wilby and Dessai, 2010; Beven, 2011; Wilby and Keenan, 2012). Despite much progress in climate modelling in recent years, there remains considerable uncertainty in (seasonal and extreme) precipitation projections from climate models, with different models projecting regional precipitation changes that can vary not just in magnitude but in sign (Hunt and Watkiss, 2011; Wilby and Dessai, 2010). Thus there is considerable uncertainty in projected impacts on water availability and flooding (e.g. Prudhomme and Davis, 2009; Kay *et al.*, 2009; Maslin and Austin, 2012; Bell *et al.*, 2012), which is a significant constraint on adaptation planning.

A particular difficulty for adaptation is distilling the complexities of the science (including uncertainty) into information more directly usable by decision-makers, to enable fuller use of the available information. It is argued by van Pelt and Swart (2011) that "formulating robust strategies will only be possible if knowledge is effectively shared between the scientific climate community and policymakers at the many relevant governance levels". Stephens *et al.* (2012) suggest that effective communication is a matter of getting the right balance between saliency, richness and robustness for different audiences.

There is a need to discourage over-reliance on traditional physical protection measures like flood barriers, with broader flood resilience being encouraged (Schelfaut *et al.*, 2011; Djordjević *et al.*, 2011). There is always the chance that physical barriers will be over-topped or fail, with potentially disastrous consequences, as when hurricane Katrina hit New Orleans in 2005 (Hallegatte, 2009). Government-implementation of large-scale technical measures can also limit or discourage alternative softer adaptation measures. For example, raising dykes, with continued development of the land behind them, reduces space for managed/temporary inundation (Haasnoot *et al.*, 2011) and proposed desalination plants and pipelines in Melbourne could discourage use-reduction, grey-water recycling and rainwater harvesting (Barnett and O'Neill, 2010). Additionally, some factors can be both barriers and incentives for adaptation. For example, flood insurance can encourage complacency and continued occupation/development of flood plains, as risk is transferred to the insurer, but could also provide financial incentives to householders to reduce their own flood risk (Wilby and Keenan, 2012).

1.2.8 Concluding remarks

The most recent IPCC report has concluded that the evidence of warming of the Earth in recent decades is unequivocal and global temperature change is likely to exceed 2 °C (IPCC, 2013). The observed warming has been associated with decreasing snow and ice cover and increasing sea levels, atmospheric humidity and rainfall intensities. In the future, extreme precipitation events are very likely to become more intense and frequent in most mid-latitude land masses and wet tropical regions. Rainfall droughts and water scarcity are likely to become more intense, as dry areas get drier and wet areas wetter. The responses of river flows are, however, more complex, these systems are very dominated by man's activities, both in terms of river impoundments and extractions and change in land use. There is also a considerable uncertainty in the response of evaporation (and hence soil moisture and river flow) to the 'fertilization effect' of increasing CO_2 concentrations (Bell *et al.*, 2011, Kay *et al.*, 2013, Prudhomme *et al.*, 2013b).

Society and business must be resilient against both climate variability and possible future climate change. Adaptation actions should be robust, flexible and/or reversible and, if possible, they should be "no-regret" actions and offer some headroom or safety margin. The high levels of uncertainty in modelling the effects of climate change on both future floods and droughts means that methods are needed for decision-making with uncertainty, for example including strategies that are 'win-win' (i.e. bring benefits even without climate change), soft rather than hard (e.g. financial rather than technical), reversible/flexible and incorporate safety margins. A particular difficulty for adaptation is distilling the complexities of the science and uncertainties into information more directly usable by decision-makers. Stakeholder participation is generally recommended during the formulation of climate change adaptation plans and has been found to enhance effectiveness. But communication remains a difficulty and must be treated with care. Many of these issues will be dealt with in more detail in later chapters.

Finally, it is clear we are living in a changed and changing world. While it is difficult to attribute an extreme event to climate change, every weather and hydrological event will be modified to an extent by increasing temperature, humidity and sea level. The question, therefore, really is: has the chance of the event been pushed over a threshold where it becomes a serious concern to society and how should society respond to these changes and the associated uncertainties?

References

Adam J.C. and Lettenmaier D.P. (2008) Application of new precipitation and reconstructed streamflow products to streamflow trend attribution in northern Eurasia. *Journal of Climate*, **21**, 1807–1828.

Allan R.P., Soden B.J., John V.O. *et al.* (2010) Current changes in tropical precipitation. *Environmental Research Letters*, **5**, 025205, doi:10.1088/1748-9326/5/2/025205.

Barnett J. and O'Neill S. (2010) Maladaptation. *Global Environmental Change*, **20**, 211–213.

Bell V.A., Kay A.L., Cole S.J. *et al.* (2012) How might climate change affect river flows across the Thames Basin? An area-wide analysis using the UKCP09 Regional Climate Model ensemble. *Journal of Hydrology*, **442–443**, 89–104, doi:10.1016/j.jhydrol.2012.04.001.

Bell, V.A., Gedney, N., Kay, A.L. *et al.* (2011) Estimating potential evaporation from vegetated surfaces for water management impact assessments using climate model output. *Journal of Hydrometeorology*, **12**, 1127–1136, doi:10.1175/2011JHM1379.1.

Beven K. (2011) I believe in climate change but how precautionary do we need to be in planning for the future? *Hydrological Processes*, **9**, 1517–1520.

Chen J., Brissette F.P., Poulin A. and Leconte (2011) Overall uncertainty study of the hydrological impacts of climate change for a Canadian watershed. *Water Resources Research*, **47**, W12509, doi: 10.1029/2011WR010602.

Christierson B.V., Vidal J. and Wade S.D. (2012) Using UKCP09 probabilistic climate information for UK water resource planning. *Journal of Hydrology*, **424–425**, 48–67.

Dai A.G., Trenberth K.E. and Qian T.T. (2004) A global dataset of Palmer Drought Severity Index for 1870–2002: Relationship with soil moisture and effects of surface warming. *Journal of Hydrometeorology,* **5**(6), 1117–1130.

Dai A., Qian T.T., Trenberth K.E. and Milliman J.D. (2009) Changes in Continental Freshwater Discharge from 1948 to 2004. *Journal of Climate*, **22**, 2773–2792.

Defra (2006) FCDPAG3 Economic Appraisal: Supplementary Note to Operating Authorities – Climate Change Impacts, October 2006, Department for Environment, Food and Rural Affairs, London, UK.

Delgado J.M., Apel H. and Merz B. (2010) Flood trends and variability in the Mekong river. *Hydrology and Earth System Sciences*, **14**, 407–418.

Deser C., Knutti R., Solomon S. and Phillips A.S. (2012) Communication of the role of natural variability in future North American climate. *Nature Climate Change*, **2**, 775–779.

Djordjević S., Butler D., Gourbesville P. *et al.* (2011) New policies to deal with climate change and other drivers impacting on resilience to flooding in urban areas: the CORFU approach. *Environmental Science and Policy*, **14**, 864–873.

Environment Agency (2011) *Adapting to Climate Change: Advice for Flood and Coastal Erosion Risk Management Authorities*. Environment Agency, September **2011**, 29 pp.

Groisman P., Knight R., Karl T.R. *et al.* (2004) Contemporary changes of the hydrological cycle over the contiguous United States: trends derived from in situ observations. *Journal of Hydrometeorology*, **5**, 64–85.

Haasnoot M., Middelkoop H., van Beek E. and van Deursen W.P.A. (2011) A method to develop sustainable water management strategies for an uncertain future. *Sustainable Development*, **19**, 369–381.

Hallegatte A. (2009) Strategies to adapt to an uncertain climate change. *Global Environmental Change*, **19**, 240–247.

Hallegatte S., Green C., Nicholls R.J. and Corfee-Morlot J. (2013) Future Flood losses in major coastal cites. *Nature Climate Change,* **3**, 802–806.

Hannaford, J. and Marsh, T.J. (2008) High-flow and flood trends in a network of undisturbed catchments in the UK. *International Journal of Climatology*, **28** (10), 1325–1338. 10.1002/joc.1643.

Hannaford, J., Lloyd-Hughes, B., Keef, C. *et al.* (2011) Examining the large-scale spatial coherence of European drought using regional indicators of rainfall and streamflow deficit. *Hydrological Processes*, **25** (7), 1146–1162. 10.1002/hyp.7644.

Hunt A. and Watkiss P. (2011) Climate change impacts and adaptation in cities: a review of the literature. *Climatic Change*, **104**, 13–49.

Huntjens P., Lebel L., Pahl-Wostl C. *et al.* (2012) Institutional design propositions for the governance of adaptation to climate change in the water sector. *Global Environmental Change*, **22**, 67–81.

IPCC (2007) *Climate Change 2007: Impacts, Adaptation and Vulnerability. Contribution of Working Group II to the Fourth Assessment Report of the Intergovernmental Panel on Climate Change,* (eds M.L. Parry, O.F. Canziani, J.P. Palutikof *et al.*), Cambridge University Press, Cambridge, UK, 7–22.

IPCC (2012) *Managing the Risks of Extreme Events and Disasters to Advance Climate Change Adaptation. A Special Report of Working Groups I and II of the Intergovernmental Panel on Climate Change* (eds C.B. Field , V. Barros, T.F. Stocker *et al.*) Cambridge University Press, Cambridge, UK, and New York, NY, USA, 582 pp.

IPCC (2013) *Working group 1 Climate Change 2013: the physical science basis. Contribution to the IPCC fifth assessment report.*

Jiang T., Kundzewicz Z.W. and Su B. (2008) Changes in monthly precipitation and flood hazard in the Yangtze River Basin, China. *International Journal of Climatology*, **28**, 1471–1481.

Kay A.L., Davies H.N., Bell V.A. and Jones R.G. (2009) Comparison of uncertainty sources for climate change impacts: flood frequency in England. *Climatic Change*, **92**, 41–63.

Kay A.L., Crooks S.M., Pall P. and Stone D. (2011a) Attribution of Autumn/Winter 2000 flood risk in England to anthropogenic climate change: a catchment-based study. *Journal of Hydrology*, **406**, 97–112, doi:10.1016/j.jhydrol.2011.06.006.

Kay A.L., Crooks S., Davies H.N., Prudhomme C. and Reynard N.S. (2011b) *Practicalities for implementing regionalised allowances for climate change on flood flows*. Report to Department for Environment, Food and Rural Affairs, Technical Report FD2648, CEH Wallingford, May 2011, 209 pp.

Kay A.L. and Jones D.A. (2012a) Transient changes in flood frequency and timing in Britain under potential projections of climate change. *International Journal of Climatology*, **32**(4), 489–502, doi:10.1002/joc.2288.

Kay A.L. and Jones R.G. (2012b) Comparison of the use of alternative UKCP09 products for modelling the impacts of climate change on flood frequency. *Climatic Change*, **14**(2), 211–230, doi:10.1007/s10584-011-0395-z.

Kay A.L., Bell V.A., Blyth E.M. *et al.* (2013) A hydrological perspective on evaporation: historical trends and future projections in Britain. *Journal of Water and Climate Change*, doi:10.2166/wcc.2013.014.

Kay A.L., Crooks S.M., Davies H.N. *et al.* (2014a) Probabilistic impacts of climate change on flood frequency using response surfaces. I: England and Wales. *Regional Environmental Change*, **14**(3), 1215–1227, doi:10.1007/s10113-013-0563-y.

Kay A.L., Crooks S.M., Davies H.N. and Reynard N.S. (2014b). Probabilistic impacts of climate change on flood frequency using response surfaces. II: Scotland. *Regional Environmental Change,* **14**(3), 1243–1255, doi:10.1007/s10113-013-0564-x.

Kundzewicz Z. W., *et al.* (2007). Freshwater resources and their management. in *Climate Change 2007: Impacts, Adaptation and Vulnerability. Contribution of Working Group II to the Fourth Assessment Report of the Intergovernmental Panel on Climate Change*, Cambridge University Press.

Kunkel K. E., Palecki M.A., Ensor L. *et al.* (2009) Trends in twentieth-century US extreme snowfall seasons. *Journal of Climate*, **22**, 6204–6216.

Maslin M. and Austin P. (2012) Climate models at their limit? *Nature*, **486**, 183–184.

Min S.-K., Zhang X., Zwiers F.W. and Hegerl G.C. (2011) Human contribution to more intense precipitation extremes. *Nature*, **470**(7334), 378–381.

Moss R.H., Edmonds, J.A., Hibbard, K.A. *et al.* (2010) The next generation of scenarios for climate change research and assessment. *Nature,* **463,** 747–756.

Murphy J.M., Sexton D.M.H., Jenkins G.J. *et al.* (2009) UK Climate Projections Science Report: Climate change projections. Met Office Hadley Centre, Exeter.

Osborne T. and Hulme M. (2004) Evidence for trends in heavy rainfall events over the UK. *Philosophical Transactions of the Royal Society London, Series A*, **360**, 1–13.

Pall, P., Aina, T., Stone, D.A. *et al.* (2011) Anthropogenic greenhouse gas contribution to flood risk in England and Wales in autumn 2000. *Nature* **470**, 382–385.

Parry, S., Hannaford, J., Prudhomme C. *et al.* (2011) *Objective drought and high flow catalogues for Europe – WATCH Technical report No 33, CEH Wallingford*, 57 pp, http://www.eu-watch.org/publications/technical-reports/2.

Peterson B.J., Holmes R.M., McClelland J.W. *et al.* (2002) Increasing river discharge to the Arctic Ocean. *Science,* **298**, 2171–2173.

Petrow T. and Merz B. (2009) Trends in flood magnitude, frequency and seasonality in Germany in the period 1951–2002. *Journal of Hydrology*, **371**, 129–141.

Piao S., Ciais, P., Huang, Y. *et al.* (2010) The impacts of climate change on water resources and agriculture in China. *Nature*, **467**, 43–51.

Pinskwar I., Kundzewicz Z.W., Peduzzi P. *et al.* (2012) Changing floods in Europe. *Changes in Flood Risk in Europe*, Special Publication No. 10 (ed. Z.W. Kundzewicz) IAHS Press, Wallingford, Oxfordshire, UK.

Prudhomme C. and Davies H.N. (2009) Assessing uncertainties in climate change impact analyses on river flow regimes in the UK. Part 2: future climate. *Climate Change*, **93**, 197–222, doi: 10.1007/s10584-008-9461-6.

Prudhomme C., Wilby R.L., Crooks S., Kay A.L. and Reynard N.S. (2010) Scenario-neutral approach to climate change impact studies: Application to flood risk. *Journal of Hydrology*, **390**, 198–209, doi:10.1016/j.jhydrol.2010.06.043.

Prudhomme C., Crooks S., Kay A.L. and Reynard N.S. (2013a) Climate change and river flooding: Part 1 Classifying the sensitivity of British catchments. *Climatic Change*, **119**(3–4), 933–948, doi:10.1007/s10584-013-0748-x.

Prudhomme C., Giuntoli I., Robinson E.L. *et al.* (2013b) Hydrological droughts in the 21st century, hotspots and uncertainties from a global multimodel ensemble experiment. *Proceedings of the National Academy of Sciences*. Doi: 10.1073/pnas.1222473110.

Prudhomme C., Kay A.L., Crooks S. and Reynard N.S. (2013c) Climate change and river flooding: Part 2 Sensitivity characterisation for British catchments and example vulnerability assessments. *Climatic Change*, **119**(3–4), 949–964, doi:10.1007/s10584-013-0726-3.

Quevauviller Ph. (2011) Adapting to climate change: reducing water-related risks in Europe. *Environmental Science and Policy*, **14**, 719–720.

Renard B., Lang, M., Bois, P. *et al.* (2008) Regional methods for trend detection: Assessing field significance and regional consistency. *Water Resources Research*, **44**.

Reynard N.S., Crooks S., Kay A.L. and Prudhomme C. (2009) *Regionalised impacts of climate change on flood flows*. Report to Department for Environment, Food and Rural Affairs, Technical Report FD2020, CEH Wallingford, November 2009, 113 pp.

Schelfaut K., Pannemans B., can der Craats I., Krywkow J., Mysaik J. and Cools J. (2011) Bringing flood resilience into practice: the FREEMAN project. *Environmental Science and Policy*, **14**, 825–833.

Serquet G., Marty C., Dulex J.P. and Rebetez M. (2011) Seasonal trends and temperature dependence of the snowfall/precipitation-day ratio in Switzerland. *Geophysical Research Letters*, **38,** L07703, doi:10.1029/2011GL046976.

Sheffield J., Wood E.F. and Roderick M.L. (2012) Little change in global drought over the past 60 years. *Nature,* **491**, 435–438.

Shiklomanov A.I., Lammers R.B., Rawlins M.A. *et al.* (2007) Temporal and spatial variations in maximum river discharge from a new Russian data set. *Journal of Geophysical Research-Biogeosciences*, **112**, G04S53, doi:c10.1029/2006JG000352.

Smith L.C., Pavelsky T., MacDonald G. *et al.* (2007) Rising minimum daily flows in northern Eurasian rivers suggest a growing influence of groundwater in the high-latitude water cycle. *Journal of Geophysical Research*, doi:doi:10.1029/2006JG000327.

St. Jacques J.-M. and Sauchyn D. (2009) Increasing winter baseflow and mean annual streamflow from possible permafrost thawing in the Northwest Territories, Canada. *Geophys. Res. Lett.*, **36,** L01401, doi:10.1029/2008GL035822.

Stahl K., Hisdel, H., Hannaford, J. *et al.* (2010) Streamflow trends in Europe: evidence from a dataset of near-natural catchments. *Hydrology and Earth System Sciences*, **14**, 2367–2382.

Stahl K., Tallaksen L.M., Gudmundsson L. and Christensen J.H. (2011) Streamflow data from small basins: A challenging test to high-resolution regional climate modeling. *J. Hydrometeor.*, **12**, 900–912.

Stahl K. and Tallaksen L.M. (2012) Filling the white space on maps of European runoff trends: estimates from a multi-model ensemble. *Hydrology and Earth System Sciences Discussions*, **9**, 2005–2032.

Stephens E.M., Edwards T.L. and Demeritt D. (2012) Communicating probabilistic information from climate model ensembles – lessons from numerical weather prediction. *WIREs Climate Change*, **3**, 409–426.

Taylor, K.E., Stouffer R.J. and Meehl G.A. (2012) An Overview of CMIP5 and the Experiment Design. *Bulletin of the American Meteorological Society*, **93**, 485–498.

Tompkins E.L., Adger W.N., Boyd E. *et al.* (2010) Observed adaptation to climate change: UK evidence of transition to a well-adapting society. *Global Environmental Change*, **20**, 627–635.

Trenberth K. and Fasullo J. (2012) Climate extremes and climate change: The Russian heatwave and other climate extremes of 2010. *Journal of Geophysical Research–Atmospheres*, **117,** D17103, doi:10.1029/2012JD018020.

van Pelt S.C. and Swart R.J. (2011) Climate change risk management in transnational river basins: the Rhine. *Water Resources Management*, **25**, 3837–3861.

Wilby R.L. and Harris I. (2006) A framework for assessing uncertainties in climate change impacts: Low-flow scenarios for the River Thames, UK. *Water Resources Research*, **42**, W02419, doi:10.1029/2005WR004065.

Wilby R.L. and Dessai S. (2010) Robust adaptation to climate change. *Weather*, **65**, 180–185.

Wilby R.L. and Keenan R. (2012) Adapting to flood risk under climate change. *Progress in Physical Geography*, **36**, 348–378.

Willett K.M., Jones P.D., Gillett N.P. and Thorne P.W. (2008) Recent Changes in Surface Humidity: Development of the HadCRUH Dataset. *Journal of Climate*, **21**, 5364–5383.

Zhang X., Zwiers, F.W., Hegerl, G.C. *et al.* (2007) Detection of human influence on twentieth-century precipitation trends. *Nature*, **448**, 461–464.

Zolina O., Simmer C., Belyaev K. *et al.* (2009) Improving estimates of heavy and extreme precipitation using daily records from European rain gauges. *Journal of Hydrometeorology*, **10**, 701–716.

1.3
Features Common to Different Hydrometeorological Events and Knowledge Integration

Barbara Zanuttigh

Department of Civil, Chemical, Environmental and Materials Engineering, University of Bologna, Bologna, Italy

1.3.1 Introduction

Hydrometeorological events are among the most common sources of natural disasters worldwide. The frequency of hydrometerological events and associated disasters is increasing as a result of many issues related both to the physical processes per se and to the social, economic, and environmental context. The aim of this section is to draw an overview of the features common to different hydrometeorological events, highlighting the key challenges for an effective implementation of risk reduction measures. These challenges include:

- integration between disaster risk management and climate change adaptation, i.e. integration of the long-term scales related to climate change effects in planning risk management strategies;
- prediction capabilities, specifically related to the inherent uncertainty of extremes, increased by the variability posed by climate change, and to the synergies among hazard sources, calling for a multi-scale and multi-hazard approach;
- set-up of a holistic approach to better understand the system exposed to the hazard, where society, economy and environment dynamically interact, affecting the overall system vulnerability in a highly nonlinear way;
- make the resilience concept operational through the integration of practical approaches to strength the reactive resilience and promote the proactive resilience of the system;

Hydrometeorological Hazards: Interfacing Science and Policy, First Edition. Edited by Philippe Quevauviller.

- increase awareness and capacity building through lessons learnt from recent experiences and through the development of a participatory approach at different levels and involving all the different actors: managers, scientists, public and policy makers;
- definition of a cascade approach from national to regional and local with identification of actors in charge in order to clarify the responsibility chain in disaster risk management;
- people-centred risk communication, to boost successful warning systems and/or reactive resilience.

Based on the synthetic concepts drawn around the challenges listed above, a roadmap towards a sustainable future is finally proposed.

1.3.2 Extreme hydrometeorological events and disasters: An increasing trend

Over the last decade the number of natural disasters, compared to 1950–1959, has increased ninefold (EM-DAT, 2011). This is generally attributed to an increase in hydrometeorological disasters (Figure 1.3.1). On average, hydrometeorological disasters account for at least 50% of the overall annual damage over all continents apart from Asia (Figure 1.3.2). Interestingly, the number of deaths attributable to natural disasters is decreasing and this may be ascribed to an improvement in both warning system technology and disaster management. On the other hand the overall number of people actually affected is increasing, together with the number of events.

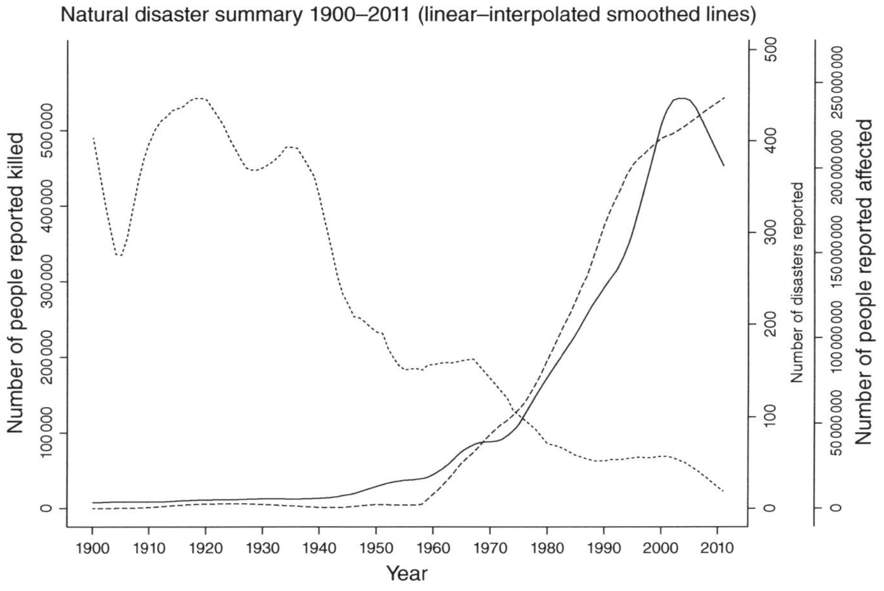

Figure 1.3.1 Trends of disasters and affected people in the period 1990–2011. *Source*: From EM-Data: The OFDA/CRED International database – www.emdat.be. See plate section for colour representation of this figure.

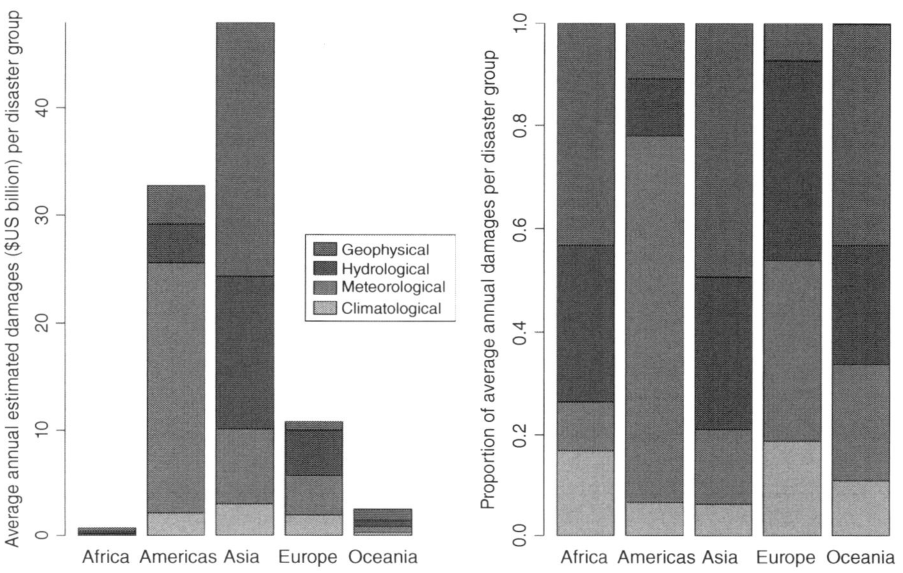

Figure 1.3.2 Average annual damage ($US billion) caused by reported natural disasters in the period 1990–2011. From EM-Data: The OFDA/CRED International database – www.emdat.be. See plate section for colour representation of this figure.

It is not possible to attribute the higher number of disasters, losses and people affected simply to improved reporting methods and media coverage. The increase in the total number of disasters is influenced by several factors. These can be divided into factors that result in greater vulnerability and those that contribute to an increase in the frequency of hazardous events.

There are several reasons for increased vulnerability. One is the growth in population, which has doubled in size from 3 to 6 billion over the last 50 years (World Bank, 2010). World population is expected to reach between 8 and 11 billion by 2050 (UNPD, 2010a) if estimated growth rates are correct. The increase in disaster impact is, however, greater than the increase in population. This indicates that there are other important factors in the equation that tend to increase overall vulnerability. The rate of urbanization is one of them. The percentage rate of worldwide urbanization has increased from 29% in 1950 to 50% in 2010 and is expected to increase to 69% in 2050 according to UN estimates (UNPD, 2010b). Population growth has also resulted in some areas becoming urbanized that were previously considered to be too much at risk from natural hazards. Today, many of the world's largest cities, that have a population of 1 million or more, are situated in either coastal or seismically active hazardous regions (Smith and Petley, 2008; Kraas, 2008).

The development of highly sensitive technologies, and therefore the increasing susceptibility of modern industrial societies to breakdowns in their infrastructure, also results in an increase in the effect that natural disasters have on society. Economic losses have increased eightfold over the last half century and insured losses by a factor of 15 (Munich Re, 2010). The losses in developed countries due to windstorms (US$90 billion) and earthquakes (US$25 billion) account for most of the rapid increase in insured losses. Insured losses resulting from flooding are surprisingly lower (in the region of US$10 billion) since they are most severe in developing countries which have lower insurance coverage (Munich Re, 2010).

In the case of disasters there is an inverse relationship between the level of development and the loss of human lives. It is estimated that approximately 85% of causalities resulting from natural disasters are from less developed countries that are home to more than 4.7 billion people. This can be attributed to several factors, such as lower quality constructions, the absence of building regulations (or the fact that they go unenforced), the construction of buildings in hazardous areas due to the absence of land use planning, lower awareness and disaster preparedness, less precise or the absence of early warning systems, the lack of evacuation planning and the lack of search-and-rescue and medical facilities.

Even if high income countries (GNI >12 000 US$ per capita) account for 65% of the overall losses due to natural disasters and low income countries (GNI < 1000 US$ per capita) account for only 3% of overall losses (World Bank, 2010), the impact on low income countries is far more serious as it may represent as much as 100% of their GNI (UNISDR, 2009a). In absolute terms, economic losses increase with the level of development. This is because the absolute value of elements-at-risk that could potentially be subject to damage during a disaster increases along with an increasing level of development. In relative terms, however, measured as a percentage of GDP, the trend is the opposite (Munich Re, 2010).

The increase in the number of natural disasters cannot, however, be explained merely by the increased exposure of the population to hazards. There is also an increase in the frequency of destructive events related to atmospheric extremes, such as floods, drought, cyclones, and landslides (EM-DAT, 2009). Over the last decade 3750 windstorms and floods were recorded, which accounted for two-thirds of all events. During the same period the number of earthquake and volcanic activity related disasters (approximately 100 per year) remained the same (Munich Re, 2010). The data seem to indicate that climate change is related to the increase in natural disasters, even if the sample period is still insufficient for it to be demonstrated with absolute confidence.

1.3.3 Integrating disaster risk management and climate change adaptation

Disasters are the result of the interaction between extreme hazardous events and vulnerable societies but disaster risk management can help reduce disaster impact. Disaster risk management can be defined as "the systematic process of using administrative decisions, organisation, operational skills and capacities to implement policies, strategies and coping capacities of the society and communities to lessen the impacts of natural hazards and related environmental and technological disasters" (UNISDR, 2009b). This comprises all forms of activities, including structural and non-structural measures to avoid (prevention) or to limit (mitigation and preparedness) adverse effects of hazards (UNISDR, 2004). Disaster risk management is aimed at disaster risk reduction, which refers to the conceptual framework of elements considered with the possibilities to minimize vulnerabilities and disaster risks within the broad context of sustainable development (UNISDR, 2004).

Over recent decades there has been a change in emphasis from 'disaster recovery and response' to 'risk management and mitigation'. The previous approach focused mainly on the hazard as being the principal cause of risk and on risk reduction by the use of physical protection measures. The newer approach concentrated on the vulnerability of communities and vulnerability reduction via preparedness and early warning systems. Later, more attention was given to the capacities of local communities and local coping strategies (Blaikie *et al.*, 1994; Pelling, 2003). The 1994 Yokohama conference brought the socio-economic aspects as a component of effective disaster prevention into the equation. It was acknowledged that social factors, such as tradition, religion, economic standing and faith in political accountability are essential elements when determining societal vulnerability. These factors need to be addressed in order to reduce the vulnerability of societies and thereby decrease the effects of natural disasters (Hillhorst, 2004). However, only local actors (public, managers) may be able to provide the necessary knowledge and understanding of local conditions that enable the accurate assessment of socio-economic factors.

The need to move away from a top-down approach for disaster management and a cycle that emphasizes reconstruction and preparedness towards a more all-embracing approach that attempts to circumvent or reduce risk prior to disasters, whilst at the same time promoting greater awareness, greater public commitment, greater knowledge-sharing and establishing partnerships in order to apply various risk reduction strategies at all levels (UNISDR, 2005) is stressed both in the 1990–2000 International Decade for Natural Disaster Reduction (IDNDR) and the subsequent International Strategy for Disaster Reduction (ISDR). This concept has been called the 'risk management cycle', or 'spiral' and is considered more practical in as much that lessons learnt from a natural disaster can encourage improvements in development planning as opposed to limiting work to the re-establishment of social conditions and the reconstruction of the built environment.

In Figure 1.3.3 the disaster cycle, together with its various components, is shown. Originally, disaster relief, recovery and reconstruction were given greater importance. This generated a mechanism in which a successive disaster would be likely to cause the same effects or even worse. More recently increased attention has been given to disaster preparedness by developing warning systems and disaster awareness programmes. Endeavours are now aiming towards disaster prevention and preparedness, thereby increasing the length of time between disasters and attenuating their effects, hence giving less importance to relief, recovery and reconstruction. The objective of disaster risk management is to lengthen this cycle and to reach the response phase only for extreme events which have a very low incidence rate.

The ultimate target is the reduction of disaster risk. This should ideally be achieved by combining both structural and non-structural procedures that centre on emergency preparedness (increased awareness, early warning systems), the incorporation of risk information into long-term land use planning and the assessment of the most cost-effective risk reduction measures. Because hazards are spatially distributed, as well as the vulnerable elements-at-risk, spatial information plays a pivotal role in the overall concept of risk management.The integration of disaster risk reduction and climate change

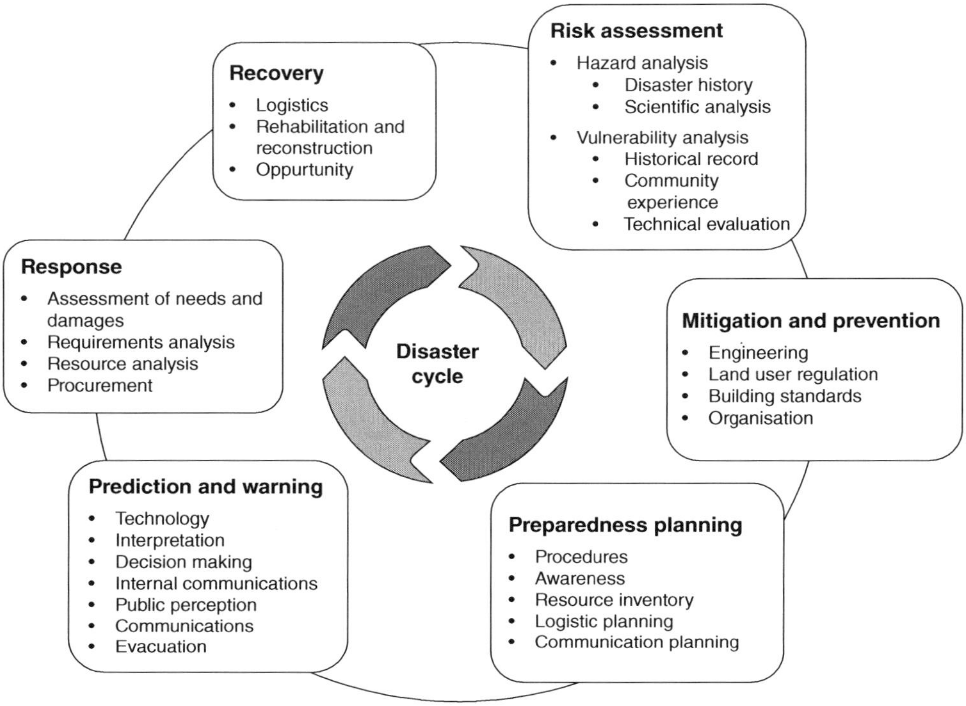

Figure 1.3.3 The "traditional" disaster cycle based on UNISDR (2009c) and IPCC (2012).

adaptation has often been difficult or impractical because they conventionally operate at different spatial scales (Birkmann and von Teichman, 2010). Disasters are usually considered to be events that occur at a particular location. Climatic change, on the other hand, is more likely to be considered as being a global or regional phenomenon. However, because the necessity for locally based climate change adaptation becomes increasingly evident (Adger, Arnell and Tompkins, 2005) and as the impact of disasters on a local scale is now acknowledged to have a more widespread impact at a larger spatial scale, this opinion is now being reconsidered. One obstacle to integration is that climate change projections fail to provide accurate local information and that adaptation strategies are generally designed for entire countries or regions (German Federal Government, 2008; Red Cross and Red Crescent Climate Centre, 2007).

The temporal scales of climate change adaptation and disaster risk reduction are also recognizably different. Traditionally, focus from the disaster community has been on relatively short term humanitarian aid, including relief and reconstruction (UNISDR, 2009c). It has been acknowledged, on the other hand, that climatic change includes long-term processes with projections extending from decades to centuries. This presents difficulties for development communities that generally focus on a shorter time span.

These temporal-scale challenges could be detected, addressed and overcome if there were greater collaboration and integration between disaster risk reduction and the climate change adaptation groups.

Fundamentally this involves an improvement in risk recognition relating to climate-related disasters in climate change adaptation and the integration of longer-term climate change risk factors into disaster risk reduction. Risk management objectives common to adaptation programmes include the:

- maintenance of functioning and healthy ecosystems,
- reduction of exposure and vulnerability of the built environment,
- strengthening of governance frameworks for climate change adaptation,
- maintenance of livelihood opportunities and diversification of options,
- reduction of human health and safety risks.

In the past many of these adaptation measures were thought of in terms of a possible contribution to achieving management objectives. The application of a climate lens means that adaptation measures are seen in terms of how they reduce the impacts and/or improve the resilience of communities and ecosystems with regards to climate change and variability, whilst at the same time promoting management objectives. This basically means taking into consideration a longer timescale and a wider range of possible variability during the planning stage.

The adaptation must be customised into a local context. This can be achieved through a comprehensive process that combines the development objectives with the issues of climatic change, whilst taking into consideration both the technical capabilities and ability of institutions together with the local community. The fundamental criteria used to decide on the best adaptation option for a specific local context are:

- Technical success: will the adaptation option be able to solve problems arising from climate change whilst at the same time fulfilling modern development or management objectives?
- Cost: what are the costs involved in implementing the adaptation measures?
- Benefits: what are the types and extent of benefits that the adaptation measure will generate and who will benefit from them?
- Implementation issues: what are the difficulties involved in the design and implementation of the option in terms of required skill level, information required and size of implementation?

Often, adaptation responses will include 'packages' of adaptation measures and, therefore, necessitate further thought when assessing multiple measures. The following may be taken into consideration when assessing and selecting the best combination of adaptation measures:

- Complementary versus substitute packages: packages should preferably include adaptation measures that are complementary. This exploits any potential synergies there may be and guarantees that some advantage is gained from every measure. If adaptation measures are interchangeable then the least costly and/or the measure that provides the most significant net benefits should be included in the package.
- Budget limitations: if there is limited funding to cover implementation costs, no-cost and low-cost measures may be the best or only option.

Unfortunately, resistance to changes in policy are not infrequent. Communities and institutions often do not feel the need for taking immediate action, especially because impacts due to climate change are cumulative and occur over long time spans. Changes to policy may be delayed because of uncertainties surrounding climate change. There are also, understandably, different approaches to the risks connected with taking action, simply because of the fact that there are uncertainties.

1.3.4 Predicting disasters: Dealing with uncertainties and scales

Extreme events pose a particular set of challenges for implementing probabilistic approaches because their relative infrequency often makes it difficult to obtain adequate data for estimating the probabilities and consequences. Even if there are statistical methods available that can estimate frequencies that are longer than the available data time series (Milly *et al.*, 2002), estimating the frequency of extreme events with any degree of reliability can be problematical or impossible merely because of their infrequency.

In the past, hazard forecasting and risk assessment systems focused on modelling atmospheric phenomenon, such as sea level, rainfall, waves, river discharge, flash flooding and so on, as separate entities. Each forecasting system consists of a linear data flow and a combination of distinct models, for example: a meteorological model which provides forcing to a shallow water flow model for the purpose of forecasting tides and surges; a wave model for predicting near-shore waves and additional models to predict subsequent overtopping and/or flooding; river flow models for forecasting pluvial and fluvial flooding. These flood modelling systems present several weaknesses (or limitations):

- the absence of inter-operability between components,
- only a single source of hazard tends to be taken into consideration,
- the absence of ensemble or data-assimilation techniques,
- the absence of tracking estimation errors applying full uncertainty analysis in order to improve the predictability skill of the probabilistic system,
- the necessity for constraining uncertainties and narrowing prediction bounds with model refinement,
- the assessment of the potential associated risk is on many occasions too simplistic and limited with respect to vulnerability and resilience,
- they assume historic/static data relating to the condition regarding landforms and defences.

In coupled models, cascading forecast uncertainty is an essential development towards improving the quality of hydrological forecasts (Cloke and Pappenberger, 2009). There is still, however, some debate as to the best methodology to quantify the total predictive uncertainty in hydrology (Beven, Smith anf Freer, 2008). There are many sources of uncertainty in the hydrometeorological forecast chain, including: the meteorological forcing, correction and downscaling procedure for meteorological predictions, previous system conditions, the observation of hydrological and meteorological networks, data assimilation methods such as discharge and soil moisture, system geometry (including flood defence structures), the likelihood of infrastructure failure (dykes or backing-up of drains), system characteristics (model parameters), and the limitations of the hydrological model to completely represent processes (for example flow processes in flood generation and routing). The importance of each individual element varies over time, depending on the dominant flow regime, and in space, due to each catchment being unique (Beven, 2000). It is also influenced by other factors, for example, the interaction between the space–time scales of the predicted event, the characteristics (area and response time) of the main catchment and the resolution of the meteorological forcing data (Thirel *et al.*, 2008). In order to track all sources of uncertainty and to evaluate both their relative influence and their total combined uncertainty, full uncertainty analysis is necessary (Pappenberger *et al.*, 2005). There is general agreement that the overall magnitude of uncertainty affects prediction quality, influences model output forecast interpretation and, ultimately, decision making (Ramos *et al.*, 2010).

A series of weather and climate forecasting systems, over recent years, have been progressively integrated into early warning systems (Glantz, 2003). This has led to an improvement in forecasting accuracy over a range of timescales and has played a role in reducing exposure to extreme climate events (Goddard *et al.*, 2009; van Aalst, 2009; Barnston *et al.*, 2010). Additionally, weather and climate information is more frequently being used for planning and climate risk management in business (Changnon and Changnon, 2010), food security (Verdin *et al.*, 2005) and health (Ceccato *et al.*, 2007; Degallier *et al.*, 2010), as well as the technology for the development of decision support tools for climate-related disaster management (van de Walle and Turoff, 2007).

Both timing and timescales are important subjects that need to be investigated further in relation to the identification and management of extreme climate and weather events, natural disasters, and adaptation strategies. The main problem when dealing with timing and timescales is that different hazards and their relative intervals of recurrence might change significantly in terms of time. At the time of writing most climate change scenarios centre around climatic change over the next one or two centuries whilst vulnerability projections often just take present socio-economic data into consideration. A major challenge for enhancing knowledge of exposure and vulnerability as key factors that determine risk requires improved data and techniques to project and identify directions and different development pathways in demographic, socioeconomic, and political trends with the same time horizon as the changes in the climate system related to physical-biogeochemical projections (Birkmann *et al.*, 2010).

Moreover, it has been demonstrated that the time dependence of risk analysis, especially if conducted at a specific point in time, is critical. More recent research emphasises that exposure, particularly that of different social groups, is a highly dynamic factor that can change not only seasonally, but also during the course of a day and even over different days of the week (Setiadi, 2011).

Additionally, variations in hazard frequency and timing during the year will have significant repercussions on the capacity of populations and ecosystems to deal with and adapt to the changes.

It is important to be aware, however, that disaster risk reduction considers risk reduction within different time frames; it incorporates short-term emergency management/response strategies and long-term risk reduction strategies, for example, building structures that resist 10 000-year earthquakes or flood barriers that resist 1000-year storm surges.

According to Turner *et al.* (2003a,b), vulnerability and resilience assessments need to take into consideration the influences on vulnerability from different scales. The practical application and subsequent analysis of these factors influencing vulnerability from different spatial scales is, however, a significant challenge and is generally not fully understood. Moreover, vulnerability analysis, especially when dealing with the identification of institutional vulnerability, must also take into consideration the various functional dimensions that natural hazards, climate change, and vulnerability as well as the administrative systems they operate on. As a general rule, current disaster management tools, measures of urban or spatial planning and water management tools (plans, zoning, norms) operate on different functional scales as compared to those of climate change. The various hazards that climate change may modify include different functional scales and cannot be sufficiently taken into consideration with only one approach. For example, because different institutions or departments normally take responsibility for policy and management of climate change and of disaster risk reduction, it becomes a challenge to develop a coherent and integrated strategy (Birkmann and von Teichman, 2010). As a consequence, it may be that mismatches in both functional and spatial scales are part of the institutional vulnerabilities that limit the capability of the governance system to respond adequately to hazards and changes brought about by climate changes.

1.3.5 Better understanding system exposure at the hazard

Vulnerability is both multi-dimensional and differential (Figure 1.3.4) meaning that it varies across physical space and among and within social groups; it is scale-dependent as regards space and units of analysis such as individual, household, region, or system; it is also dynamic because characteristics and driving forces transform over time. Because both vulnerability and exposure are not fixed an understanding of their trends is essential in order to reduce uncertainties.

The last decade has seen various attempts, methods and tools developed to comprehend, evaluate and manage risks (Wisner *et al.*, 2004; Birkmann, 2006, 2011a,b; the FP 6 FLOODsite project, www.floodsite.net). Progress has been made in the following:

- characterizing hazards and modelling their occurrence,
- assessing the vulnerability of the system affected (Vogel and O'Brien, 2004; Vogel *et al.*, 2007),
- determining long-term impacts by using scenario analyses of risk drivers (Thorne, Evans and Penning-Rowsell, 2007; Nicholls *et al.*, 2008).

If the correlation between hazards and their impact across both natural and human systems is to be understood, it is essential to establish a common view of the event under

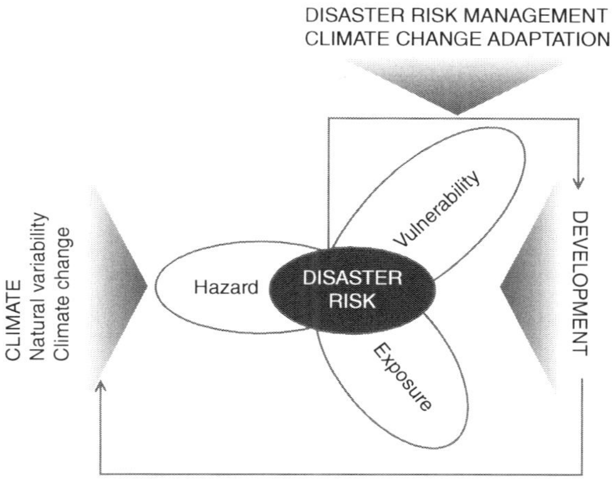

Figure 1.3.4 Illustration of the vulnerability concept, based on IPCC (2012).

investigation, especially where an objective decision-relevant assessment and response is required. A clear methodological approach, conceptual model and analytical framework are essential in order to develop such a view.

There are essentially the following four approaches to understanding vulnerability and its causes, embedded in political economy, social-ecology, vulnerability, disaster risk assessment, as well as adaptation to climate change.

1. The pressure and release model. This model is more widely used in social science-related vulnerability research. It puts greater emphasis on social conditions and principle causes of exposure giving rise to unsafe conditions, rather than the hazard. This approach connects vulnerability to unsafe conditions that associates local vulnerability to wider national and global shifts in the political economy of resources and political power. (Blaikie *et al.*, 1994; Wisner *et al.*, 2004).
2. The social ecology perspective. This emphasizes the necessity to concentrate on linked human-environmental systems (Turner *et al.*, 2003a,b). It gives importance to the ability of societies to transform nature and also the implications of environmental change for social and economic systems. The exposure and susceptibility of a system can only be sufficiently understood if these coupling processes and interactions are taken into consideration.
3. Holistic perspectives. These attempt to go beyond technical modelling to incorporate a wider and more comprehensive explanation of vulnerability. They differentiate between exposure, susceptibility and societal response capacities as causes or factors of vulnerability (Cardona, 2004; IDEA, 2005; Birkmann, 2006; Carreño, Cardona and Barbat, 2007). The feedback loop which underlines the fact that vulnerability is dynamic is a core of this type of approach and is the main driver and determinant of current or future risk.
4. Vulnerability definitions. Different vulnerability definitions and concepts have been developed and discussed in the context of climate change adaptation. One of the most famous definitions is the one reflected in the IPCC Fourth Assessment

Report. Here vulnerability is described as a function of exposure, sensitivity, and adaptive capacity (amongst others Füssel and Klein, 2006; Füssel, 2007; O'Brien *et al.*, 2008). This approach differs from the understanding of vulnerability from the disaster risk management point of view because it takes into account the rate and magnitude of climate change. Here, the concept of vulnerability includes external environmental factors of shock or stress. Therefore, this approach takes into consideration the magnitude and frequency of potential hazard events when dealing with vulnerability due to climate change. This approach also differs in that it addresses long-term trends and stresses, rather than current shock forecasting. This is something not explicitly excluded from disaster risk management approaches but is rarely considered.

The effectiveness of disaster risk management and adaptation to climate change is hindered by the lack of a comprehensive conceptual framework that facilitates common multidisciplinary risk evaluation (Cardona, 2004). The option for anticipatory disaster risk reduction and adaptation exists because risk is a latent condition that indicates potential future adverse effects (Lavell, 1999). If disaster risk management is understood as a social process it allows for a shift in emphasis away from responding to the disaster event towards an understanding of disaster risk (Cardona, 2011). This necessitates knowledge about how human interactions with the natural environment bring about new hazards, and how people, property, infrastructure, goods and the environment are exposed to potentially damaging events. Additionally, it requires a thorough understanding of the vulnerability of people and their livelihoods, including the allocation and distribution of social and economic resources that can work for or against the achievement of resistance, resilience and security (ICSU-LAC, 2010). Generally speaking, there is confidence that although hazard events are usually considered the cause of disaster risk, vulnerability and exposure are the key factors that determine it. Furthermore, unlike the hazard, vulnerability and exposure can frequently be influenced by policy and practice, even in the short to medium term. Disaster risk management and adaptation strategies have, therefore, to address mainly these same risk factors (Vogel and O'Brien, 2004; Birkmann, 2006; Leichenko and O'Brien, 2008).

One approach that is being increasingly used which helps to understand how different parts of a system influence one another within the whole is "systems thinking". This method views 'problems' as part of the overall way a system functions. It attempts to anticipate any unintended consequences by looking at the whole rather than reacting to the specific problem. The Source–Pathway–Receptor–Consequence (SPRC) approach supplies a simple, widely applicable conceptual model that has been widely used in flood risk projects (e.g. Thorne *et al.*, 2007). It can be applied at a relevant scale in order to provide insight into the way the flood system functions. The identification of relevant drivers of, and responses to changes which affect the consequences of extreme events for the flood system are linked to the use of the SPRC. Both changes in drivers and the identification of responses can affect flood system resilience; it is applicable to human, ecological and geomorphological aspects and processes. Recent events such as Hurricane Katrina have shown the value of such resilience and hence the interest in how to maintain and promote it (Heinz Center and CERES, 2009).

For both human populations and natural systems, physical exposure can have a considerable impact on vulnerability, but it is the lack of resilience that is frequently the most important factor that generates hotspots of human vulnerability when built-up areas are exposed to natural and anthropogenic hazard. Considering the recent and anticipated increases in damage from extreme events, the insurance industry is making greater use of catastrophe models. These include event generation (storm magnitude and frequency), hazard simulation (wind stresses, surge heights, waves), damage modelling (scale of structural damage), and financial modelling (costs). Further progress has been made in moving from the traditional cost-benefit analysis to assessments that combine monetary, social and natural science criteria. Different methodologies for risk assessments and risk indices have been recently reviewed. This has resulted in their merger and subsequent development for the integration of results (see e.g. the FP 7 THESEUS project, www.theseusproject.eu). Key research in the field of disaster risk indices has been conducted by the World Bank (natural disaster hotspots) and UNISDR (mortality index in GAR2009, natural disaster atlas, 2012 in cooperation with Maplecroft).

1.3.6 Resilience: From concept to operation

The UN International Strategy for Disaster Reduction (UNISDR), with particular reference to natural hazards, defines the term resilience as being "the capacity of a system, community or society to resist or to change in order that it may obtain an acceptable level in functioning and structure" (UNISDR, 2009b).

This is determined by the extent to which the social system is capable of organising itself and its ability to increase capacity for learning and adaptation, including the capability to recover from a disaster. In addition, it states that "the motivation to invest in disaster risk reduction is first and foremost a human, people-centred concern. It is about improving standards of safety and living conditions with an eye on protection from hazards to increase resilience of communities". It maintains that it should be possible to increase the resilience of communities to disasters and that current trends of environmental degradation and dealing with hazard management can be reversed through the adapted, sustainable and integrated management of natural resources, including reforestation projects, proper land use, and well-planned settlements.

Nevertheless, there is still an insistent commitment to using the concept and the general acceptance that resilience is good and must be encouraged remains unquestioned, regardless of the potential risks to society (cf. Handmer and Dovers, 1999; Adger, 2000). Even after many years of academic debate, the problem still remains of transforming the concept into an operational tool for policy and management purposes. Resilience needs to have a specific operational definition in order for it to become a tool for practical policy or management.

If resilience is to be considered a useful attribute of natural and social systems or not will depend on the definition of the concept. Even if traditional definitions that assume some equilibrium state are considered by some to be out-dated, they still attract considerable attention when resilience is mentioned. It is undisputable that systems are in a continuous state of change and that it is impossible to "bounce back" to the original state after a disaster. Moreover, once a system has been struck by a disaster, it follows that its original state was one in which it was vulnerable to the disaster. It would be undesirable, therefore, to return to the original as the system would be just as vulnerable as before. It is not surprising, therefore, that the most important development over recent years is the increasing recognition of the fact that human and ecological systems are interlinked. It is also recognised that their resilience relates to the way the systems function or interact, as opposed to the stability of their components or their capability to maintain or return to some equilibrium state. Subsequent definitions of resilience therefore focus on the functioning of systems, including their self-organising capacity. It is much more advantageous to interpret resilience in this way: it facilitates and contributes to the process of recovery after a disaster. A resilient system would therefore be less likely to suffer a severe long-term impact following a disaster. Nevertheless, this type of resilience does not contribute to the prevention of disasters or reduce their immediate impacts.

A recent interpretation of resilience includes the degree to which a system can develop and increase its learning and adaptation capacity (Carpenter *et al.*, 2001). The capacity for learning and adaptation is obviously a desirable attribute, even if few people would perhaps instinctively associate the capacity to increase it with resilience. This interpretation of resilience relates to adaptive management and adaptive capacity: two concepts that have their own literature and interpretations, but are perhaps more appropriate for policy and management.

Resilience of social-ecological systems consists of three distinct dimensions (Carpenter *et al.*, 2001):

- the level of disturbance a system can absorb whilst remaining within the same state or domain of attraction,
- the extent to which the system is capable of self-organisation,
- the extent to which the system can develop and increase the capacity for learning and adaptation.

The first two are what Handmer and Dovers (2007) refer to as "reactive resilience", which enables what is known in natural hazards literature as "coping" (e.g., Corbett, 1988) and what the climate change community calls "autonomous adaptation" (e.g., Carter *et al.*, 2007). Both attributes, to a greater or lesser degree, can be measured and monitored. There still remain, however, questions regarding the relationship between natural system and social system resilience that still have to be fully investigated.

The third attribute refers to "proactive resilience" (Handmer and Dovers, 2007), i.e. the adaptive capacity that includes the ability to prepare for and plan for hazards, as well as to the ability to implement technical measures before, during and after a hazard event. Adaptive capacity is defined as the ability to plan, prepare for, facilitate and implement adaptation options. Economic wealth, technology and infrastructure, information, knowledge, skills, the nature of institutions, commitment to equity and social capital are all factors that determine a country's or community's adaptive capacity to hazards (Smit *et al.*, 2001). Consequently, most industrialised countries have higher adaptive capacities than developing countries.

Proactive resilience connects the analysis of present and future hazardous conditions to the evaluation of specific strategies for improving the capacity for disaster prevention and preparedness. More importantly, it would provide hazard management groups with an instrument that is similar to resilience with regards to its relationship to vulnerability but which offers greater application potential. This is especially relevant when attempting to move away from disaster recovery to disaster prevention and preparedness. Hazard and risk reduction and adaptation strategies are effectively an on-going and iterative process that includes information development, awareness raising, planning, design, implementation and monitoring (Klein *et al.*, 2007).

Five generic approaches to anticipatory adaptation can be identified as follows:

- Increasing the ability of physical infrastructure to withstand the impacts. One approach, for example, would be to extend the "disturbance" range that a system can withstand; another would be to modify a system's tolerance to loss or failure.
- Increasing the flexibility of potentially vulnerable systems that are managed by humans. This could include allowing for mid-term adjustments in management practices, including changes in use or location.
- Enhancing the adaptability of vulnerable natural systems. This could involve reducing stresses due to hazardous effects, or removing barriers to the migration/displacements of vulnerable elements.
- Reversing trends that increase vulnerability. This could range from reducing human activity in vulnerable areas to preserving natural systems that protect against hazards.
- Improving public awareness and preparedness. This could include informing the public about the risks and possible consequences of hazards, as well as setting up early warning systems for extreme events.

To summarise, a society that relies on reactive resilience endeavours to strengthen existing conditions and make the current system resistant to change. A society that develops proactive resilience, on the other hand, accepts that change is inevitable and tries to create a system that is capable of adjusting to it.

1.3.7 Learning from experience

Previous experience with disaster risk management under uncertainty, or where long return periods for extreme events prevail, can inform effective risk reduction, response, and preparation, as well as disaster risk management strategies in general (Lonsdale *et al.*, 2008). Climate variability occurs over a broad range of timescales and therefore documentation is often available that details previous attempts to manage and adapt to climate-related risk that are relevant to risk management under climate change. These efforts provide a starting point for learning from the evaluation of responses, interventions and recovery from previous impacts. Even if efforts to incorporate learning into the management of weather and climate related risks have not always succeeded, such adaptive approaches provide a possible model for longer-term efforts. The process of learning from past experience encourages participation of a wide range of groups, such as institutions, the public, risk managers, scientists and policy makers.

Data relating to disasters, their impact on populations and their cost are essential for disaster risk management. A number of organisations now collect disaster related information, but on different scales and with different objectives.

The Centre for Research on the Epidemiology of Disasters (CRED) has been maintaining an Emergency Events Database (EM-DAT, 2009) since 1988. In order to be included in the EM-DAT database disasters have to meet certain criteria: they must have caused at least 10 casualties, 100 or more people should have been affected, a declaration of emergency should have been issued or there should have been a call for external help.

Reinsurance companies also collect data on disaster impacts. The Munich Re data base for natural catastrophes NatCatSERVICE, for example, includes more than 28 000 records in relation to material and human loss events worldwide (Munich Re, 2010) and SwissRe maintains a similar disaster event database (SIGMA). The data, however, are not publicly available.

A new disaster database named Glidenumber (2010) has been started by the Asian Disaster Reduction Center (ADRC). In this database each disaster receives a unique identifier and a number of relevant attributes.

There are also many other disaster databases that collect data on a national level or that collect data on a specific type of hazard. DisDAT is a service launched by the Global Risk Identification Program (GRIP) and the Centre for research in Epidemiology of Disasters (CRED) which aggregates all publicly available disaster databases from different countries (GRIP, 2010). It contains 60 registered disaster databases, of which 13 are global.

Several problems are associated with the use of disaster statistics for hazard and risk assessment from the above-mentioned sources. Because data is collected from a variety of sources, official disaster statistics, such as those in the EM-DAT, have data standardization issues. Often data are only linked to the main type of disaster, whereas associated disasters, for example a landslide that is triggered by an earthquake or tropical storms, are classified under the triggering event. In this way they are not reported as actual disasters. Accurate data relating to the number of people affected are frequently difficult to obtain. This is because it entails a subjective decision of the extent to which people need to be affected in order to enter the statistics in the database. The problem with data collected by insurance companies is that they are collected for a specific purpose, that is related to the coverage of the insurance premiums, which may influence the values and the events that are reported. If disaster information is collected at local level (e.g. DesInventar) it is usually more complete because it also includes small magnitude/high frequency events. The global coverage of such databases is, however, limited. One of the main limitations of the use of disaster databases for natural hazard and risk assessment is that there is no proper geo-reference for the reported events (Verelst, 1999).

Recent landfalls of extreme hydrometeorological events have resulted in a number of disasters. A list with a critical synthesis is reported in Table 1.3.1. Despite a by far non-exhaustive number of events reported in the table, we believe that the highlighted weak and strong points in disaster risk management can be of general interest for many other hydrometeorological events. Most of these cases illustrate how disasters are related to:

- lack of public awareness of the risk level;
- still poor representation of multiple-hazard cases;
- missing pro-active mentality among the population.

Table 1.3.1 Synthetic information of some recent hydrometeorological events (in the period 2004–2013) that can be considered relevant as lessons learnt for disaster risk management practice.

Storm, year, place	Physical drivers	Impact	Weak/strong points of DRM
Bangladesh flood, 2004	Heavy rainfall within and outside the country, combined with monsoon and peak flows of the major rivers	Death of about 800 people Flooded around 75% of the country 36M homeless	Insufficient river management Lack of disaster preparedness at local level, with provision of rescue resources, emergency funding mechanisms Insufficient information management and contingency planning
Hurricane Katrina 2005, United States	Rainfall (up to 25.4 mm h^{-1}), wind (up to 140 mph) and storm surge (till 902 mb)	The costliest natural disaster in the history of the United States. Death of 1833 people. Total property damage was estimated at \$81 billion, over 233 000 km^2. 3M people without electricity.	Early warning timely issued Teams of volunteers Catastrophic failure of the levee system in New Orleans Mismanagement and lack of leadership in the relief efforts in response to the storm and its aftermath in Louisiana, especially in organising evacuation
Britta storm 2006, Germany	High wind speeds, extreme storm surge (return period 100 years) and high waves (up to 17 m)	Damage to dunes, buildings and infrastructures In Hamburg only undefended areas were flooded.	Careful design of banks and defences
United Kingdom Floods, 2007	Rainfall (up to 25.7 mm h^{-1}), flash flooding and extensive flood	Death of 13 people, hundreds rescued. Damages up to £6M. –48 000 affected homes	Warnings properly issued Unclear governmental responsibilities (4 departments) Need for reinforced flood banks
Xynthia storm 2010, France	Storm surge and high tide	Death of 47 people. Total damages: €2.5billion 200 km of dykes, sand dunes, transport, houses, M houses lost power supply.	Wind correctly forecast, inappropriate alerts, no flood warning

(*continued overleaf*)

Table 1.3.1 *(continued)*

Storm, year, place	Physical drivers	Impact	Weak/strong points of DRM
Pakistan flood, 2010	Heavy monsoon rains, which affected the Indus River basin	One-fifth of Pakistan's total land area underwater. Affected about 20M people. Death toll of close to 2000. Total damage €35billion.	Sluggish and disorganised response to the floods from the government Non-prompt international aids (a.o. 10Mpeople drunk unsafe water)
Becky storm 2011, Spain	Waves up to 8 m, falling tide	3 injured in Spain. Damages: €4M only in Galicia. 20 tons of sand on Santander seafront	In place warning system
Liguria Flash flood 2011, Italy	High rainfall rate, storm	Death of 6 people. Damage to houses, buildings, infrastructures ~€1 billion	Rainfall correctly forecast, warning issued in advance but ignored
Hurricane Sandy, 2012, United States	High winds (80 mph), exceptionally high spring tide	Death of 149 people. Overall US estimated loss: \$20 billion in property damage and \$10–30 billion in lost business.	Alerts properly issued Shutdown of transportation Evacuation of most threatened areas
Eastern Europe floods, 2013, Germany, Czech, Austria	High rainfall rate and exceptionally long raining period, record levels of soil moisture	Death of 21 people Estimated losses for the insurance industry \$3.5–4.5 billion (SwissRe)	Alerts properly issued Evacuation of about 70 000 people Teams of volunteers placed sand bags for bank reinforcing

and call for:

- more efficient warning systems and better risk communication;
- explicit chains of responsibility;
- more strategic planning of resistant flood defences and maintenance of existing mitigation measures.

Other cases illustrate effective forecasting, warnings, evacuations and precautionary measures which limited the loss of lives, but also showed that residual risk cannot be avoided when coping capacity is overwhelmed, leading to high damages and disruption of human activities.

1.3.8 Risk governance: Responsibility and participation

Research generally demonstrates that to effectively mainstream both adaptation and disaster risk management on a national level involves 'whole of government' coordination across different levels and sectors of governance, including the involvement of a broad range of stakeholders (Few *et al.*, 2006; Thomalla *et al.*, 2006; OECD, 2009).

Governments, despite these evident interdependences, have generally attempted to manage these issues within their respective sectors. This means that environment or energy authorities and scientific institutions are normally responsible for climate change adaptation, while disaster risk management authorities may belong to a variety of national government departments and national disaster management offices (Sperling and Szekely, 2005; Thomalla *et al.*, 2006; Prabhakar, Srinivasan and Shaw, 2009). Progress in planning for adaptation and developing and implementing strategies within government agencies is usually dependent on political commitment, institutional capacity and, in some cases, on enabling legislation, regulations and financial support (Few *et al.*, 2006; OECD, 2009; Krysanova *et al.*, 2010).

Research undertaken on a national level indicates that it may be important to clearly identify a lead for disaster and climate risk reduction efforts where that lead has influence on budgeting and planning processes (Few *et al.*, 2006; OECD, 2009). In order to develop and strengthen their responsible institutions, countries and regions, in some cases, may be able to expand on phases of increased awareness and greater attention to disaster risk (Few *et al.*, 2006; Krysanova *et al.*, 2010).

Additional vulnerabilities for communities exposed to environmental change can be the result of conflicts between formal and informal governance or governmental and non-governmental strategies and norms. Flood protection measures often demonstrate the conflicts between formal and informal strategies. For example, local people might be willing to use resources to cope with an increasing number of flood events (e.g., adapting livelihoods and production patterns) whereas formal adaptation strategies, particularly in developing countries, tend to give priority to structural measures (e.g. dyke systems or relocation strategies) that have severe consequences on the vulnerability of communities that are dependent on local ecosystems, such as fishing and farming (see Birkmann, 2011a,b). Conflicts between formal and informal or governmental and non-governmental management systems and norms are an important issue that leads to an increase in vulnerability and reduces the adaptive capacity of the overall system (Birkmann *et al.*, 2010).

For risk governance to be effective it needs to involve a wide range of stakeholder groups in the process of exchanging, integrating, and sharing knowledge and information. These groups may include scientists, policymakers, private companies, non-governmental organizations, media, educators and the public. Sustainability science promotes the interactive co-production of knowledge between experts and other actors, based on transdisciplinarity (Jasanoff, 2004; Pohl *et al.*, 2010) and social learning (Pelling *et al.*, 2008; Pahl-Wostl, 2009).

In view of the long term and large-scale nature of coastal risks, the development of prevention, mitigation and preparedness strategies should adopt a hierarchical planning approach. This should range from a top down national (master) planning level to the analysis of individual flood-prone areas and the consideration of specific strategies and measures within these areas. The master plan should facilitate communication and coordination between coastal managers, contingency planners and crisis managers. Moreover, it should guarantee the continuity of coastal risk management policies, protection measures and the related operational procedures (i.e. financing, institutional arrangements, legal rights/obligations, operational responsibilities).

So how is this coordination to be brought about and legitimized in pluralist, democratic societies? The plural actors of society should be an integral part of the decision-making process, i.e. inclusive governance should be pursued (Renn, 2008). Inclusive governance assumes that all stakeholders have something to contribute to the risk governance process. It also assumes that mutual communication and exchange of ideas, assessments, and evaluations leads to an improvement in the final decisions as opposed to obstructing the decision-making process or compromising the quality of scientific input and the legitimacy of legal requirements.

Inclusive governance, in relation to preparedness towards natural hazards, requires that there has been a significant or clear attempt to:

- involve representatives from all relevant actor groups,
- permit all actors to participate actively and constructively in the discussion,
- co-design the framing of the (emergency response) problem or the issue in an interchange with the different groups,
- promote a common understanding about the framing of the problem, potential solutions and their probable consequences (based on the expertise of all participants),
- oversee a decision-making forum that provides equal and fair opportunities for all participants to voice their opinion and to express their preferences,
- establish a relationship between the participatory bodies of decision-making and the political implementation level.

Evidence shows that if these conditions are achieved, the actors not only develop faith in their own competence but start to trust each other and gain confidence in the process of risk management. This is especially true for the local and regional level where the participants are familiar with each other and have more immediate access to the issue. It is much more difficult, however, to reach consensus and build up trust on highly complex and transgressional subjects such as extreme floods. Being inclusive and open to social groups does not, therefore, always guarantee constructive cooperation by participants. Some actors, for example, may reject the framing of the issue and decide to withdraw and others may benefit from the collapse of an inclusive governance process.

Inclusive governance (Renn, 2008) also needs to address the second part of the decision-making process (i.e. reaching closure on a set of options that are chosen for further consideration, whilst rejecting others). "Closure" does not signify having the final word on a community preparedness plan or a regulation. It simply represents the product of deliberation (i.e. the agreement that the participants reached). The problem is that the more actors, viewpoints, interests and values there are, the more difficult it is to reach either a consensus or some other kind of joint agreement. It is important, therefore, to establish common rules both for generating and evaluating evidence and for dealing with conflicting values. In order to do this a second set of criteria is required to evaluate the process by which the closure of debates (either final or temporary) is brought about, as well as the quality of the decision or recommendation that is generated through the closure procedure.

The quality of the closure process can be assessed by answering the following issues:

- Have all arguments been treated correctly? Have all claims been fairly and accurately tested against commonly agreed validation standards?
- Has all relevant evidence been collected and processed, in accordance with the actual state-of-the-art knowledge?
- Was systematic, experiential and practical knowledge and expertise adequately included and processed?
- Were all interests and values considered, and was there a significant effort to arrive at fair and balanced solutions?
- Were all normative judgements made clear and thoroughly explained? Were normative statements derived from accepted ethical principles or legally prescribed norms?
- Was every effort made to preserve the plurality of lifestyles and individual freedom and to restrict the realm of collectively binding decisions to those areas in which binding rules and norms are essential and necessary to produce the wanted outcome?

When looking at the issues of outcome, additional criteria need to be addressed (Renn, 2008). They have been discussed in political science and governance literature for a considerable time and include:

- Effectiveness: Does the preparedness plan achieve the required effect?
- Efficiency: Does the option achieve the desired effect with the least resource expenditure?
- Minimisation of external side effects: Does the option infringe on other valuable goods, benefits or services, such as community development, public health, environmental quality, social cohesion and so on? Does it reduce the efficiency and acceptance of the governance system itself?
- Sustainability: Does the option contribute to the overall goal of sustainability? Does it assist in sustaining vital ecological functions, economic prosperity and social cohesion?

- Fairness: Does the option burden the subjects of regulation in a fair and equitable manner?
- Political and legal implementability: Is the option compatible with legal requirements and political programmes?
- Ethical acceptability: Is the option morally acceptable?

The quality of the participation process directly affects the potential benefits resulting from inclusive governance.

It is of paramount importance to treat the time and effort of the participants as extra resources that need to be handled with care and respect. The participation process should encourage the various actors to contribute in those areas in which they feel they are competent and in which they can offer input that will improve the quality of the final product.

1.3.9 Risk communication

Risk communication is a complex cross-disciplinary field that involves reaching different audiences in order to make risk comprehensible. According to the IRGC Risk Governance Framework (Renn, 2005), communication is needed from the framing of the risk situation to the implementation and monitoring of measures (see Figure 1.3.5).

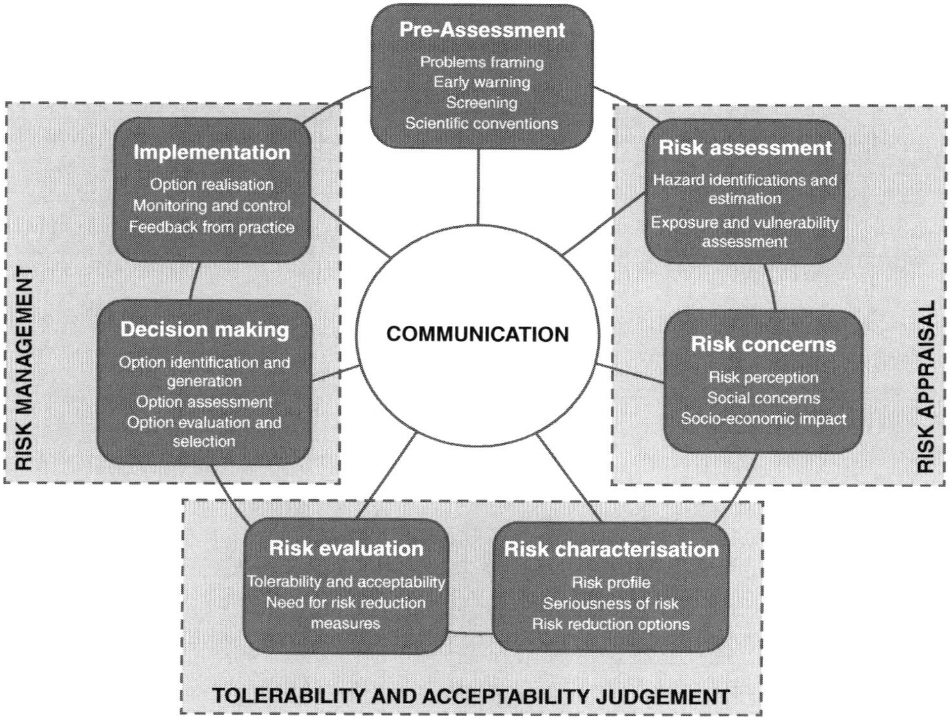

Figure 1.3.5 Risk communication and the five elements of risk governance based on (IRGC, 2009).

Social scientists, in the field of natural hazards, maintain that communication is vital before, during and after a hazard event (Steinführer *et al.*, 2009). Or, from a managerial point of view, communication transcends and links the four main hazard cycle phases: prevention/preparation, warning, emergency response and recovery/reorganisation.

Failures in risk communication have been uncovered in past disasters, such as Hurricane Katrina in 2005 or the Pakistan floods in 2010 (German Committee for Disaster Reduction, 2011). In the context of early warning or climate change adaptation, weak and insufficient risk communication as well as the loss of trust in government can be viewed as key components of institutional vulnerability.

Risk communication is strengthened by the explicit characterization of uncertainty. Attention must therefore be given to initiatives that encourage or reinforce the participation of forecasting experts in the hydrological forecasting chain. For the communication of uncertain forecasts it is essential to have face-to-face forecast briefings that focus on sharing the knowledge of how forecasters interpret, describe and perceive the model output forecasted scenarios, together with continuous technical training. One should also encourage training for the users so that flood forecasters and users can work together to define and agree on the level of uncertainty acceptable for each community or specific problem. Importance should also be given to studies that take into consideration the forecast requirements of different types of users, with different forecast formulation requirements, tolerance to false alarms and optimum probability thresholds for the detection of critical events.

More information and communication about vulnerability factors, dynamic temporal and spatial changes of vulnerability, and the coping and response capacities of societies or social-ecological systems at risk are required in order to better understand the ways in which disasters are framed. Frames are shaped by knowledge of and underlying views of the world (Schon and Rein, 1994). They arerelated to the organization of knowledge that people have about their world in relation to their underlying attitudes toward key social values (e.g., nature, peace, freedom), their ideas of agency and responsibility (e.g., individual autonomy, corporate responsibility) and their judgments about reliability, relevance, and weight of competing knowledge claims (Jasanoff and Wynne, 1997).

'Early warning' requires information interventions into an environment in which much about vulnerability is already assumed. In this respect, risk communication is not exclusively linked to a top-down communication process. Effective risk communication requires that communication be recognized as a social process. This means that risk communication also deals with local risk perceptions and the local framing of risk. Risk communication, therefore, also functions as a tool to increase local knowledge and needs (bottom-up approach). Effective risk communication achieves the process of informing people at risk about both the key determinants of their particular risks and of impending disaster risk (early warning). It also involves different stakeholders in defining a problem and the identification of solutions (see van Aalst *et al.*, 2008).

Public engagement and social change are fundamental to reducing risk. Communications that include social, interpersonal, physical, environmental and policy factors help achieve this (Brulle, 2010). A participatory approach emphasizes the need for several communication channels that lead to credibility, trust, and cooperation (Frumkin and McMichael, 2008), which are particularly important in high-stress situations such as those associated with climate extremes. Approaches of this nature include participatory video production (Baumhardt *et al.*, 2009), engagement with community leaders or opinion leaders in accessing social networks (Maibach, Roser-Renouf and leiserowitz, 2008), visualizing methods such as mapping, cartographic animations and graphic representations (Shaw, Sharma and Takeuchi, 2009). Overall risk communication should:

- design integrative messages that cover cognitive, experiential as well as normative dimensions of human behaviour,
- make the context and relevance of a specific given source of hazard explicit,
- be targeted to specific behaviours and pay close attention to the psychological determinants of the behaviours that need to be changed.

Additionally, communication is more effective when risk information does not exceed the capacity for coping, thereby reinforcing resilience (Fritze *et al.*, 2008). Research suggests that focusing on the personal risk of specific damages can be a key element in motivating interest and attitude change (Leiserowitz, 2007). Risk messages often tend to vary in threat specificity, guidance specificity, repetition, consistency, certainty, clarity, accuracy, and sufficiency (Mileti and Peek, 2002).

Public interest, leadership and acceptance are all a necessary part of climate change adaptation strategies as well as disaster risk reduction. Both previous and current research indicate that the general public still lacks important knowledge about climate change in all important areas: it is perceived as a non-urgent, temporally and geographically distant threat. In order to close this gap more efficiently, communication campaigns need to pay closer attention to the convergence of knowledge and attempt to integrate information about the causes, consequences and solutions to climate change. For example, there has been little methodical research on message effectiveness in prompting local action according to differing characteristics, such as the precision of message dissemination, penetration into normal activities, message specificity, message distortion, rate of dissemination over time, receiver characteristics, sender requirements and feedback (NRC, 2006). Mass media are often closely linked to political awareness and are governed by their own journalistic rules and priorities; that also means that little space for alternative frames of communicating climate change is provided by the mass media (Carvalho, 2005; Olausson, 2009). Boykoff and Boykoff (2007) conclude that this process might also result in an informational bias, especially towards the presentation of events instead of a comprehensive analysis of the problem. The acceptance and admission of the limits of knowledge about the future is therefore an important aspect of improving risk communication and the respective knowledge base (Birkmann and von Teichman, 2010).

1.3.10 A roadmap towards a sustainable future

Responding to disasters and enhancing resilience to extreme hydrometeorological events needs a holistic, participatory and inter-disciplinary approach, where science is embedded in the social, cultural and economic context in which communities live.

The approach should focus on how to operationalise the concept of resilience by the analysis, advance, development, testing and integration of a wide range of complementary adaptation and disaster risk mitigation approaches and tools to face extreme hydrometeorological events and their key research challenges.

The roadmap to develop such an approach therefore builds on the four main pillars identified by the disaster risk mitigation and adaptation strategy (IPCC, 2012):

1. Risk Identification and Early Warning
2. Policy and Governance
3. Awareness and Education
4. Mitigation and Preparedness.

Such a roadmap should be considered as part of a medium-term vision (Figure 1.3.6), which calls for a tailored-process leading to the two main recipients, public (i.e. citizens) and institutional bodies (i.e. managers and policy makers), as a whole, to progressively achieve a better resilience to disasters.

These recipients are prompted into a paradigm shift that is driven by the learning process around experience (case studies). While boosting this transforming process, this vision recognises that avoidance of future risk is the most cost-effective solution and, therefore, there is the need to urgently address the challenge to reduce exposure and vulnerability and increase resilience capacity. The vision comprises the enhancement of short term disaster risk reduction measures, such as better identification risks, including impact assessment and uncertainties; improved forecasting, prediction and early warning capabilities and optimised design of mitigation and preparedness measures.

The roadmap can be schematised through the vertical and horizontal reading keys. Along the 'non-technological' learning axis (i.e. the horizontal axis), institutional bodies move from more disaggregated and fragmented actions towards a real partnership, i.e. clear responsibility chain and interaction, through the pillar 'Policy and Governance'.

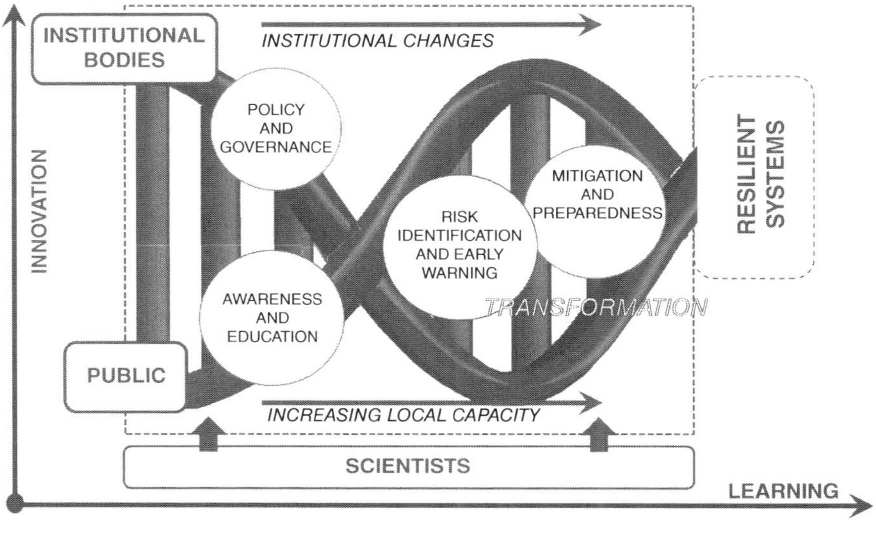

Figure 1.3.6 The sustainability road-map to face the consequences of hydrometeorological events: resilience is achieved through capacity building and institutional transformation, boosted by participatory learning process and scientific innovation. The strategy is supported by the four pillars of disaster risk management.

Public at large transforms from feeling the disaster to adapting and facing the disaster, shifting progressively from a reactive to a proactive behaviour, through the 'Awareness and Education' pillar. Along the 'technological' innovation axis (i.e. the vertical axis), specific scientific and technological frameworks – within the 'Risk Identification and Early Warning' and the 'Mitigation and Preparedness' pillars – both address key emerging research challenges and promote and enhance a functional connection of the two recipients. This road-map recognises that the human factor is essential in assessing and managing risk (i.e., risk is strongly dependent on social perception and disaster risk reduction measures on risk communication strategies). This will lead to a strong advance of existing tools and solutions coherently structured and harmonised into people-centred strategy.

The following four main cross-cutting concepts support the roadmap by allowing both the horizontal (concepts 1, 4) and the vertical (concepts 2, 3) reading of the strategy.

Concept 1-Integrative participatory framework. The effective management of climate-related disaster risks requires the active involvement and cooperation of multiple actors (national/regional governments, the private sector, research bodies, civil society, and community-based organizations and communities) playing differential but complementary well-defined roles across spatial and temporal scales (Wisner, 2011). Learning from past and recent extreme events supports the development of a participatory framework where all stakeholders have something to contribute to the process of risk governance and mutual communication. Exchange of ideas, assessments, and evaluations improve the final decisions and the quality of research hence moving institutional bodies to a real partnership.

Concept 2-Prioritising management actions through a disaster-impact perspective. Determining protection levels and designing corresponding management solutions have been often based on the "principle of equal hazard" (i.e. every area should be safe from, say, a 100-year flood). While this may be equitable, the need is to turn the disaster management approach on risk – and not hazard – in order to achieve a disaster-impact prioritised planning. However, this approach can be effective only if an enhanced and integrated assessment of vulnerability and resilience is achieved to create a readily understandable link between the theoretical concepts and the day-to-day decision-making process through:

- the experience gained in disasters;
- the development of a participatory approach embracing local communities, end users, managers and key authorities;
- the systematic collection of data to improve the knowledge and support the implementation of risk transfer measures;
- the mainstream use and verification of existing vulnerability indicators and the development of new ones where gaps still exist, especially focused on resilience of society, economies and environment and on the results from innovative combined monitoring/observation/ modelling chains.

Concept 3-Portfolio of people-centred solutions. The social construction of risk, awareness and preparedness are crucial to make disaster risk management plans effective. Yet, when facing extremes, very little knowledge and information exist in terms of the dynamic interactions between risk, individual and institutional awareness of this risk and individual and collective preparation against receiving it. Central issues here comprise the improved understanding, awareness, and people-centred mitigation options of risk. Hence, alternative strategies of protection and management need to be based on a logical and coherent mix of measures. The cost-benefit assessment of the relevant impacts of the alternative strategies has to be carried out for different scenarios in order to deal with the various (large) uncertainties. The evaluation should not so much aim at selecting the 'best' strategy within a specific scenario, but rather at identifying the most 'robust' strategy, showing an acceptable performance across all relevant scenarios.

Concept 4-Participatory, resilience-centric approach to disasters. Resilience-centric approaches offer four key contributions for living with extremes:

- providing a holistic framework to evaluate hazards in coupled social-ecological systems;
- putting emphasis on the capacities to deal with hazard or disturbance;
- helping to explore options for dealing with uncertainty and future changes; and
- identifying enabling factors to create proactive responses (Obrist, Pfeiffer and Henley, 2010).

These approaches require active and conscious participation at both community and policy maker levels and can be achieved only through the promotion of reflective social learning –from reaction to transformation – and increasing capacity – from feeling to facing the disaster. They also involve being prepared for surprises – those events beyond the lived experience or occurring very infrequently: as Aristotle said: "It is probable that the improbable will happen".

References

Adger W.N. (2000) Social and ecological resilience: are they related? *Progress in Human Geography*, **24**(3), 347–364.

Adger W.N., Arnell N.W. and Tompkins E.L. (2005) Successful adaptation to climate change across scales. *Global Environmental Change*, **15**(2), 77–86.

Barnston, A.G., Li, S., Mason, S.L. *et al.* (2010) Verification of the first 11 years of iri's seasonal climate forecasts. *Journal of Applied Meteorology and Climatology*, **49**, 493–520.

Baumhardt, F., Lasage, R., Suarez, P. and Chadza C. (2009) Farmers become filmmakers: Climate change adaptation in Malawi. *Participatory Learning and Action*, **60**, 129–138.

Beven K.J., Smith P.J. and Freer J.E. (2008) So just why would a modeler choose to be incoherent? *Journal of Hydrology*, **354**(1–4), 15–32.

Beven K.J. (2000) Uniqueness of place and process representations in hydrological modelling. *Hydrology and Earth System Sciences*, **4**(2): 203–213.

Birkmann J. (2006) Measuring vulnerability to promote disaster-resilient societies: conceptual frameworks and definitions. in *Measuring Vulnerability to Natural Hazards: Towards Disaster Resilient Societies* (ed. J. Birkmann). United Nations University Press, Tokyo, Japan, pp 9–54.

Birkmann J. (2011a) First and second-order adaptation to natural hazards and extreme events in the context of climate change. *Natural Hazards*, **58**(2), 811–840, doi:10.1007/s11069-011-9806-8.

Birkmann J. (2011b) Regulation and coupling of society and nature in the context of natural hazards. in *Coping with Global Environmental Change, Disasters and Security* (eds H.G. Brauch, U. Oswald Spring, C. Mesjasz *et al.*) Springer, Berlin, Germany, pp 1103–1127.

Birkmann, J. and von Teichman K. (2010) Integrating disaster risk reduction and climate change adaptation: key challenges – scales, knowledge, and norms. *Sustainability Science*, **5**(2), 171–184.

Birkmann, J., Garschagen, M., Kraas, F. and Quang N. (2010) Adaptive urban governance: new challenges for the second generation of urban adaptation strategies to climate change. *Sustainability Science*, **5**(2), 185–206.

Blaikie, P., Cannon, T., Davis. I., and Wisner B. (1994) *At Risk: Natural Hazards, People's Vulnerability and Disasters*. Routledge, London.

Boykoff, M.T. and Boykoff J. (2007) Climate change and journalistic norms: a case-study of US mass-media coverage. *Geoforum*, **38**, 1190–1204.

Brulle R. (2010) From environmental campaigns to advancing the public dialogues: Environmental communication for civic engagement. *Environmental Communication*, **4**(1), 82–98.

Cardona O.D. (2004) The Need for rethinking the concepts of vulnerability and risk from a holistic perspective: A necessary review and criticism for effective risk management. in *Mapping Vulnerability: Disasters, Development and People* (eds G. Bankoff , G. Frerks and D. Hilhorst eds.) Earthscan Publishers, London, UK, pp 37–51.

Cardona O.D. (2011) *Disaster risk and vulnerability: Notions and measurement of human and environmental insecurity*. in *Coping with Global Environmental Change, Disasters and Security – Threats, Challenges, Vulnerabilities and Risks* (eds H.G. Brauch , U. Oswald Spring, C. Mesjasz *et al.*) Springer Verlag, Berlin, Germany, pp 107–122.

Carpenter, S.R., Walker, B.H., Anderies, J.M. and Abel N. (2001) From metaphor to measurement: resilience of what to what? *Ecosystems*, **4**(8), 765–781.

Carreño, M.L., Cardona, O.D. and Barbat A.H. (2007) A disaster risk management performance index. *Journal of Natural Hazards*, **41**(1), 1–20.

Carter, T.R., Jones, R.N., Lu, X. *et al.* (2007) New assessment methods and the characterisation of future conditions. in *Climate Change 2007: Impacts, Adaptation and Vulnerability. Contribution of Working Group II to the Fourth Assessment Report of the Intergovernmental Panel on Climate Change* (eds M.L. Parry , O.F. Canziani, J.P. Palutikof *et al.*). Cambridge University Press, Cambridge, UK, pp 133–171.

Carvalho A. (2005) Representing the politics of the greenhouse effect: discursive strategies in the British media. *Critical Discourse Studies*, **2**(1), 1–29.

Ceccato, P., Ghebremeskel, T., Jaiteh, M. *et al.* (2007) Malaria stratification, climate, and epidemic early warning in Eritrea. *American Journal of Tropical Medicine and Hygiene*, **77**, 61–68.

Changnon D. and Changnon S.A. (2010) Major growth in some business related uses of climate information. *Journal of Applied Meteorology and Climatology*, **49**, 325–331.

Cloke H. and Pappenberger F. (2009) Ensemble flood forecasting: a review. *Journal of Hydrology*, **375**(3–4), 613–626.

Corbett J. (1988) Famine and household coping strategies. *World Development*, **16**(9), 1099–1112.

Degallier, N., Favier, C., Menkes M. *et al.* (2010) Toward an early warning system for dengue prevention: modeling climate impact on dengue transmission. *Climatic Change*, **98**, 581–592.

German Committee for Disaster Reduction (ed.) (2011) *Adaptive Disaster Risk Reduction*. Enhancing Methods and Tools of Disaster Risk Reduction in the light of Climate Change. DKKV Publication Series, no.43, Bonn.

EM-DAT (2009), (2011). *The OFDA/CRED International Disaster Database*, Université Catholique de Louvain, Brussels, Belgium. http://www.emdat.be

Few, R., Osbahr, H.,Bouwer, L.M. *et al.* (2006) *Linking Climate Change Adaptation and Disaster Risk Management for Sustainable Poverty Reduction* (Synthesis report). Vulnerability and Adaptation Resource Group, Washington, DC.

Fritze, J., Blashki, G., Burke, S. and Wieseman J. (2008) Hope, despair and transformation: Climate change and the promotion of mental health and well-being. *International Journal of Mental Health Systems*, **2**, 1–12.

Frumkin, H. and McMichael A.J. (2008) Climate change and public health: Thinking, communicating, acting. *American Journal of Preventive Medicine*, **35**(5), 403–410.

Füssel, H.-M.and Klein R.J.T. (2006) Climate change vulnerability assessments: an evolution of conceptual thinking. *Climatic Change*, **75**, 301–329.

Füssel H.M. (2007) Vulnerability: A generally applicable conceptual framework for climate change research. *Global Environmental Change*, **17**, 155–167.

German Federal Government (2008) *German Strategy for Adaptation to Climate Change*. Adopted by the German Federal Cabinet on 17 December 2008, The Federal Government, Berlin, Germany, www.bmu.de/files/english/pdf/application/pdf/das_gesamt_en_bf.pdf.

Glantz M. (2003) *Early warning systems: Dos and don'ts. Report of Workshop on Early Warning Systems*, Shanghai, China, 20–24 October 2003, ISBN 978-0756744953.

Glidenumber (2010). *Disaster database*. ADRC. See www.glidenumber.net

Goddard, L., Aitchellouche, Y., Baethgen, W. *et al.* (2009) *Providing Seasonal-to-Interannual Climate Information for Risk Management and Decision Making*. White Paper presented at the World Climate Conference 3 in Report of the World Climate Conference 3, Report No. 1048, World Meteorological Organization, Geneva, Switzerland.

GRIP (2010) *DISDAT the disaster data collection initiative portal. Global Risk Identification Program* (GRIP). http://www.gripweb.org/

Handmer J. and Dovers S. (2007) *The Handbook of Disaster and Emergency Policy and Institutions*. Earthscan, London, UK, 192 pp.

Heinz Center and CERES, (2009) *Resilient Coasts: A Blueprint for Action*. http://www.ceres.org/resources/reports/resilient-coasts-blueprint-for-action-2009.

Hilhorst D. (2004) Unlocking domains of disaster response. in *Mapping Vulnerability: Disasters, Development and People* (eds G. Bankoff, G. Frerks and D. Hilhorst) Earthscan Publications, London, pp 52–66.

ICSU-LAC (2010) *Science for a Better Life: Developing Regional Scientific Programs in Priority Areas for Latin America and the Caribbean. Vol. 2, Understanding and Managing Risk Associated with Natural Hazards: An Integrated Scientific Approach in Latin America and the Caribbean* (eds O.D. Cardona , J.C. Bertoni, A. Gibbs *et al.*). ICSU Regional Office for Latin America and the Caribbean, Rio de Janeiro, Brazil.

IDEA (2005) *Indicators of Disaster Risk and Risk Management – Main Technical Report. English and Spanish edition*, National University of Colombia/Manizales, Institute of Environmental Studies/IDEA, Inter-American Development Bank, Washington, DC, 223 pp.

IPCC (2012) *Managing the Risks of Extreme Events and Disasters to Advance Climate Change Adaptation. A Special Report of Working Groups I and II of the Intergovernmental Panel on Climate Change* (eds C.B. Field , V. Barros, T.F. Stocker *et al.*) Cambridge University Press, Cambridge, UK, and New York, NY, USA, 582 pp.

IRGC (2009) *Risk Governance Deficits: An Analysis and an Illustration of the Most Common Deficits in Risk Governance*. International Risk Governance Council, Geneva, Switzerland, 92 pp.

Jasanoff S. and Wynne B. (1997) *Handbook of Science and Technology Studies*, Sage Publications, London, UK, 828 pp.

Jasanoff S. (2004) *States of Knowledge: The Co-Production of Science and Social Order*. Routledge, London, UK.

Klein, R.J.T., Eriksen, S., Næss, L.O. *et al.* (2007) Portfolio screening to support the mainstreaming of adaptation to climate change into development. *Climatic Change*, **84**, 23–44.

Kraas F. (2008) Megacities as Global Risk Areas. *Urban Ecology*, **V**, 583–596.

Krysanova, V., Dickens, C., Timmerman, J. *et al.* (2010) Cross-comparison of climate change adaptation strategies across large river basins in Europe, Africa and Asia. *Water Resources Management*, **24**, 4121–4160.

Lavell A. (1999) Environmental degradation, risks and urban disasters. issues and concepts: Towards the definition of a research agenda. in *Cities at Risk: Environmental Degradation, Urban Risks and Disasters in Latin America* (ed. M.A. Fernandez) A/H Editorial, La RED, US AID, Quito, Ecuador, pp 19–58.

Leichenko, R.M. and O'Brien K.L. (2008) *Environmental Change and Globalization, Double Exposures*. Oxford University Press, New York, NY.

Leiserowitz, A. (2007) Communicating the risks of global warming: American risk perceptions, affective images and interpretive communities. in *Creating a Climate for Change* (eds S. Moser and L. Dilling) Cambridge University Press, New York, NY, pp. 44–63.

Shaw, R., Sharma, A. and Takeuchi, Y. (2009) Indigenous Knowledge and Disaster *R*isk Reduction: From Practice to Policy. Nova Science Publishers, Hauppauge, NY, pp. 409.

Lonsdale, K.G., Downing, T., Nicholls, R. *et al.* (2008) Plausible responses to the threat of rapid sea-level rise for the Thames estuary. *Climatic Change*, **91**(1–2), 145–169.

Maibach, E., C. Roser-Renouf and Leiserowitz A. (2008) Communication and marketing as climate change - intervention assets: A public health perspective. *American Journal of Preventive Medicine*, **35**(5), 488–500.

Mileti D.S. and Peek L.A. (2002) Understanding individual and social characteristics in the promotion of household disaster preparedness. in *New Tools for Environmental Protection: Education, Information, and Voluntary Measures* (eds T. Dietz , T. and P.C. Stern) The National Academies Press, Washington, DC, pp 125–140.

Milly, P.C.D., R.T. Wetherald, K.A. Dunne and Delworth T.L. (2002) Increasing risk of great floods in a change climate. *Nature*, **415**, 514–517.

Munich Re (2010) *Topics Geo: Natural catastrophes 2009 – Analyses, assessments, positions*. Munich Re, Munich, Germany, www.munichre.com.

Nicholls, R.J., Wong, P.P., Burkett, V. *et al.* (2008) Climate change and coastal vulnerability assessment: scenarios for integrated assessment. *Sustainability Science*, **3**, 89–102.

NRC (National Research Council) (2006) *Completing the Forecast: Characterizing and Communicating Uncertainty for Better Decisions Using Weather and Climate Forecasts*. Committee on Estimating and Communicating Uncertainty in Weather and Climate Forecasts, National Research Council, The National Academies Press: Washington, D.C.; 124 pp. www.nap.edu.

O'Brien, K., Sygna, L., Leinchenko, R. *et al.* (2008) Disaster Risk Reduction, Climate Change Adaptation and Human Security. GECHS Report 2008:3, Global Environmental Change and Human Security, Oslo, Norway.

Obrist, B., Pfeiffer, C. and Henley R. (2010) Multi-layered social resilience: a new approach in mitigation research. *Progress in Development Studies*, **10**(4), 283–293.

OECD (2009) *Policy Guidance on Integrating Climate Change Adaptation into Development Co-operation*. Organisation for Economic Cooperation and Development, Paris, France.

Olausson U. (2009) Global warming – global responsibility? Media frames of collective action and scientific certainty. *Public Understanding of Science*, **18**, 421–436.

Pahl-Wostl C. (2009) A conceptual framework for analysing adaptive capacity and multi-level learning processes in resource governance regimes. *Global Environmental Change*, **19**(3), 354.

Pappenberger F., Beven K.J., Hunter M.N. *et al.* (2005) Cascading model uncertainty from medium range weather forecasts (10 days) through a rainfall-runoff model to flood inundation predictions within the European Flood Forecasting System (EFFS). *Hydrology and Earth System Sciences* **9**(4), 381–393.

Pelling M. (2003) *The Vulnerability of Cities. Natural disasters and Social Resilience*, Earthscan Publications, London, 256 pp.

Pelling, M., High, C., Dearing, J. and Smith D. (2008) Shadow spaces for social learning: a relational understanding of adaptive capacity to climate change within organisations. *Environment and Planning A*, **40**(4), 867–884.

Pohl, C., Rist, S., Zimmermann, A. *et al.* (2010) Researchers' roles in knowledge co-production: experience from sustainability research in Kenya, Switzerland, Bolivia and Nepal. *Science and Public Policy*, **37**(4), 267–281.

Prabhakar, S.V.R.K., Srinivasan, A. and Shaw R. (2009) Climate change and local level disaster risk reduction planning: Need, opportunities and challenges. *Mitigation and Adaptation Strategies for Global Change*, **14**(1), 7–33.

Ramos, M.H., Mathevet, T., Thielenc J. and Pappenbergerd F. (2010) Communicating uncertainty in hydro-meteorological forecasts: mission impossible? *Meteorol. Appl.*, **17**, 223–235, DOI: 10.1002/met.202.

Red Cross and Red Crescent Climate Centre (2007) *Red Cross, Red Crescent Climate Guide*. Red Cross, The Hague, The Netherlands.

Renn O. (2005) *Risk Governance: Towards an Integrative Approach. International Risk Governance Council Report*, White Paper N.1, http://www.irgc.org/IMG/pdf/IRGC_WP_No_1_Risk_Governance__reprinted_version_.pdf.

Renn, O. (2008) *Risk Governance – Coping with Uncertainty in a Complex World*. Earthscan, London, UK.

Schon D.A. and Rein M. (1994) *Frame Reflection: Toward the Resolution of Intractable Policy Controversies*. BasicBooks, New York, NY, USA, 247 pp.

Setiadi N. (2011) Daily mobility – excursus Padang, Indonesia. in: *Early Warning in the Context of Environmental Shocks: Demographic Change, Dynamic Exposure to Hazards, and the Role of EWS in Migration Flows and Human Displacement* (eds D. Chang Seng and J. Birkmann) Migration and Global Environmental Change, Foresight SR4b, Vulnerability Assessment, Risk Management and Adaptive Planning Section, United Nations University for Environment and Human Security, Government Office for Science, pp 41–45.

Smit, B., Pilifosova, O., Burton, I. *et al.* (2001) Adaptation to climate change in the context of sustainable development and equity. in: *Climate Change 2007. Impacts, Adaptation and Vulnerability. Contribution of Working Group II to the Fourth Assessment Report of the Intergovernmental Panel on Climate Change* (eds M.L. Parry , O.F. Canziani, J.P. Palutikof *et al.*) Cambridge University Press, Cambridge, UK, pp 877–912.

Smith, K. and Petley D.N. (2008) *Environmental Hazards. Assessing Risk and Reducing Disaster*. Taylor & Francis, London.

Sperling F. and Szekely F. (2005) *Disaster risk management in a changing climate, vulnerability and adaptation*. Discussion paper prepared for the world conference on disaster reduction – reprint with addendum on conference outcomes. Vulnerability and Adaptation Resource Group, Washington, DC.

Steinführer, A., Kuhlicke, C., De Marchi, B. *et al.* (2009). *Local communities at risk from flooding Social Vulnerability, Resilience and Recommendations for Flood Risk Management in Europe*, www.floodsite.net.

Thirel G, Rousset-Regimbeau F, Martin E. and F. Habets (2008) On the impacts of short-range meteorological forecasts for ensemble stream flow predictions. *Journal of Hydrolometerology* **9**(6), 1301–1317.

Thomalla, F., Downing, T., Spanger-Siegfried, E. *et al.* (2006) Reducing hazard vulnerability: towards a common approach between disaster risk reduction and climate adaptation. *Disasters*, **30**(1), 39–48.

Thorne, C., Evans E. and Penning-Rowsell E. (eds) (2007) *Future Flooding and Coastal Erosion Risks*, Thomas Telford, London, 350 pp.

Turner, B.L., Kasperson, R.E.,Matson, P.A. *et al.* (2003a) A framework for vulnerability analysis in sustainability science. *Proceedings of the National Academy of Sciences*, **100**(14), 8074–8079.

Turner, B.L. II,, Matson, P.A.,McCarthy, J.J. *et al.* (2003b) Illustrating the coupled human-environment system for vulnerability analysis: Three case studies. *Proceedings of the National Academy of Sciences*, **100**(14), 8080–8085.

UNISDR (2004) *Living With Risk: A Global Review of Disaster Reduction Initiatives*. United Nations International Strategy for Disaster Reduction, Geneva, Switzerland.

UNISDR (2005) *Reducing Disaster Risk: A Challenge for Development*. United Nations, International Strategy for Disaster Reduction, Geneva, Switzerland, http://www.undp.org/bcpr/disred/rdr.htm.

UNISDR (2009a). *Economic damages: share of GDP, by natural disaster and country*. United Nations, International Strategy for Disaster Reduction, Geneva, Switzerland. http://www.unisdr.org/disasterstatistics/top50.htm.

UNISDR (2009b) *Terminology on Disaster Risk Reduction*. United Nations International Strategy for Disaster Reduction, Geneva, Switzerland. http://www.unisdr.org/files/7817_UNISDRTerminologyEnglish.pdf.

UNISDR (2009c) *Second Global Platform on Disaster Risk Reduction*, Geneva, June 2009: Concluding Summary by the Platform Chair. United Nations International Strategy for Disaster Reduction, Geneva, Switzerland.

UNPD (2010a) *World Population Prospects: the 2008 Revision. United Nations Population Division*. United Nations, Department of Economic and Social Affairs, New York, http://esa.un.org/unpd/wpp2008/index.htm.

UNPD (2010b) *World Urbanization Prospects: the 2009 Revision*. United Nations Population Division. United Nations, Department of Economic and Social Affairs, New York, http://esa.un.org/unpd/wup/index.htm.

Van Aalst, M., Maarten, K., Cannon, T. and Burton I. (2008) Community level adaptation to climate change: The potential role of participatory community risk assessment. *Global Environmental Change: Human and Policy Dimensions*, **18**(1), 165–179.

van Aalst M.K. (2009) Bridging timescales. in *World Disasters Report 2009. Focus on Early Warning, Early Action*. IFRC, Geneva, Switzerland, pp 68–93.

Van de Walle B. and Turoff M. (2007) Emergency response information systems: Emerging trends and technologies. *Communications of the ACM*, **50**(3), 29–31.

Verdin, J., CFunk, C., Senay, G. and Choularton R. (2005) Climate science and famine early warning. *Philosophical Transactions of the Royal Society B*, **360**(1463), 2155–2168.

Verelst L. (1999) *Recommendations for Geo-referencing EM-DAT Data Base*, Consultancy Report prepared for the Asia FIVIMS Project and the CRED, Rome.

Vogel C. and O'Brien K. (2004) *Vulnerability and Global Environmental Change: Rhetoric and Reality*. AVISO 13, Global Environmental Change and Human Security Project, Ottawa, Canada, http://hdl.handle.net/10625/39859.

Vogel, C., Moser, S. C., Kasperson, R. E. and Dabelko, G. D. (2007) Linking vulnerability, adaptation, and resilience science to practice: Pathways, players, and partnerships. *Global Environmental Change*, **17**, 349–364.

Wisner B. (2011) Are we there yet? Reflections on integrated disaster risk management after ten years. *Journal of Integrated Disaster Risk Management*, **1**(1), doi:10.5595/idrim.2011.0015.

Wisner, B., Blaikie, P., Cannon, T. and Davis I. (2004) *At Risk, Natural Hazards, People's Vulnerability and Disasters*. Routledge, London, UK.

World Bank 2010: *Natural Hazards, UnNatural Disasters: The Economics of Effective Prevention. International Bank for Reconstruction and Development/World Bank*, Washington, DC, 254 pp., gfdrr.org/gfdrr/nhud-home.

1.4 Science and Policy Interfacing

Philippe Quevauviller[1]

Department of Hydrology and Hydrological Engineering, Vrije Universiteit Brussels, Brussels, Belgium

1.4.1 Introduction

Most policies require a sound scientific evidence base to be developed and implemented at the most appropriate level (international, national, regional, local). Links among science and policy are, however, not always straightforward and require specific actions to ensure a proper transfer of information among the (science or policy) knowledge providers. It may be organised on an international scale, a salient example of which is the way the Climate Change community is managing to efficiently transfer inputs from scientists to policy-makers through the Intergovernmental Panel on Climate Change (IPCC) process (see Chapter 2.2). In relation to hydrometeorological extreme events, science–policy interactions are also organised on a UN scale through the International Strategy for Disaster Reduction (see Chapter 2.1) and on a European scale through various policies which are discussed in Chapter 2.3. From the various exchanges among scientists and policy-makers is growing clear evidence that science–policy interfacing requires a strategic framework which has to be structured and developed. This chapter provides some insight into the concept of science–policy interfacing, which is relevant to hydrometeorological extreme events.

1.4.2 Taking account of the knowledge base

Various policy settings described in Part 2 of this book provide platforms that enable building up communication and best practice exchanges among actors involved in prevention management of hydrometeorological extreme events (policy implementers, technology providers, the scientific community, industrial stakeholders, NGOs etc.). In this area, research strives to develop knowledge to better understand extreme events

[1] The views expressed in this chapter are purely those of the author and may not in any circumstances be regarded as stating a formal position of any EU institution

Hydrometeorological Hazards: Interfacing Science and Policy, First Edition. Edited by Philippe Quevauviller.

(in particular floods, droughts and storms) at scales that are relevant to decision-making (preferably linked to policy-making), including their socio-economic dimensions, in order to improve the management and mitigation of risks and uncertainties. A wide range of research projects funded at international, EU or national levels thus contribute to building up the knowledge base which is indispensable to develop and implement policies. Examples of such projects are given in this book.

1.4.3 Concept of science and policy interfacing

Experiences gathered from exchanges among scientists and policy-makers often tend to show that efforts are needed for presenting results of research and demonstration projects in a form that is easily usable by policy-makers, and for highlighting policy-related research needs in a way that may be tackled by scientists. It is recognised that the way research results are considered by the policy-making community is not straightforward, mainly for political reasons and difficulties in integrating the latest research developments in legislation. The difficulty is enhanced by the fact that the policy-making community is probably not defining its role as "client" sufficiently well. In other words, the dialogue and communication are far from being what one would hope to ensure an efficient flow of information (Quevauviller, 2010). The concept of "science–policy interface" is not new: efforts have been made in the past ten years in the water policy sector to support the implementation of the Water Framework Directive (European Commission, 2000) and its parent legislation such as the Flood Directive (European Commission, 2007a) as well as policy trends in the drought sector (European Commission, 2007b) and adaptation to climate change (European Commission, 2009), see Chapter 2.3 for additional details. To make it operational, such an interface should include:

- A screening phase for evaluating needs linked to specific policy milestones (this may concern gathering of background scientific information or tailor-made research, or demonstration of existing methods, tools, guidelines etc.).
- A mechanism to identify those research projects which are most relevant with respect to policy steps, and which might be "validated" through demonstration or capacity-building activities with the aim of being disseminated and applied at the most appropriate level (international, EU, national, regional or local).
- An exchange platform enabling both scientists and policy-makers to discuss the corresponding research and policy agendas from the very beginning in order to ensure a more structured communication at all appropriate levels of policy formulation, development, implementation and review.

A mechanism of science–policy interfacing should offer more than just fostering dissemination and application of research to respond to policy needs. It should actually strive to establish strong links among different research and demonstration funding mechanisms and thematic policies, in particular for combining the implementation of results of successfully completed research or demonstration projects with the implementation of related policy steps. In addition, it should be anticipated that specific skills should be used/developed by project coordinators (or dedicated partners) in order for the required projects to deliver appropriate scientific support to policies. Since most of the research projects are gathering different scientific disciplines (physics, chemistry, natural sciences, economics, social sciences etc.) and different socio-economic sectors (managers, policy-makers, enterprises, NGO, ...), project coordinators with 'translating' expertise (see Quevauviller, 2010) are key resources to drive these projects. In addition, since hydrometeorological extreme events are global problems but occur at local level, such an interface is needed at a regional/local level as well as on an international scale.

1.4.4 Matching research with policy needs

1.4.4.1 Type of research

It is not always possible to establish a clear distinction between "basic" and "applied" research. Also the timing aspect (short-, medium- and long-term) is intimately linked to the way research instruments are being operated. The identification of research needs matching policies is of course fed by advances in scientific knowledge, but is also directly influenced by the evolution and requirements of policies. The needs for ensuring coincidence of research and policy agendas may depend upon the stage of development of the policy in a given thematic area. In this respect, one may distinguish three different categories of needs in the sector relevant to hydrometeorological extreme events, depending on timing considerations:

- *Short-term* (~1–2 years): Needs are basically concerning accessibility of research knowledge required for the development of policies on a short-term basis. Timing is not adapted to develop new types of research (unless very specific needs are identified, which may be sorted out in a 6–12 months period). Policy development also requires an efficient and user friendly access to background scientific information and archives. In this context, the time needed for the design, approval and operation of *ad hoc* calls for research proposals (e.g. such as the ones leading to research funding under the EU Framework Programmes) makes it difficult to respond to short-term research needs, that is a specific need expressed at a given time will rarely be met through a project selected under a call for proposals the year after. Therefore, to date such needs may only be tackled in a timely fashion through either studies or possibly national research programmes.
- *Medium-term* (~2–5 years): The timing of medium-term research is adapted to responses of needs expressed in relation to the implementation agenda of well-defined policies (representing a "stable platform" for building strong partnerships among policy implementers, the scientific community and various stakeholders). Such platforms exist at EU level, for example the so-called Common Implementation Strategy for the Water Framework Directive (which includes expert groups on floods and droughts), and are also in place at international level, for example through the UN International Strategy for Disaster Reduction, and at national or regional levels under the responsibilities of governmental ministries/agencies. The most efficient response for enabling "medium-term" research to support policies is when a detailed description of research needs is made by policy-makers, leading to projects which are closely followed-up by them as well

as other stakeholders. Typically, research projects running over a 2–3 years period may fulfil medium-term research needs focusing for eample on well-defined policy milestones.

- *Long-term* (~5–10 years): Scientific progress in this respect supports either policy milestones which are clearly identified at a 10-years horizon, or the review process of a given legislation. Long-term research needs may be linked to the development of a programme of measures which have to be operational 5–10 years later. It may also concern the review process of technical requirements detailed in relevant annexes of a given directive (see Chapter 2.2).

1.4.4.2 Short analysis of drawbacks

The definition of the research funding programmes in principle includes topics that are aimed to respond to policy needs (but not only these, i.e. research should not be seen as only directed to support policies). This is the case of ranges of EU-funded projects that have been tailor-made to support policies over the past decades (Quevauviller, 2011). While the principle of designing research programmes that may match policy needs is well recognised, the practice may face some drawbacks, for example research projects may be designed according to clear technical specifications which may not anticipate the outcome and orientations of political decisions regarding specific policies (not defined at the starting time of the projects). This has sometimes resulted in results which although they had potential were not fit for policy developments. Another drawback is the insufficient coordination among different actors, which often hampers a full integration of scientific inputs from research projects into the policy discussions; this is linked to the lack of a clear "science–policy interface" mechanism. Indeed, interactions among policy and scientific communities are rarely coordinated in a systematic way and rather function on an *ad hoc* basis (based on links among "individuals" rather than "institutions"). On the other hand, various policy departments have different formal and informal contact points coming from numerous research and "policy" communities with discrete interests. These communities are often also competing internally (e.g. competition within the research community, or between responsible national authorities/local risk managers), which creates an obstacle to the pursuit of effective collaboration.

1.4.5 Research–policy interactions

1.4.5.1 Interactions with the scientific community

At the start of projects which have been identified as relevant to any policies (including those relevant to hydrometeorological extreme events), there is certainly a need to clarify policy issues by describing the aims, milestones, technical challenges to the research project consortia, so that they understand what are the policy expectations over the duration of their project. These exchanges of information/knowledge rarely occur, which may lead to divergent directions being taken by the projects in comparison to policy orientations. Regular contacts among scientists and policy-makers are essential during the lifetime of projects. These interactions are generally recognised as useful by both communities, but in reality they are not designed to meet expected targets

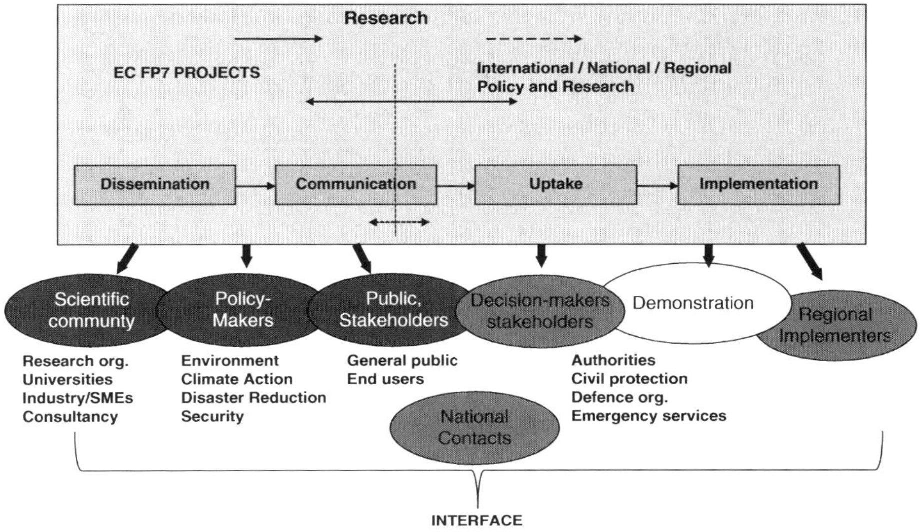

Figure 1.4.1 Different stakeholders require different types of communication.

(sharing of information, take-up of research outputs, recognition of policy needs etc.). This is directly due to a lack of a proper interface and facilitating mechanism. Scientists strive to involve policy-makers in the same way as they interact with their fellow scientists, that is by inviting policy representatives to scientific conferences or workshops and expecting them to participate as experts, following highly technical discussions. This is overlooking both the fact that policy-makers are rarely scientific experts and that they often do not have the resources to travel abroad and participate in meetings lasting one or several days. On the other hand, scientists are not trained to communicate in a concise way and to adapt their communication to different communities (see Figure 1.4.1).

Frequently, scientists participating in policy-makers' meetings (typically meetings of Member States representatives in charge of the implementation of a policy in a given field) present a full lecture of scientific facts and figures, with few or no references to the policy which they claim to support; in policy meetings, presentations are usually restricted to short interventions (5–10 minutes) to highlight very specific issues related to policy items, which does not correspond to the way scientists communicate ... both situations lead to misunderstanding and frustration, and create the feeling that the "others" are not making any effort to understand each other's viewpoints, while the problem is rather due to the formatting of the information. One solution to improve communication among the scientific and policy communities would be to rely on "mediators" who would predigest information and format it in a way which matches expectations from the different actors.

1.4.5.2 Synthesis needs

Linked to the difficulties of interaction, the most critical issue is the way information is "digested". On the one hand, the scientific information is published in a rather technical way which is not easily usable for policy purposes; improvements have been noticed in the last few years with the attempt to produce "policy briefs" within research projects, but these briefs still often remain barely accessible for policy use. On the other hand, policy information is almost never communicated in a way which highlights where and how science could contribute to the policy process. This creates gaps which are essentially due to a lack of efficient communication bridging and which is certainly the weakest link of the science–policy chain. Indeed, only a small percentage of research projects are known by policy implementers, which illustrates the need to improve awareness about research outputs but also to encourage policy actors to reflect on research needs linked to their portfolios. This may be translated into needs to carry out synthesis works in the form of "policy digests" (addressed to the scientific community from the policy implementer's side) and "science digests" (prepared by the scientific community for the policy implementers – scientists claiming that they are producing "policy briefs" often overlook the fact that they are not producing policy recommendations but rather highlight the possible scientific contribution to the policy implementation cycle, they should hence speak about "science–policy" or "science-into-policy" briefs).

1.4.5.3 Exchange platforms

As said above, participation of policy officers in research project meetings may not be practicable due to a lack of resources. Other ways have to be found to regularly hold joint meetings focusing on specific themes, starting from the formulation of research topics and interactions from the very start of research projects. This is already happening in the Water Framework Directive (Quevauviller, 2010, 2011) and should be systematised.

1.4.6 Conclusions

This chapter highlights some aspects of science-policy interfacing which are common to all fields, including the hydrometeorological one. Difficulties experienced to date stem from the fact that there is no sufficient streamlining of information from for example the scientific community to policy decision-makers. In this respect, efforts are on-going in the framework of various initiatives to examine how an efficient and operational 'science–policy interface' could be developed in support of the implementation of the Water Framework Directive and other water policies which are relevant to hydrometeorological extreme threats. The ultimate aim is to develop such an interface in a way that it could meet the demands of different levels of users (e.g. policy-makers, industry etc.) and stakeholders (e.g. the scientific community, academia etc.), ensuring an efficient dissemination and use of research results. This interfacing goal is ambitious and involves many different actors, hence its complexity as it considers not only the "human side" with different mentalities but also many different disciplines/sectors with their specific practices and jargons (Quevauviller, 2010). At present, this framework remains to be practically developed and implemented in Europe and is a task to be tackled for the forthcoming decade. The challenge over the next few years will hence be to establish operational links among different pieces of this complex puzzle, constituting a general interfacing mechanism within the timeframe of the Horizon2020 work programme (2014–2020).

As a concluding remark, it should be recalled that policy orientations rely on scientific evidence. In this respect, the efficient use of science represents an increasing challenge for the scientific and policy-making community, the private sector, NGOs, citizens' associations and professional organisations. The need to improve the role that science plays in environmental policy-making has been widely debated over the last few years; in particular, the need to ensure better links between policy needs and research programmes, with enhanced coordination regarding programme planning, project selection and management, and mechanisms for knowledge transfer, to ensure that outputs from research programmes really do contribute to policy development, implementation and review. This issue has been discussed in depth in the water sector at European Union level within the last ten years, underlining the need to develop a conceptual framework for a science–policy interface related to water which would gather together various initiatives and knowledge. This is discussed in depth in the literature (Quevauviller, 2010) and is illustrated in Figure 1.4.2, showing the necessary links between research recommendations or tools and "users" (policy-makers, stakeholders, water managers) and the need to ensure a "memory" of scientific information (facilitated by various dedicated websites), demonstration of the applicability of the research and dissemination through appropriate communication and "translation" of the scientific information. This closes the present chapter in underlining that the bridge between policy and research is not trivial and deserves more attention from all actors concerned.

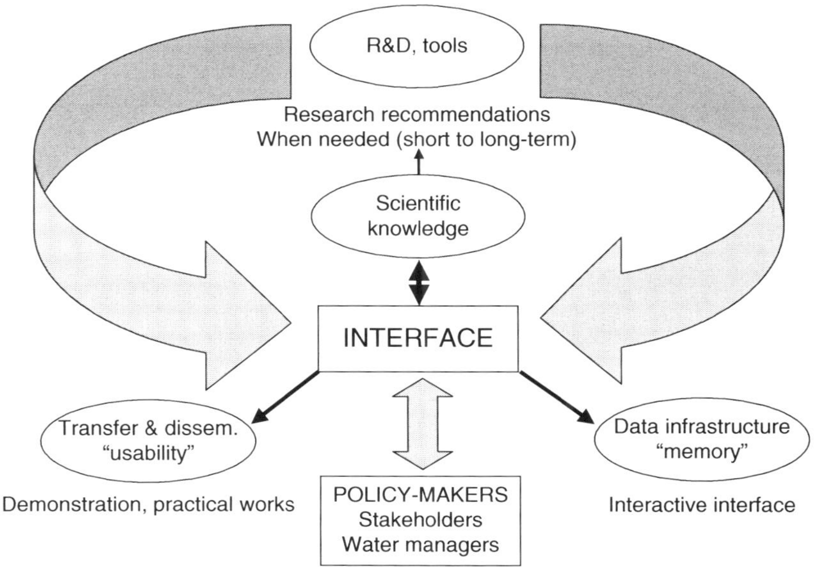

Figure 1.4.2 Needs of an interface between science and policy. *Source:* Adapted from Quevauviller, 2010. Reproduced by permission of The Royal Society of Chemistry.

References

European Commission (2000) Directive 2000/60/EC of the European Parliament and of the Council of 23 October 2000 establishing a framework for Community action in the field of water policy, *Official Journal of the European Communities*, L 327, 22.12.2000, 1.

European Commission (2007a) Directive 2007/60/EC of the European Parliament and of the Council of 23 October 2007 on the assessment and management of flood risks, *Official Journal of the European Communities*, L 288, 6.11.2007, 27.

European Commission (2007b) *Communication to the European Parliament and the Council – Addressing the challenge of water scarcity and droughts in the European Union*, COM/2007/04141 final.

European Commission (2009) *White Paper, Adapting to climate change: Towards a European framework for action*, COM(2009) 147 final.

Quevauviller Ph., Ed. (2010) *Water System Science and Policy Interfacing*, RSC Publishing, Cambridge, ISBN: 978-1-84755-861-9, 430 pp.

Quevauviller Ph. (2011) Adapting to climate change: Reducing water-related risks in Europe – EU policy and research considerations, *Environmental Science and Policy*, **14**(7), 722.

Part Two
Policy Settings

2.1 When Science Meets Policy: Enhancing Governance and Management of Disaster Risks

Demetrio Innocenti

University of Antwerp, Institute of Development Policy and Management (IOB), Antwerp, Belgium; The United Nations Office for Disaster Risk Reduction (UNISDR), Brussels, Belgium

2.1.1 Science and disaster risk management

Without a solid scientific basis there is not disaster risk management. Managing disaster risks relies on scientific research and technological advances. This is the case for all the risks posed by natural hazards: geophysical, hydrological, climatological and meteorological alike.

The progress of technology in the past five decades has allowed the possibility of setting accurate forecasting systems for a number of weather-related hazards. Improvements in computing power coupled with research in climate modelling have permitted a detailed understanding of the atmospheric system and the development of reliable early warning systems.

The progress in the technology in forecasting capacities and early warning systems has allowed the saving ofthousands of lives of those exposed and vulnerable to weather-related hazards such as floods and typhoons. The 2011 Global Assessment Report on Disaster Risk Reduction (GAR11) reported how mortality risk from floods and tropical cyclones diminished worldwide in the period 1990 to 2010. This result was mainly driven by a fall in causalities due to tropical cyclones in East Asia and the Pacific region where are concentrated about 44% of the exposed population (UNISDR, 2011b).

This remarkable outcome was achievable for a series of factors. On the one hand, the socio-economic situation of many South East Asian countries has improved dramatically in the past 30 years. The United Nations Development Programme (UNDP) reported that the average Human Development Index (HDI) for East Asia and the

Hydrometeorological Hazards: Interfacing Science and Policy, First Edition. Edited by Philippe Quevauviller.

Pacific region was 0.383 in 1980 and increased to 0.643 in 2010 (UNDP, 2013). On the other hand, applied science was the main driver in the reduction of vulnerability to weather-related hazards. Along with improving economic and development conditions, it allowed enhancements in early warnings, preparedness and response which positively reduced the mortality risk.

Technological progress jointly with effective risk communication and increased capacities of disaster risk management authorities have provided good examples of how science makes a difference in reducing disaster risk. A well-documented case is the Ganges–Brahmaputra delta flood warning system developed in Bangladesh (Webster, 2013). Bangladesh invested significantly in developing a reliable forecast system that can provide information with a good degree of certainty up to ten days in advance. In parallel, the Bangladeshi National Flood Forecast and Warning Centre invested in dissemination and appropriate risk communication through the expansion of its district-based networks, which became able to effectively advise the population on the most appropriate mitigation measures, including early harvesting and provision of medicines, food and drinkable water.

No less relevant is the role of science in preventing and mitigating geophysical risks. Although there is a general scientific consensus that earthquakes cannot be predicted (Jordan *et al.*, 2011), in the last decades improvements in the available scientific knowledge on seismic risks have been of vital importance in saving lives and economic assets. When seismic risk assessments and mapping are coupled with modern civil engineering and the capacity to develop and apply sound building codes, even the impact of the most intense earthquake can be radically reduced.

Often the 2010 Haiti and Chile earthquakes are examples mentioned to show the different impact that seismic events can have when countries have invested differently in disaster risk reduction. There are several studies on the aspects which have influenced the opposite outcomes of the two earthquakes (Kefer, 2010; Rojas, 2010). Chile's earthquake had an estimated magnitude of 8.8 on the Richter scale, Haiti's only 7.0. In terms of energy released the earthquake in Chile was approximately five hundred times greater than that in Haiti and yet, despite this, the number of reported casualities in Chile was 486, a fraction of the 222 517 fatalities in Haiti (UNOCHA, 2010).

The impact of disasters, in turn, might provide lessons that can be used by science to inform policies and different decision makers on how to handle prevention, preparedness and response for future disasters. The 2010 eruption of the Eyjafjallajokull volcano in Iceland caused US$4.7 billion in global GDP losses as a consequence of the ash cloud which paralysed the European airspace (Oxford Economics, 2010). The post-disaster analysis allowed an understanding of all the shortcomings in preventing and mitigating the impact of the volcanic ash cloud which could have been predicted well in advance, allowing time to prepare a contingency plan and avoid huge economic losses (Sammonds, 2010).

From the examples mentioned above we can start to draw some implications of the role of science in managing disaster risks:

1. Research and technology are the backbone of applied disaster risk management. There would be less possibilities of mitigating disaster risks if today's technologies did not provide instruments and tools to forecast climatological and meteorological phenomena, handle terabytes of data for climate modelling, apply Geographical

Information Systems (GIS) to risk mapping and use innovative engineering and material science to build seismic resilient infrastructures.
2. Science and technology alone cannot be effective in reducing risks if they do not operate in a conducive environment, which requires collaboration between policy-makers and scientists, appropriate legislation, strong institutions with capacities and clear mandates to implement and enforce policies and laws and ultimately promote awareness of disaster risk and consequent uptake of responsibilities among communities, households and individuals.
3. Research and development in science and technology have to be a key area for public investment for governments and financial institutions at different levels. Technological improvements have been the key to achieving major results in mortality reduction and will continue to be the pivotal driver for reducing vulnerabilities to natural hazards in the years to come.
4. Technological access and transfer has a prominent role in increasing the disaster resilience of developing countries.

2.1.2 Knowledge-based policy

Before discussing the role of policy and its interconnections with science in disaster risk management, it is useful to understand what the word policy means. From one of the most-known English Dictionaries (also publicly available on the internet) the Merriam-Webster, we get the following definition of policy: "*a definite course or method of action selected from among alternatives and in light of given conditions to guide and determine present and future decisions*".

Can the above definition of policy apply in the context of disaster risk management?

In the first place, it is essential to reflect on the historical struggle in creating a standard terminology in disaster risk management. Academics and practitioners have used words such as "risk", "disaster risk", "disaster risk management" or "disaster risk reduction" meaning different concepts. Further, within the United Nations system, different agencies dealing with disaster risk management have adopted different definitions. In recent years, and following the adoption of the HFA, the United Nations Office for Disaster Risk Reduction (United Nations International Strategy for Disaster Reduction, UNISDR) proposed a standard terminology which is nowadays likely to be the most widely referred to and the one which this chapter uses (UNISDR, 2009). How diverse communities of practitioners and scholars have differently approached the definitions of risk and disaster risk have had consequences on how the framing of disaster risk itself remained far from being unequivocal. Framing disaster risk can vary according to specific interests and perspectives and it is different for an insurance regulator, an urban planner, a climatologist or a community leader in a developing country.

These difficulties in framing the concept are reflected as well in the definition of disaster risk management policy, as each actor/stakeholder might intend policy as a set of planned actions only from his/her own point of view (if we use the definition from the dictionary quoted above).

Moreover, framing disaster risk from the perspective of one discipline can lead to shortcomings that undermine policies and interventions. For example, in hydrometeorology, technical scientists approach early warning systems through a linear method:

observation, data collection, modelling, analysis and generation of the warning to be disseminated through the competent authorities. Within this process there is nonetheless a lapse in the "demand side" or, in other words, how the warnings are received and elaborated by the final users. Are the recipients of the warnings, such as homeowners or farmers, taking actions upon receiving them? Or are there other factors that influence the decisions of those individuals on how and when to react to such warnings (Basher, 2013)? Empirical practice rather suggests that a farmer would first consult with his/her neighbours and community leaders before deciding on a costly early harvest or other risk mitigation measures.

Along with the technical institutional capacities to develop early warning systems, it is necessary to plan how the warnings will be communicated to the final users. This implies a role for social research which can profile the users and, based on their cultural, socio-economic and other context specific variables, inform policies on how warnings can be effectively communicated. Social and technical disciplines need, therefore, to work together to support appropriate policy decisions. This is the underlying concept of good "risk governance".

Therefore, the way in which we define policy in disaster risk management has to be tailored to the context and based on a holistic perspective. A possible inclusive definition can be the following:

Disaster Risk Management policy: a plan of action in which a set of coherent decisions – based on a comprehensive and holistic situation-analysis – works in synergy and coordination towards the common goal of building nations' and community resilience to natural and technological hazards.

2.1.3 The science–policy interface in practice

The European Commission (EC) and UNISDR have worked closely together over the past four years to ensure that within the framework of the EC-supported European research projects there is a clear linkage between science and policy in the fields of climate adaptation and disaster risk management.

Between 2009 and 2011 three international workshops were organised[1] by the two organisations in Brussels (Belgium) and proceedings have been published (European Commission, 2010). These workshops have provided the venue for coordinators of relevant pan-European research projects and senior European policy-makers (those that at the national level are responsible for the implementation of the HFA and the national DRR agenda) to exchange views on each other's perspectives. This ignited a debate between the "demand side" seeking evidence-based knowledge (the policy-makers) and the "supply side", those who can provide research findings to sustain sound disaster risk policies (the research coordinators).

Those workshops were a practical example of what it means to create an interface between science and policy and allowed an understanding of the barriers to interfacing the two as well as the possible solutions and opportunities. The Science–Policy Interface (SPI) is still a relevant issue within the projects supported by the EC, especially in

[1] Proceedings and presentations available at: http://www.preventionweb.net/english/professional/trainings-events/events/v.php?id=10785 (2009) http://www.preventionweb.net/english/professional/trainings-events/events/v.php?id=14411 (2010); http://www.preventionweb.net/english/professional/trainings-events/events/v.php?id=22160 (2011).

the area of climate-related risks. In November 2012, during the FloodRisk Conference in Rotterdam (Netherlands), the EC organised two specific workshop sessions on the SPI and further discussed the SPI barriers and solutions with the European scientific community involved in climate risks management[2].

The outcomes of those workshops led to the identification of the following SPI issues:

- Policy-makers needs from the scientific community:
  - scientific evidence relevant to identifying priorities of actions for formulating new policies or correcting the existing ones,
  - information that is understandable to non-experts and
  - research reachable within the context of limited resources in which national disaster management institutions have to operate.
- One of the main barriers to the use of research findings is often the lack of institutional mechanisms that allow coordination between researchers and policy-makers. Mechanisms such as the National Platforms for Disaster Risk Reduction (NP)[3] can provide the venue for informing disaster policy formulation on the basis of research findings and shape research planning in a way that finds relevant application in practice.
- Communication is at the core of an effective interface between science and policy. EC-supported research projects in disaster risk management are an example of good practice since in recent years most of them include clear communication plans to policy-makers within their envisaged outputs. This result has been achieved thanks to dialogue between policy-makers, donor institutions and academia which has become more frequent in recent years.

Risk communication is a crucial business in disaster risk management. Besides the aspects highlighted above on the role of social science in assuring that risk is well understood and followed by actions, there is also another implication attached to it: the political one. Disaster risk reduction practitioners have long argued that low investments in prevention and risk reduction measures are (also) a consequence of how disasters are communicated through the media (especially TV and journals). Journalism tends to privilege visible headlines when an intensive disaster strikes and widely covers the emergency and early reconstruction phases. Consequently politicians are inclined to be active in sustaining an emergency response rather than prevention and risk reduction measures as there is an immediate return in visibility and the public's attention. In addition, to be "tagged" by media as unresponsive in managing a national crisis can cost political figures and their parties the possibility of being re-elected. There are documented cases on how disaster response might have affected political elections (Wisner *et al.*, 2011). Recently, this argument was also discussed in the United States by some newspapers (Cohen, 2012) in the aftermath of the 2012 "superstorm Sandy" and what

---

[2] For more information: http://www.floodrisk2012.net/news-archive.asp.

[3] The National Platforms for Disaster Risk Reduction are nationally owned mechanisms for coordination at the national level on issues related to climate adaptation and disaster risk. Their establishment fully complies with the action set by the HFA (priority 1) and their mandate, composition and functional mechanism are country specific. A list of the National Platforms established worldwide to date can be found at http://www.preventionweb.net/english/hyogo/national/list/?pid:23&pih:2.

the media's description of the response meant for the re-election of the current US President.

The political attention to the response side, driven by visibility issues, can also influence the attitude of donor countries involved in international financing assistance. Studies tracking the Organisation for Economic Cooperation and Development (OECD) Development Assistance Committee (DAC) indicate that only 4% of humanitarian aid expenditure is invested in disaster prevention and risk reduction while the vast majority of DAC donors invest less than 1% of their Official Development Assistance (ODA) in disaster risk reduction (Sparks D., 2012). These figures are still below those indicated by the second session of the Global Platform for Disaster Risk Reduction (GPDRR) in 2009 (see Box 2.1.1) which set as financial targets for ODA spent in disaster risk reduction to 10% of humanitarian relief, 10% of post-disaster reconstruction and recovery and 1% of all national development funding and development assistance (GPDRR 2009, Chair's Summary, paragraph 19, page 4)[4]. Social and communication studies can, therefore, play a vital part in influencing and driving the media and the general public towards a culture that politically rewards prevention and risk reduction measures rather than mainly emergency response. Cases of national and local administrations that successfully invest in prevention, adaptation and risk reduction, avoiding human and economic impact when disaster occurs, will find the deserved visibility in the media for voters' appreciation. Yet, this cannot happen if there are no capacities built for media professionals to document disasters "through a different lens" – to quote the title of guidelines that the United Nations (UNISDR, 2011a) developed for journalists and reporters to approach the subject of disaster risk reduction.

Box 2.1.1 The Global Platform for Disaster Risk Reduction

The Global Platform for Disaster Risk Reduction, which takes place every two years, is the global forum for accelerating world-wide momentum on disaster risk reduction. As the primary gathering for the world's disaster risk community, it brings together Governments, UN, international regional organisations and institutions, NGOs, scientific/academic institutions and the private sector. It is mandated by the United Nations General Assembly (A/RES/62/192) to:

- assess progress made in the implementation of the Hyogo Framework for Action
- enhance awareness of disaster risk reduction
- share experience and lessons from good practice, and
- identify remaining gaps and recommend targeted action to accelerate national and local implementation.

Source: PreventionWeb; *http://www.preventionweb.net/english/professional/contacts/profile.php?id=3388*

Viewed from the opposite direction, disasters can influence politics and, consequently, policies. The 2011 Great East Japan Earthquake and the resultant Fukushima

[4] Global Platform 2009. Outcome Document: Chair's Summary of the Second Session Global Platform for Disaster Risk Reduction, Geneva Switzerland. Available at: http://www.preventionweb.net/files/10750_GP09ChairsSummary.pdf.

Daiichi nuclear accident impacted the overall global policy approach to nuclear industry (Joskow, 2012). While just before Fukishima nuclear power was enjoying a global reconsideration (driven by the commitments to reduce CO_2 emissions pledged by many OECD countries), following the 2011 fall out, major nuclear power producers such as Germany and Switzerland decided to reduce their share of nuclear energy production. Germany shut down 8 out of 17 of its power plants, the Switzerland Federal Council recommended to phase out nuclear energy production by 2034. Even Japan, following an investigation report published in September 2012 on the Fukushima accident announced its intention to phase out its nuclear energy programme by 2040, however, given the economic consequence of this decision it later announced that it would not pursue this commitment (Joskow, 2012).

An even more radical impact of disasters on national "politics" was argued by Cavallo (2010) who discussed how the 1972 earthquake that devastated Managua in Nicaragua might have played a role in the fall of Anastasio Somoza. Public anger rose when, instead of helping to rebuild the city, Somoza allegedly used relief money to help pay for his National Guard luxury homes, while the homeless were delivered poorly constructed wooden shacks.

2.1.4 Evidence-based disaster risk policies

In recent years the attention of international organisations dealing with disaster risk management has moved towards the economic argument as a main driver for positively influencing policies addressing governance of risks.

A remarkable effort was the release of the 2010 study from the World Bank and the United Nations "*Natural Hazards, UnNatural Disasters – The Economics of Effective Prevention*" (World Bank–United Nations, 2010). This publication was accompanied by positive remarks from several Nobel Prize winners, such as Kenneth Arrow, Gary Becker, Elinor Ostrom, Thomas Shelling, Amartya Sen and Robert Solow. among several others notable economists, policy-makers and disaster risk managers. This study was generally recognised as a pioneering attempt to make a strong economic case for influencing disaster management policies in upscaling disaster risk reduction at the core of national public investment portfolios. It provided practical examples of what has worked in integrating disaster risk reduction in public investments, taking into consideration several cases drawn from different scientific disciplines and practical engineering solutions applied to structural measures (infrastructures).

The World Bank–UN report findings triggered solution-oriented recommendations for policy-makers. One of them is to promote advocacy for investing in disaster risk reduction systematically accounting for disaster losses. This was a core recommendation of the 2011 Global Assessment Report on Disaster Risk Reduction (GAR11; UNISDR, 2011b) which described the examples of several countries, especially in Latin America, which since the 1990s have invested in building their own national disaster databases on economic losses. The economic argument to stimulate governments considering disaster prevention as a priority in their investments' agenda was an outcome priority point of the 3rd and 4th sessions of the Global Platform for Disaster Risk Reduction in 2011

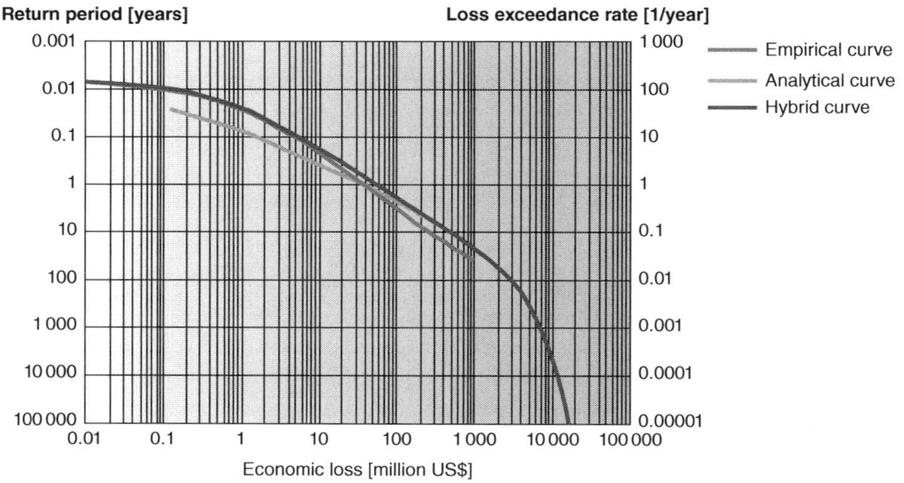

Figure 2.1.1 Hybrid risk model: loss exceedance curve for Colombia. *Source:* Data from ERN-AL in the UNISDR 2011 Global Assessment Report on Disaster Risk Reduction. See plate section for colour representation of this figure.

and 2013. The 2011 Chair Summary[5] of the 3rd session of the Global Platform specifically called upon all governments to invest in the development of nationally owned (and validated) databases on disaster losses (paragraph 8.4, page 3).

GAR11 provides examples of scientific applications based on the analysis of the data on economic losses which can be employed for sustaining national disaster risk management policies and investment decisions. From the databases on disaster losses, models on disaster risk can be drawn. Figure 2.1.1 (ERN-AL., 2011) is an example of a "hybrid model".

The figure shows an exceedance curve which informs on the expected level of maximum losses that a country (in this case Colombia) can expect from disasters in a given period (in this case one year). This model integrates the information from national analytical risk assessments (probabilistic models derived from the analysis of the location of economic assets and their exposure and vulnerability to natural hazards) with the empirical curve that can be derived from a study of the historical records on economic losses recorded in the national database. The resulting hybrid curve, which merges the two, has proved to be more accurate in estimating the expected economic losses. While the probabilistic curve is more accurate in estimating the possible losses associated with intensive disasters such as earthquakes or typhoons, the empirical curve is more accurate in estimating the losses associated with extensive risks such as recurrent floods which are normally underestimated in the probabilistic model.

In this specific example Colombia can expect to lose around US$30 million every year (loss exceedance rate around 1) or US$1 billion every 30 years because of disasters associated with natural hazards.

---

[5] Available at http://www.preventionweb.net/files/20102_gp2011chairssummary.pdf.

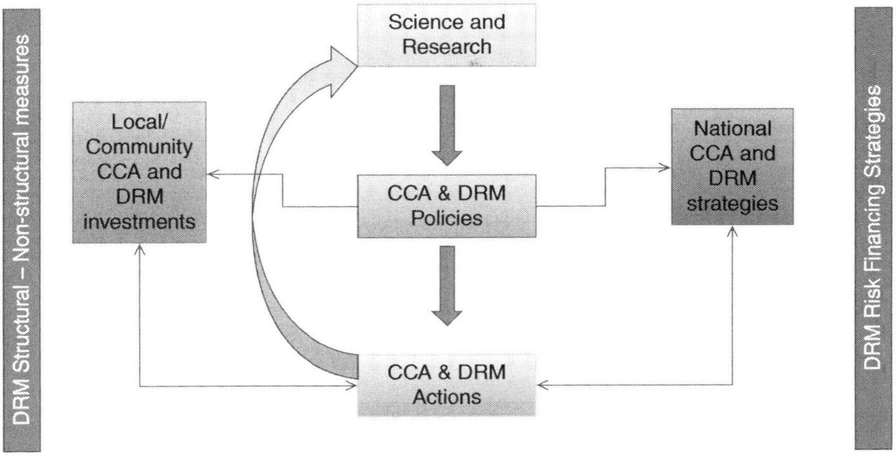

Figure 2.1.2 From science to policy actions.

This scientific research on risk modelling has implications for policies and politics. Vis-à-vis policies, it offers the possibility to take well-informed decisions concerning the most appropriate disaster risk management options at local and national level. Concerning politics it implies a "no excuse" situation for not considering disaster risk reduction a priority when planning resources and investments.

Making the real costs of disasters visible to policy-makers is not the only objective of research in disaster risk management and climate change adaptation. Research has also to provide management options (Figure 2.1.2). Systematic evaluation of climate change adaptation and disaster risk reduction policies and investments is fundamental in informing research on what has worked and what has not. Lessons learned from past disasters can improve research models and methods and contribute further to building resilience informing future policies.

Tailoring disaster risk management options to national and local situations implies that social and natural scientific disciplines inform on the most cost effective disaster policies, and that policies are able to translate these inputs into practical risk reduction actions. Structural and non-structural adaptation and risk reduction measures are (mainly) to be planned at the local level, as they must be context specific to be effective, whereas disaster risk financing strategies, such as the appropriate level of reserve funds, market-based risk transfer tools (insurance/reinsurance) and capital market instruments (cat-bonds), mainly involve the national level if not the regional as in the case of the Caribbean Catastrophe Risk Insurance Facility (CCRIF) (Suarez, 2011).

Regional approaches to disaster insurance can be effective at regional level for small and medium sized economies. In the Western Balkans the World Bank, the Swiss Cooperation, the Global Environmental Facility and UNISDR in partnership with the private sector are promoting the establishment of the South Eastern Europe and Caucasus Catastrophe Risk Insurance Facility (SEEC CRIF). The SEEC CRIF intends to play a key role in reducing the level of economic and fiscal exposure to disasters in South Eastern Europe and Caucasus. The facility is established as a regional catastrophe risk pool owned by countries and it is managed by Europa Re. Europa Re is a service company whose purpose is to promote disaster insurance/reinsurance among its shareholder governments (Box 2.1.2). The Europa Re case shows the importance of creating public–private partnerships in the field of disaster risk management.

Box 2.1.2 Europa Reinsurance Facility Ltd (Europa Re)

Europa Re is a catastrophe (cat) risk insurance services and reinsurance company owned by countries of Southeast Europe. It provides homeowners, farmers, enterprises, and the government of Southeastern Europe with access to affordable catastrophe and weather risk insurance through local insurers, who can also reinsure with the Facility.

Europa Re was created to address the very low levels of catastrophe and weather risk insurance penetration in the SEE region. Preparation of the Facility began in 2008 with the financial and technical support of the World Bank, the Global Environment Facility (GEF), the Swiss Secretariat for Economic Affairs (SECO) and the United Nations Office for Disaster Reduction (UNISDR).

The Europa Re mission is to educate consumers about the exposure of their properties and businesses to weather and catastrophe risks; to improve and standardize the credit quality of catastrophe insurance products; to enable insurance companies to sell complex weather and catastrophe risk insurance products; and to help governments and insurance regulators to enact regulatory and policy reforms conducive to the development of catastrophe and weather risk markets.

Source: Adapted from Europa Re website: http://www.europa-re.eberlesystems.ch/.

2.1.5 Climate research and disaster economics: Two scientific pillars of governance of disaster risks

Climate change adaptation and disaster risk reduction have often been associated in recent years. The two fields share the common goal of building resilience of societies and the environment to present and emerging risks, though from different perspectives and, traditionally, from different communities of practitioners. International organisations, such as the United Nations Framework Convention on Climate Change (UNFCCC) and UNISDR had a key role in setting the policy discussions on the linkages between the two. A boost to promote the importance of mainstreaming disaster risk reduction measures into climate change adaptation strategies at the national level came from the outcomes of UNFCCC Conference of Parties (COPs) meetings. In particular from COP 12 (2006) which set the Nairobi Work Programme on impacts, vulnerability and adaptation to climate change, COP 13 (2007) which adopted the Bali Action Plan and the establishment of an ad hoc Working Group on Long-term Cooperative Action (AWG-LCA) and the COP 16 (2010) when the Cancun Adaptation Framework (CAF) was endorsed by the parties.

The CAF makes specific reference to the HFA as a guiding tool for the disaster risk reduction measures which have to be undertaken as part of climate adaptation measures: "*Enhancing climate change related disaster risk reduction strategies, taking into consideration the Hyogo Framework for Action, where appropriate, early warning systems, risk assessment and management, and sharing and transfer mechanisms such as insurance, at the local, national, subregional and regional levels, as appropriate*" (par. 14e page 5) (UNFCCC, 2010).

The ground work of the UNFCCC in climate change, both mitigation and adaptation, requires the highest standards in scientific methodologies. The reliability of the data and the scientific soundness are for the Parties of the Convention a critical element to overcome political pressures attached to climate change policies. This is understandable given the enormous vested interests that several groups of stakeholders (especially in the energy sector) have at stake in instruments such as the Clean Development Mechanism (CDM) and the regulation of greenhouse gas emissions.

The International Panel on Climate Change (IPCC) scientific work is the pivotal driver for encouraging bold steps in climate change policies and investments and their reports are most influential in guiding international climate change agreements and climate policy formulation. The upcoming 5th Assessment Report will have a major impact on the 2015 Global Climate Change Agreement.

Specifically on the linkage between climate change adaptation and disaster risk reduction, the 2007 IPCC 4th Assessment Report influenced the way of approaching the two subjects. The 4th assessment report suggests that there is evidence that climate change is impacting the frequency and severity of climate-related hazards (IPCC, 2007). The scientific analysis on the connection between climate change, hazards and disaster impact was further investigated in the 2012 IPCC *Special Report on Managing the Risk of Extreme Events and Disasters to Advance Climate Change Adaptation (SREX)*. The SREX brings for the first time, at the highest scientific level, different communities of climate change experts and disaster practitioners to provide scientific evidence of the impact of climate on hazards while promoting adaptation and risk reduction measures to cope with them.

The IPCC reports have been influential for several policy documents at international, regional and national level. For example the 2013 EU Climate Change Adaptation Strategy and the 2009 EU Strategy for Supporting Disaster Risk Reduction in Developing Countries refers to the scientific evidence of the IPCC reports and their proposed actions as viable policy options.

Evidence-based policies are also sustained by the economic argument as discussed in the previous sections of this chapter. Once the magnitude of disasters is fully documented, the analysis can illuminate policies and guide governance of public investments in a way that considers disaster risk reduction as a pillar of sustainable investments and economic growth.

Nonetheless, data and economic risk models are not the only field in which economics can influence policies in disaster risk management. The 2013 Global Assessment Report on Disaster Risk Reduction (GAR13) is a ground-breaking study which highlights the linkages between disaster economics, public policies and private corporate investment decisions. It shows, through a series of examples, how these three components are in our globalised economy totally interconnected.

GAR13 presents an overview of what globalisation meant in terms of accumulating disaster risks, especially in those developing countries which in the past two decades have experienced a rapid and often unsustainable economic growth. Globalisation moved a massive flow of Foreign Direct Investments (FDI) in developing economies in the areas most exposed to risks, such as the East cost of China and South East Asia.

While this positively triggered the private sector contribution to employment and growth of those developing economies, it often externalised disaster risks (UNISDR, 2013a).

The United Nations Conference on Trade and Development (UNCTAD) estimated that by the end of 2013 the FDI will reach globally US$1.8 trillion and further increase in future years, 44% of which (approximately US$777 billion) will flow into low and middle income countries (UNCTAD, 2012). International and national policies will be the determining factor in the way those massive economic resources will result in long-term development opportunity or rather, increase the stock of disaster risks of those emerging economies.

There is an emerging need to rethink public risk governance towards a model that considers the role of the private sector at the centre of disaster risk reduction policies. Disasters affect the private sector not only in terms of directly damaging their privately owned assets and infrastructures but also in terms of disrupting production as a consequence of power cuts and damage to publicly owned power plants and networks and infrastructures such as bridges and roads. Therefore, the involvement of the private sector in contributing to reducing disaster risks is not only a matter of avoiding the "free rider" effect, when the costs of accumulated risks resulting from private investments are discounted into public goods, but also a crucial element for promoting business continuity and countries' economic competiveness in the global markets.

The link between disasters and national economic competitiveness is becoming more and more evident when analysing the impact of recent major disasters. Some authors (Haraguchi, 2012) (Ye, 2012) documented how the Chao Praya River floods in Thailand in October 2011 affected Japanese car makers (Nissan, Honda and Toyota) supply chain, stopping or slowing the production in other parts of the world. This was due to the specific spare parts manufactured in those factories hit by the Thai floods. They also showcased how this localised disaster was able to influence the entire global IT industry as the price of hard disks (which were heavily produced in factories located in the flooded areas) rose steeply following the floods. The same argument has been discussed in the aftermath of the 2011 Great East Japan Earthquake. The production of electronic components for cars carried out in the tsunami's affected factories hit the automobile production capacity in Thailand (20% production decrease), the Philippines (18% decrease) and Malaysia (6% decrease) (Ye, 2012). The 2012 Emilia earthquake in Italy had similar consequences on the price of parmesan cheese and disruption of bio-medical supplies which were heavily produced in the area affected by the earthquake (Jordan, 2012).

2.1.6 Conclusions

This chapter has discussed the multiple scientific roots nurturing disaster risk management and specifically disaster risk reduction. Natural science disciplines, such as geology, hydrology, meteorology, climatology and ecology among others, are coupled with social science and economics to create the scientific framework for disaster risk reduction.

Disaster risk management has emerged as a relatively new research area. This meanss that its knowledge base, though rapidly growing, is still limited, as well as its methodologies and standardisations (both in methods and definitions). With regard to the specific area of disaster risk reduction, some authors consider it as a potential new field of science as a robust body of knowledge is rapidly being developed around it (Basher, 2013). This knowledge can be further structured to develop academic graduate and postgraduate courses which have the potential to contribute significantly to countries' disaster resilience, shaping well-trained future professionals and experts.

The intention of this chapter was to give an overview of why it is important to invest in science and research which contributes to climate change adaptation and disaster risk reduction policies. There is no good governance of disaster risk if there is not a solid scientific basis that informs decision makers on the most appropriate and cost-efficient solutions.

It was highlighted how climate change adaptation and disaster risk reduction have to deal with uncertainty. While this is not an issue for science per se, uncertainty is a concern when it comes to advising policy-makers who need solid evidence to justify their decisions in front of several stakeholders and, not least, their tax payers. While scientific progress and technological innovations will likely contribute in the future to mitigating uncertainty, there is a contingent need to tackle the way in which both researchers and policy-makers interact. Enhancing the synergy between the two communities in a "non-business as usual" environment in which their respective roles and responsibilities are redefined is essential to make national and local governance of disaster risk more effective.

From the examples reported in the previous sections, there is evidence that this shift is already happening. Within the scientific community, complementary disciplines, such as natural and social science, are increasingly being integrated (such as in the example of risk communication and early warning systems). This is also the case for policy-makers, as documented by the mid-term review of the implementation of the HFA, who are increasingly looking at disaster prevention and risk reduction as an investment rather than a cost (UNISDR, 2011c, 2013b) and rely, more than ever, on the advice of their scientific community.

This chapter underlined, with specific examples, the results of an integrated science–policy approach in terms of enhanced effectiveness of disaster risk governance and emphasised the crucial role that some national mechanism, such as the National Platforms, can have to address the existing communication gaps between science and policy and the part that international organisations such as the EC and the UN can play in providing venues and means of cooperation at international level to promote a dialogue between science and policy.

It was stressed that the research areas which are currently most in need of knowledge production are likely to be disaster loss data and multi-hazards risk assessments. For example, risk modelling has a direct impact on how governments can define meaningful disaster risk financing strategies through a mix of public–private partnerships, as seen in the case of the insurance–reinsurance sector and cat-bonds.

Finally, this chapter looked at two specific research areas: climate science and disaster economics. This more in-depth look does not imply that those disciplines are considered more relevant than other areas in disaster risk management. They were specifically investigated as, internationally, policy frameworks currently under discussion, such as the post-2015 HFA (or HFA 2), are considering them of particular relevance for policy guidance.

This contribution reaffirmed that an effective science–policy nexus is, and will continue to be, the strategic element of successful governance of disaster risk and, in general, development policies: building resilient societies is the underlying pillar for any sustainable development intervention.

References

Basher R. (2013) *Science and Technology for Disaster Risk Reduction: A Review of Application and Coordination Need*. UNISDR, Geneva, Switzerland.

Cavallo E.E. (2010) Catastrophic natural disasters and economic growth. Inter-American Development Bank Working Paper Series No. IDB-WP-183.

Cohen J. (2012, October 31). http://www.washingtonpost.com/blogs/the-fix/wp/2012/10/31/wapo-abc-tracking-poll-high-marks-for-president-obama-on-hurricane-sandy-response/. Retrieved from Washingtonpost.com.

ERN-AL., Consortium Evaluación de Riesgos Naturales – América Latina. (2011) Probabilistic modelling of disaster risk at global level: Development of a methodology and implementation of case studies. Phase 1A: Colombia, Mexico, Nepal. UNISDR, Geneva, Switzerland.

European Commission (2010) *International workshop climate change impacts and adaptation: reducing water-related risks in Europe – scientific and policy report*, . European Commission, DG Research and Innovation, Brussels.

Haraguchi M.A. (2012) *Flood Risks and Impacts. Future Research Questions and Implications to Private Investment Decisions-Making for Supply Chain Networks*. UNISDR, Geneva, Switzerland.

IPCC (2007) *Fourth Assessment Report, Working Group I, Summary for Policymakers*. United Nations, Bonn, Germany.

IPCC (2012) *Managing the Risks of Extreme Events and Disasters to Advance Climate Change*. Cambridge University Press, New York.

Jordan T.H., Chen, Y-T, Grisparini, P. *et al.* (2011) Operational earthquake forecasting. State of knowledge and guidelines for utilization. *Annals of Geophysics*, 54(4), doi: 10.4401/ag-5350, available at http://www.annalsofgeophysics.eu/index.php/annals/article/view/5350, accessed 6 June 2014.

Joskow P.A. (2012) The Future of Nuclear Power After Fukushima. *MIT Center for Energy and Environmental Policy Research (CEEPR) Working paper 2012-001*.

Kefer P.N. (2010) *Earthquake propensity and the politics of mortality prevention,*. Washington DC, USA: Policy Research Working paper 4952, The World Bank.

Oxford Economics (2010) *The Economic Impact of Air Travel Restrictions Due to Volcanic Ash*. Oxford University, Oxford.

Rojas F.L. (2010) An overview of building codes and standards in Chile. in *The Structural Design of Tall and Special Buildings* (ed. M.L. Mehrain) John Wiley & Sons Inc. New York, pp. 823–934.

Sammonds P.M. (2010) *Volcanic Hazard from Iceland: Analysis and Implications of the Eyjafjallajökull Eruption*. UCL Institute for Risk and Disaster Reduction, London.

Sparks D. (2012) *Aid Investments in Disaster Risk Reduction – Rhetoric to Action*. Global Humanitarian Assistance, Bristol.

Suarez P.A.-B. (2011) *Insurance related instruments for disaster risk reduction*. United Nations, Geneva, Switzerland.

UNCTAD (2012) *World Investment Report 2012: Towards a New Generation of Investment Policies*. United Nations, Geneva, Switzerland.

UNDP (2013) *Human Development Report - The Rise of the South: Human Progress in a Diverse World*. UNDP, New York.

UNFCCC (2010) *Report of the Conference of the Parties on its sixteenth session, held in Cancun from 29 November to 10 December 2010 - Addendum Part Two: Action taken by the Conference of the Parties at its sixteenth session*. United Nations, Bonn, Germany.

UNISDR (2009) *Terminology on Disaster Risk Reduction*. United Nations, Geneva, Switzerland.

UNISDR (2011a) *Disaster through a different lens, behind every effect, there is a cause*. United Nations, Geneva, Switzerland.

UNISDR (2011b) *Global Assessment Report on Disaster Risk Reduction – Revealing Risk, Redefining Development*. United Nations,Geneva.

UNISDR (2011c) *Hyogo Framework for Action 20052015, Building the Resilience of Nations and Communities to Disasters, Mid-Term Review 2010–2011*. United Nations, Geneva, Switzerland.

UNISDR (2013a) *From Shared Risk to Shared Value – The Business Case for Disaster Risk Reduction. Global Assesment Report on Disaster Risk Reduction*. UNISDR, Geneva, Switzerland.

UNISDR (2013b) *Implementing the Hyogo Framework for Action in Europe, Regional Synthesis Report 2011–2013*. UNISDR Europe, Brussels.

UNOCHA (2010) *Haiti situation report 19*. United Nations, New York.

Webster P. (2013) Improve weather forecasts for the developing world. *Nature*, **493**, 17–19.

Wisner, B. Kent, G., Carmalt, J. *et al.* (2011) *Political Will for Disaster Reduction: What Incentives Build It, And Why Is It So Hard To Achieve?* UNISDR, Geneva, Switzerland.

World Bank–United Nations (2010) *Natural Hazards, UnNatural Disasters – The Economics of Effective Prevention*. The World Bank, Washington DC.

Ye L.A. (2012) *The Impacts of Natural Disasters on Global Supply Chain*. Bangkok, Thailand: Asia-Pacific Research and Training Netwrok on Trade. ARTNet Working Paper Series N. 115 / June 2012. ESCAP.

2.2 Hydrometeorological Extremes and the Science–policy Interface: IPCC

Zbigniew W. Kundzewicz

Institute for Agricultural and Forest Environment, Polish Academy of Sciences, Poznan, Poland
Potsdam Institute for Climate Impact Research, Potsdam, Germany

2.2.1 Introduction

The Intergovernmental Panel on Climate Change (IPCC) is an international body of intergovernmental nature. It was set up in 1988 by the World Meteorological Organization (WMO) – a specialist intergovernmental agency – and the United Nations Environment Programme (UNEP). The Panel is open to all Member Countries of the UNO and WMO and now the membership of the IPCC is almost 200.

The initial task for the IPCC as outlined in the UN General Assembly Resolution 43/53 of 6 December 1988, endorsing establishing of the IPCC, was to prepare a comprehensive review and recommendations with respect to the state of knowledge of the science of climate change: the social and economic impact of climate change and possible response strategies and elements for inclusion in a possible future international convention on climate. Nowadays, the IPCC's task is broader: "... to assess on a comprehensive, objective, open and transparent basis the scientific, technical and socio-economic information relevant to understanding the scientific basis of risk of human-induced climate change, its potential impacts and options for adaptation and mitigation".

The Panel has not conducted research nor has it monitored climate-related data or parameters. The IPCC has periodically reviewed and assessed the most recent scientific, technical and socio-economic information produced worldwide, relevant to the understanding of climate change. The Panel has been providing the world governments with a clear scientific view on the current state of knowledge in climate change, its observed and potential impacts and options for mitigation of and adaptation to climate change.

Hydrometeorological Hazards: Interfacing Science and Policy, First Edition. Edited by Philippe Quevauviller.

The IPCC is organized in three Working Groups and a Task Force on National Greenhouse Gas Inventories. If necessary, Task Groups and Steering Groups may be established by the IPCC to consider a specific issue. The Working Groups I, II, and III deal, respectively, with:

- the physical science basis of climate change,
- climate change impacts, adaptation and vulnerability, and
- mitigation of climate change.

The present chapter refers to hydrometeorological extremes and the science–policy interface in the IPCC context. After delivering information on the IPCC in a nutshell, special attention is given to the selected key messages conveyed by the IPCC Special Report on Managing the Risks of Extreme Events and Disasters to Advance Climate Change Adaptation (SREX), cf. IPCC (2012). Although the author of this chapter has been involved in different roles in many IPCC publications[1], including the SREX Report, this chapter is *not* an IPCC report. Whereas the author believes that his conclusions are congruent with the IPCC material, the views expressed are those of the author only, and do not represent, in any way, an IPCC report or supplement.

It is hoped that this chapter encourages the readership to seek useful information in the wealth of IPCC products.

2.2.2 IPCC at the interface of science and policy

The IPCC is a truly unique body at the interface of science and policy. These two areas interact in different ways. The IPCC mantra reads: "*policy relevant but not policy prescriptive*", that is IPCC works should be neutral with respect to policy, although they may deal with scientific, technical and socio-economic factors relevant to the application of particular policies. The Panel does not tell governments what should be done. Indeed, the form "should" does not belong in the language used in IPCC documents addressed to governments.

In the process leading to preparation of IPCC deliverables, representatives of science do assess relevant information available in the literature and prepare draft deliverables (assessment reports, special reports, and technical papers) that are subject to detailed and comprehensive review by experts. Reviewers do their best to render IPCC deliverables objective, balanced, comprehensive and transparent.

[1] The author has been collaborating with IPCC since 1994. He was a contributing author of Chapter 14 (Water resources management) in the IPCC WG II Second Assessment Report SAR) published in 1995. In the IPCC WG II Third Assessment Report (2001) he was a coordinating lead author of Chapter 13 (Europe), a lead author of the Technical Summary and a member of the core writing team for the Summary for Policymakers and the Synthesis Report (WG I, II, III) and of the Summary for Policymakers of the Synthesis Report. He was a lead author of the IPCC Technical Paper on Climate Change and Biodiversity (TP V), 2002. In the IPCC WG II Fourth Assessment Report (2007) he was a coordinating lead author of Chapter 3 (Freshwater resources and their management), a lead author of the Technical Summary and a member of the core writing team for the Summary for Policymakers and the Synthesis Report (WG I, II, III) and of the Summary for Policymakers of the Synthesis Report. He was a coordinating lead author of the IPCC Technical Paper on Climate Change and Water (TP VI), 2008. In the IPCC Special Report on Managing the Risks of Extreme Events and Disasters to Advance Climate Change Adaptation (SREX), published in 2012, he was a coordinating lead author of Chapter 4 (Changes in Impacts of Climate Extremes: Human Systems and Ecosystems). In the work on the IPCC Fifth Assessment Report, he plays the role of Review Editor of the WG II Chapter 3 (Freshwater resources) and the Synthesis Report.

The IPCC has delivered, on a regular basis, the most comprehensive scientific reports about climate change produced worldwide – the Assessment Reports. In addition, the Panel has also responded to the needs expressed by the UNFCCC (United Nations Framework Convention on Climate Change) and other bodies for information on particular scientific and technical matters. Along with regular assessment reports, the IPCC has produced several Special Reports on various topics of broad interest, as well as Technical Papers and Methodology Reports. It has also produced guidelines in order to help UNFCCC parties prepare national greenhouse gas inventories.

All IPCC reports must be endorsed by an IPCC Working Group during its Plenary Session. There are three different levels of endorsement of different categories of IPCC deliverables – approval, adoption, and acceptance:

1. Summaries for Policymakers of the Reports are "approved", i.e. the material is subject to line by line discussion and agreement.
2. Synthesis Reports and overview chapters of Methodology Reports are "adopted", that is endorsed section by section.
3. Endorsement by "acceptance" means that the material has not been subject to line by line nor section by section discussion and agreement, but nevertheless presents a comprehensive, objective and balanced view of the subject matter.

By endorsing the IPCC reports in one of the three ways specified above, governments acknowledge the authority of their scientific content.

Representatives of governments of IPCC Member Countries participate in the process of review of IPCC deliverables and in the Plenary Sessions of IPCC and their Working Groups, where main decisions about the IPCC work programme are taken and deliverables are endorsed. The Panel meets approximately once a year in Plenary Sessions attended by Government Representatives of all member countries (officials and experts from relevant Ministries, Agencies and Research Institutions from Member Countries and from observer organizations). The Members of the IPCC Bureau, including the Chair, are also elected during the Plenary Sessions of the Panel by representatives of IPCC Member Countries. The IPCC Bureau comprises the IPCC Chair (at present – Mr Rajendra K. Pachauri), the IPCC Vice-Chairs, the Co-Chairs and Vice-Chairs of the Working Groups and the Co-Chairs of the Task Force.

Members of the Bureau, whose task is to provide scientific and technical support to the Chair of the IPCC and the Co-Chairs of the Working Groups, are selected on the basis of their qualifications. There is also a need for adequate geographical composition of the IPCC Bureau, so that Members represent the different regions of the world. A Working Group is led by two or more co-chairs (typically, one or more from a developing country and one from a developed country). The government of the developed country Co-Chair of a Working Group hosts (and provides financial support to) a Technical Support Unit (TSU), assisting the Working Group. Likewise, a TSU may be established to support the preparation of the Synthesis Report.

IPCC Member Countries can have IPCC Focal Points identified by the relevant authorities in the country, or else – by default – all IPCC-relevant matters are addressed to the Ministries of Foreign Affairs.

Representatives of the scientific community, most of whom are nominated by the governments of their countries, are involved in the preparation of IPCC reports.

Coordinating lead authors (CLAs) and lead authors (LAs) for IPCC reports are selected by the relevant Working Group Bureau from a pool of experts proposed by governments and participating organizations, and other experts known through their works. The composition of the group of CLAs and LAs aims to reflect a range of scientific, technical and socio-economic views and expertise; adequate geographical representation and gender balance. Attempt is also made to nominate a mixture of experts with and without previous experience in IPCC. The CLAs (usually two or more per chapter: one or more from a developing country and one or more from a developed one) coordinate the content of the chapter they are responsible for, while the LAs create the content of the chapter. If appropriate, contributing authors (CAs) may be invited to provide technical information on specific subjects.

The most numerous set of individuals participating in the process of creation of IPCC deliverables are expert reviewers. There are thousands of experts from all over the world who review IPCC draft deliverables. During the first review round of an IPCC Report by experts, a first order draft (FOD) is circulated to competent experts. During the second review round, a second order draft (SOD) of the IPCC Report and a first draft of the summary for policymakers (SPM) are distributed to all governments of IPCC Member countries and to the reviewers involved in the expert review.

The review process of IPCC deliverables is overseen by the review editors (REs), who ensure that all substantive expert and government review comments are given adequate consideration by the author teams. REs advise LAss on how to handle contentious and/or controversial issues and ensure that genuine controversies are adequately reflected.

Review is a very essential part of the IPCC process, ensuring an objective assessment of current information, reflecting a range of views and expertise. Because of its scientific and intergovernmental nature, the IPCC embodies a unique opportunity to provide rigorous and balanced scientific information to decision makers.

A central IPCC Secretariat, hosted at WMO headquarters in Geneva, supports the work of the IPCC and liaises with Governments. It is supported financially by IPCC parent agencies: WMO and UNEP.

2.2.3 Evolution of IPCC over 25 years

As stated in Section 2.2.1, the task of the IPCC has evolved during the 25 years of existence of the Panel. One unchanging element is that every few years, the Panel publishes an actual Assessment Report, a truly iconic product, consisting of three parts (each prepared by one Working Group of the IPCC). In addition, as a result of multiple stages of the process of "distillation", two shorter documents are prepared, namely the Technical Summary and – even shorter – the Summary for Policymakers, accompanying every comprehensive volume of the IPCC Assessment Report. This means that each of three parts of AR4 consists of three separate documents: Assessment Report proper, Technical Summary and Summary for Policymakers. Since the TAR, there is also an aggregate deliverable, produced by all three Working Groups and drawing from each of the three volumes of particular Working Group reports – the Synthesis Report (longer Synthesis Report proper and Summary for Policymakers of the Synthesis Report).

The participation of the scientific community in the work of the IPCC has substantially grown, in terms of the number of authors and contributors involved in writing and reviewing the reports, the geographical distribution of the authors, and the topics covered by the reports.

The first four IPCC Assessment Reports were published, respectively, in 1990, 1995, 2001, and 2007. The complete set of IPCC publications is freely available in electronic form on the official IPCC internet portal (www.ipcc.ch), as well as being published in book form by Cambridge University Press.

The scientific evidence brought up by the First Assessment Report (FAR) of IPCC in 1990 unveiled the importance of climate change as a topic deserving broad attention and a political platform among countries and paved the way to the UNFCCC. This key international treaty to reduce global warming and cope with the consequences of climate change, was opened for signature at the United Nations Conference on Environment and Development (UNCED), also called Earth Summit, held in Rio de Janeiro in June 1992. Over 100 countries sent their heads of state or government to the Earth Summit in Rio. The IPCC Second Assessment Report (SAR) of 1995 provided important material used by climate policy negotiators in the run-up to adoption of the Kyoto Protocol in 1997. The Fourth Assessment Report (AR4), "Climate Change 2007" gained huge international visibility and interest and helped the world understand the ongoing climate change. At the end of 2007, months after acceptance and publication of AR4, the Intergovernmental Panel on Climate Change was honoured with the Nobel Peace Prize, shared with Albert (Al) Arnold Gore Jr, for efforts to build up and disseminate greater knowledge about man-made climate change, and to lay the foundations for the measures that are needed to counteract such change.

Over 3500 experts from more than 130 countries contributed to the IPCC's Fourth Assessment Report (AR4) released in 2007 (over 450 LAs, including CLAs, over 800 CAs and over 2500 expert reviewers providing over 90 000 review comments).

The writing and review of IPCC reports and other publications is done in accordance with the detailed procedures for the preparation of the various types of IPCC material that have been regularly reviewed and revised over the years. The procedures address all steps leading to the preparation of IPCC material starting with the scoping process, nomination process and selection of authors, preparation of drafts by the writing teams, the review by experts and governments and finally the approval, adoption and acceptance process in plenary sessions. They also contain a description of the tasks of authors, reviewers and review editors.

Statements attributing climate change to anthropogenic impacts, underpinned by results of climate modelling, play a crucial role in the assessments made by the Panel. The attribution statements have evolved with time in the warming world, in consecutive IPCC assessment reports. The First Assessment Report (FAR) and the Second Assessment Report (SAR) noted, respectively, "little evidence of detectable anthropogenic influence on climate" in 1990 and a "discernible human influence on climate" in 1995. In the light of accumulated evidence, the attribution statements became stronger in the last three IPCC assessment reports as the time series of atmospheric concentration of greenhouse gases and global temperature continued to increase and reached new heights. The Third Assessment Report (TAR) of IPCC stated in 2001 that "most of the observed warming over the last 50 years is *likely* to have been due to the increase in greenhouse gas concentrations", while the Fourth Assessment Report (AR4) published in 2007, conveyed the message that "most of the observed increase in globally averaged temperatures since the mid-20th century is *very likely* due to the observed increase in anthropogenic greenhouse gas concentrations". The qualifiers "*likely*" and "*very likely*" used in the last two statements should be understood in terms of probabilities in excess of 66% and 90%, respectively.

The Fifth Assessment Report (AR5) of IPCC WG I, published in September 2013, conveys an even stronger attribution statement: "It is extremely likely that more than half of the observed increase in global average surface temperature from 1951 to 2010 was caused by the anthropogenic increase in greenhouse gas concentrations and other anthropogenic forcings together." An additional term extending the calibrated uncertainty language used in IPCC AR5 was "*extremely likely*" that should be understood in terms of a probability in excess of 95%.

Following the "black PR" related to IPCC in the media (so called "climategate" and "glaciergate") in 2009/2010, the UN Secretary-General, Mr. Ban Ki-moon, and the IPCC Chair, Mr. Rajendra K. Pachauri, asked the InterAcademy Council (IAC) to carry out an independent review of IPCC processes and procedures. The IAC responded positively to that request, carried out a thorough review and issued its report and recommendations which have been considered in detail by the IPCC. In reaction to the IAC review, the Panel introduced a number of changes in the following areas: governance, review process, characterizing and communicating uncertainty, communication and transparency. Several new procedures were introduced and the existing ones were strengthened.

2.2.4 IPCC SREX messages in a nutshell

The IPCC Special Report on Managing the Risks of Extreme Events and Disasters to Advance Climate Change Adaptation (SREX), approved and accepted by the IPCC in November 2011, is indeed a very special deliverable of the Panel. Noting that extreme weather and climate events interacting with exposed and vulnerable human and natural systems can lead to disasters, this Special Report explores the physical as well as social dimensions of weather- and climate-related disasters, considering opportunities for managing risks at various spatial scales. The SREX report is of strong relevance to the present book, which is devoted to prevention of hydrometeorological extreme events and interfacing science and policy.

The SREX report assesses the scientific literature on a range of issues – from the relationship between climate change and hydrometeorological extreme events ('climate extremes') to the implications of these events for society and sustainable development. The assessment concerns the interaction of climatic, environmental and human factors that can lead to impacts and disasters, options for managing the risks posed by impacts and disasters, and the important role of non-climatic factors.

Since the risks of hydrometeorological extremes cannot be fully eliminated, disaster risk management and adaptation to climate change is important. This focusses on reducing system exposure and vulnerability and increasing resilience to changing risks. Adaptation and risk management can embrace improving risk preparedness systems as well as response and recovery, risk transfer and risk sharing.

The SREX report integrates perspectives from several historically distinct communities dealing with climate science, climate impacts, adaptation to climate change and disaster risk management. Each of these communities brings in different viewpoints, approaches, goals, cultures and vocabularies, providing important insights into the status of the knowledge base and its gaps. Several key assessment findings come from the interfaces among these communities.

The SREX report is a product of two IPCC Working Groups (I and II, see Section 2.2.1). It addresses, for the first time in the history of the IPCC, how integrating expertise in climate science, disaster risk management and adaptation can inform discussions on reducing and managing the risks posed by hydrometeorological extreme events in a changing climate. The report evaluates the role of climate change in altering characteristics of extreme events. It assesses experience with a wide range of adaptation and management options to reduce exposure and vulnerability and improve resilience to hydrometeorological extremes.

The SREX report consists of nine chapters. The first two chapters, of an introductory nature, deal with Climate Change: (1) New Dimensions in Disaster Risk, Exposure, Vulnerability, and Resilience, and (2) Determinants of Risk: Exposure and Vulnerability, respectively. Chapters three and four deal with Changes in Climate Extremes and Their Impacts, in reference to: (3) Natural Physical Environment, and (4) Human Systems and Ecosystems, respectively. Chapters five to seven deal with management of risk from climate extremes at three spatial scales: (5) Local, (6) National and (7) International. The latter chapter also discusses integration across scales. The penultimate chapter eight has the title "Toward a Sustainable and Resilient Future", while the last chapter presents Case Studies that illustrate specific extreme events and their impacts in different parts of the world, as well as a range of risk management activities.

In the SREX report, as well as in several other IPCC deliverables, it is attempted to accurately convey the authors' evaluation of the degree of uncertainty of key statements, following the guidance encapsulated in Mastrandrea *et al.* (2010). Author teams applied expert judgment, using two metrics to communicate the degree of uncertainty in key findings: qualitative confidence in the validity of a finding based on evaluation of the underlying scientific evidence and agreement; and quantified measures of uncertainty expressed as probabilities. Confidence in the validity of a finding is based on the type, amount, quality and consistency of evidence and the degree of agreement, while levels of confidence are expressed qualitatively through five qualifiers: *very low*, *low*, *medium*, *high* and *very high*. This guidance framework includes summary terms to describe the available evidence: *limited*, *medium*, or *robust*, and the degree of agreement: *low*, *medium*, or *high*. Where quantified measures of uncertainty in a finding can be expressed probabilistically (based on statistical analysis of observations or model results or expert judgment), a finding can include calibrated likelihood language. The likelihood terms and their corresponding probability ranges are as follows: *virtually certain*, 99–100% probability; *very likely*, 90–100% probability; *likely*, 66–100% probability; *about as likely as not*, 33–66% probability. Likewise, *unlikely* means 0–33% probability; *very unlikely*, 0–10% probability; and *exceptionally unlikely*, 0–1% probability. This calibrated uncertainty language was extended in AR5 by inclusion of the term "extremely likely" (95–100% probability), (cf. Section 2.2.3).

The report assesses how exposure and vulnerability to weather and climate events determine impacts and the likelihood of disasters (disaster risk). It evaluates the influence of natural climate variability and anthropogenic climate change on climate extremes and other weather and climate events that can contribute to disasters, as well as the exposure and vulnerability of human society and natural ecosystems. It also considers the role of development in trends in exposure and vulnerability, implications for disaster risk, and interactions between disasters and development. It examines how disaster risk management and adaptation to climate change can reduce exposure and vulnerability to weather and climate events and thus reduce disaster risk and increase resilience to the risks that cannot be eliminated.

Throughout the SREX report, observed or projected changes are marked with a qualifier related to confidence. However, it has to be stressed that assigning "low confidence" in observed or projected changes in a specific extreme neither implies nor excludes the possibility of changes in this extreme.

The character and severity of impacts from hydrometeorological extremes depend not only on the extremes themselves but also on system exposure and vulnerability to these extremes that may vary across temporal and spatial scales. Hydrometeorological extremes, exposure, and vulnerability are influenced by a wide range of factors, including cultural, institutional, governance and environmental factors, anthropogenic climate change, natural climate variability, and socioeconomic development, as illustrated in Figure 2.2.1, stemming from SREX.

Exposure and vulnerability are key determinants of disaster risk and of impacts when risk is realized. For example, a tropical cyclone can have very different impacts depending on where and when it makes landfall. Similarly, a heatwave can have very different impacts on different populations, depending on their vulnerability.

Extreme impacts can result from individual extreme hydrometeorological events or from a compounding of events or their impacts that may not be extreme when considered independently (e.g. drought, coupled with extreme heat and low humidity, can increase the risk of wildfire). Extreme impacts may also result from non-extreme events where system exposure and vulnerability are high.

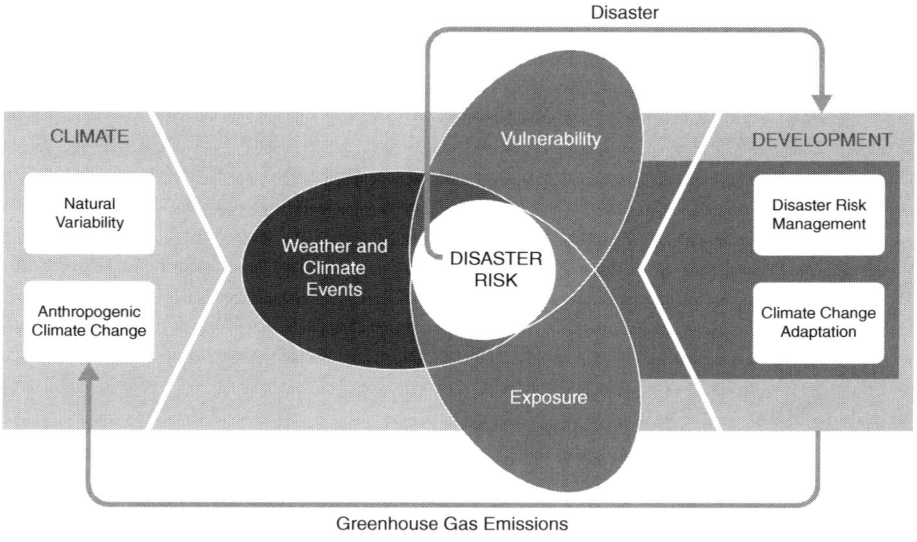

Figure 2.2.1 Illustration of the core concepts of the SREX report.

A changing climate may lead to changes in the frequency, intensity, spatial extent, duration and timing of extreme hydrometeorological events. The SREX report provides information on observed and projected changes in hydrometeorological events. In some parts of the world, increases in some extreme hydrometeorological events have been observed and further increases are projected for the future.

2.2.4.1 Observations

Confidence in observed changes in extremes depends on the quantity and quality of data records and the availability of studies analysing these data, which vary across regions and for different extremes. Since truly extreme events are rare (this statement is indeed a kind of tautology), even if a long time series of records is available, there are only few data available to make assessments regarding changes in their magnitude, frequency or intensity. The more rare the event the more difficult it is to identify long-term changes.

Observations gathered over the last 60 years demonstrate evidence of significant changes in some hydrometeorological extremes. As intuitively expected, cold extremes are getting less frequent, while warm extremes are getting more frequent in comparison to the reference period. As stated in SREX, it is *very likely* (90–100% probability) that there has been an overall decrease in the number of cold days and nights, and an overall increase in the number of warm days and nights, at the global scale, that is, for most land areas with sufficient data. At the regional scale, the qualifiers of likelihood differ for different regions.

The SREX report states that there have been statistically significant trends in the number of heavy precipitation events in some regions (Seneviratne *et al.*, 2012). It is *likely* (66–100% probability) that more of these regions have experienced increases than decreases, although there are strong regional and subregional variations in these trends.

There is *low confidence* in any observed long-term increases in tropical cyclone activity (i.e., intensity, frequency, duration). It is *likely* that there has been a poleward shift in the main Northern and Southern Hemisphere extratropical storm tracks.

There is *low confidence* in observed trends in small spatial-scale phenomena, such as tornadoes and hail, because of data non-homogeneities and inadequacies in monitoring systems.

There is *medium confidence* that some regions of the world have experienced more intense and longer droughts, in particular in southern Europe and West Africa, but in some regions droughts have become less frequent, less intense or shorter (e.g. in central North America and northwestern Australia).

Climate-driven observed changes in the magnitude and frequency of floods at regional scales are less evident, because the available instrumental records at gauge stations are limited in space and time, and because of confounding effects of changes in land use and flood defences.

It is *likely* that there has been an increase in extreme coastal high water related to increases in mean sea level.

Records show that global economic losses from hydrometeorological extremes have increased but there is large spatial and inter-annual variability. Observed changes in climate extremes reflect the influence of anthropogenic climate change in addition to natural climate variability, with changes in exposure and vulnerability influenced by both climatic and non-climatic factors. Generally, increasing exposure of people and economic assets (resulting from settlement patterns, urbanisation, and socioeconomic changes) has been the major cause of long-term increases in economic losses from hydrometeorological extremes (Handmer *et al.*, 2012). Hence, reduction in exposure and vulnerability are core common elements of adaptation and disaster risk management.

Long-term trends in economic disaster losses adjusted for wealth and population increases have not been attributed to climate change, but a role for climate change has not been excluded. Total absolute economic losses (including insured losses) from natural disasters are higher in developed countries, while relative economic losses (expressed as a proportion of gross domestic product, GDP) are higher in developing countries. Deaths from natural disasters occur much more in developing countries. For example, from 1970 to 2008 more than 95% of deaths from natural disasters occurred in developing countries. Loss estimates are lower-bound estimates because many impacts are difficult to value and monetise. For instance, impacts on the informal or undocumented economy as well as indirect economic effects are generally not counted in reported estimates of losses. In 2001–2006, losses amounted to about 1% of GDP for middle-income countries, while this ratio has been about 0.3% of GDP for low-income countries and less than 0.1% of GDP for high-income countries. In small exposed countries, particularly small island developing states, losses expressed as a percentage of GDP have been particularly high, exceeding 1% in many cases and 8% in the most extreme cases, averaged over both disaster and non-disaster years for the period 1970–2010 (IPCC, 2012).

There is evidence that changes in some extremes (e.g. warming of extreme daily minimum and maximum temperatures, intensification of extreme precipitation at the global scale, increasing extreme coastal high water due to an increase in mean sea level) are a result of anthropogenic influences, including increases in atmospheric concentrations of greenhouse gases. However, attribution of single extreme events to anthropogenic climate change is challenging.

2.2.4.2 Projections

The SREX report reviews and assesses projections of hydrometeorological extremes. It concludes that the confidence in projecting changes in the direction and magnitude of hydrometeorological extremes depends on multiple factors, such as the type of extreme (also the amount and quality of observational data, the level of understanding of the underlying processes and the reliability of their simulation in models), the region and the season. Projected changes in hydrometeorological extremes in the coming two to three decades generally do not diverge strongly under different emissions scenarios, but the projected changes over this time frame are relatively small compared to natural climate variability.

Models project substantial warming in temperature extremes by the end of the twentyfirst century. The SREX report concludes that it is *virtually certain* (99–100% probability) that increases in the frequency and magnitude of warm daily temperature extremes and decreases in cold extremes will occur throughout the twentyfirst century on a global scale. It is *very likely* that heat waves will increase in length, frequency and/or intensity over most land areas. For assumed A1B and A2 emissions scenarios, a 1-in-20 year hottest day is *likely* to become a 1-in-2 year event (i.e., occurring, on average, every year or every second year, i.e. 10–20 times more frequently than in the control period) by the end of the twentyfirst century in most regions, except in the high latitudes of the Northern Hemisphere, where it is *likely* to become a 1-in-5 year event. Under the B1 scenario, a 1-in-20 year event is *likely* to become a 1-in-5 year event (and a 1-in-10 year event in Northern Hemisphere high latitudes). The 1-in-20 year extreme daily maximum temperature (i.e., a value that was exceeded on average only once during the period 1981–2000) will *likely* increase by about 1 to 3 °C by the mid-twentyfirst century and by about 2 to 5 °C by the late twentyfirst century, depending on the region and emissions scenario (based on the B1, A1B, and A2 scenarios), cf. Seneviratne *et al.* (2012).

It is *likely* that the frequency of heavy precipitation or the proportion of total rainfall from heavy precipitation will increase in the twentyfirst century over many areas, in particular in the high latitudes and tropical regions, and in winter in the northern mid-latitudes. There is *medium confidence* that, in some regions, increases in heavy precipitation will occur despite decreases in total precipitation in those regions. Based on a range of emissions scenarios (B1, A1B, A2), a 1-in-20 year annual maximum daily precipitation amount is *likely* to occur more frequently, becoming a 1-in-5 to 1-in-15 year event by the end of the 21st century in many regions. In most regions the higher emissions scenarios (A1B and A2) lead to a stronger projected decrease in return period (Seneviratne *et al.*, 2012).

It is *likely* that the average maximum wind speed of tropical cyclones will increase throughout the coming century, although possibly not in all ocean basins, and that overall the global number of tropical cyclones will either decrease or remain essentially unchanged. There is *medium confidence* in a projected poleward shift of extratropical storm tracks and in reduction of the number of extratropical cyclones averaged over each hemisphere.

There is *medium confidence* that droughts will intensify in the twentyfirst century in some seasons and areas (such as southern Europe and the Mediterranean region, central Europe, central North America, Central America and Mexico, northeast Brazil, and southern Africa), due to reduced precipitation and/or increased evapotranspiration.

Overall there is *low confidence* at the global scale regarding climate-driven changes in magnitude or frequency of river-related flooding, due to limited evidence and because the causes of regional changes are complex. However, there is *medium confidence* (based on physical reasoning) that projected increases in heavy rainfall in some areas would contribute to increases in local flooding.

It is *very likely* that an average sea level rise will contribute to upward trends in extreme coastal high water levels.

Low-probability, high-impact changes associated with the crossing of poorly understood climate thresholds cannot be excluded in the future.

Extreme hydrometeorological events have (and will continue to have) greater impacts on sectors with closer links to climate, such as water, agriculture and food security, forestry, health and tourism. However, climate change is only one of the drivers of future changes and is not necessarily the most important driver. At the local scale, the main drivers of future increases in economic losses due to some climate extremes can be socioeconomic in nature. Future changes in exposure, vulnerability and climate extremes resulting from natural climate variability, anthropogenic climate change and socioeconomic development can alter the impacts on natural and human systems and the potential for disasters.

Uncertainty in projection of climate change and its impacts is high and should be reduced by addressing knowledge gaps (e.g. via enhanced observation and research), to help in designing effective adaptation and risk management strategies. There is *low confidence* in specific projections of changes in indices of hydrometeorological extremes. However, even in the absence of increased risk due to changes in the climate system, disaster risk will generally increase as exposure continues to rise.

2.2.4.3 Managing the risk of hydrometeorological extremes

One of important messages conveyed by the SREX report is that experience with disaster risk management and climate change adaptation provides a knowledge base for informing effective approaches to prepare for and respond to extreme climate- and weather-related events and disasters. Effective risk management generally involves a portfolio of risk-reducing actions, ranging from incremental improvements in governance and technology (e.g. improving infrastructure) to transformational changes (involving alterations of fundamental system attributes). Some measures (so-called no-regret or low-regret measures), such as development of early warning systems, changes in land use planning, sustainable land management, ecosystem management, improvements in health surveillance, water supplies and drainage systems, development and enforcement of building codes, better risk communication between decision makers and local citizens, and better education and awareness make sense under a broader range of future climates. Many measures for managing current and future risks have additional benefits and help address other development goals, such as improving peoples' livelihood and well-being, and conserving biodiversity. Post-disaster recovery and reconstruction provide an opportunity for reducing the risks posed by future weather- and climate-related disasters. However, short-term measures to protect people from immediate risks can increase future risks. For instance, improvements in levees may encourage further development in flood plains (maladaptation). The most effective adaptation and disaster risk reduction actions are those that offer development benefits in the relatively near term, as well as reductions in vulnerability over the longer term.

Risk management works best when tailored to local circumstances, hence combining local knowledge with additional scientific and technical expertise can help communities reduce their risk and adapt to climate change. There is room for improved integration across a range of scales.

The interactions among climate change mitigation, adaptation and disaster risk management may have a major influence on resilient and sustainable pathways.

2.2.5 Final remarks – AR5 is there

At the time of drafting this chapter, the IPCC was working on its Fifth Assessment Report (AR5), planned to be released in September 2013 (Working Group I) and in March and April 2014 (Working Groups II and III, respectively). The plenary sessions of the IPCC Working Group I and Working Group II approved and accepted the SPM of the WG I AR5 and WG II AR5 from 23 to 28 September 2013 and from 25 to 29 March 2014, respectively. There is no doubt that the AR5 conveys important and more recent information for the context of the present book, on top of the material contained in the IPCC Special Report on Managing the Risks of Extreme Events and Disasters to Advance Climate Change Adaptation (SREX), published in 2012 (IPCC, 2012). The AR5 assesses many more recent references, providing an important update of the SREX material.

Acknowledgements

The author has drawn from a wealth of IPCC products and particularly from the IPCC Special Report on Managing the Risks of Extreme Events and Disasters to Advance Climate Change Adaptation (SREX).

References

Handmer J., Honda, Y., Kundzewicz *et al.* (2012) Changes in impacts of climate extremes: human systems and ecosystems in *Managing the Risks of Extreme Events and Disasters to Advance Climate Change Adaptation. A Special Report of Working Groups I and II of the Intergovernmental Panel on Climate Change* (eds C.B. Field, V. Barros, T.F. Stocker *et al.*). Cambridge University Press, Cambridge and New York, pp. 231–290.

IPCC (2012) *Managing the Risks of Extreme Events and Disasters to Advance Climate Change Adaptation. A Special Report of Working Groups I and II of the Intergovernmental Panel on Climate Change* (eds C.B. Field, V. Barros, T.F. Stocker *et al.*). Cambridge University Press, Cambridge and New York, 582 pp.

Mastrandrea M.D., Field, C.B., Stocker, T.F. *et al.* (2010) *Guidance Note for Lead Authors of the IPCC Fifth Assessment Report on Consistent Treatment of Uncertainties*. Available at http://www.ipcc.ch.

Seneviratne S.I., Nicholls, N., Easterling, D. *et al.* (2012) Changes in climate extremes and their impacts on the natural physical environment in *Managing the Risks of Extreme Events and Disasters to Advance Climate Change Adaptation. A Special Report of Working Groups I and II of the Intergovernmental Panel on Climate Change* (eds C.B. Field, V. Barros, T.F. Stocker *et al.*) Cambridge University Press, Cambridge and New York, pp. 109–230.

2.3

A Snapshot of EU and International Policies Relevant to Hydrometeorological Events

Philippe Quevauviller[1]

Department of Hydrology and Hydrological Engineering, Vrije Universiteit Brussels, Brussels, Belgium

2.3.1 Introduction – A complex policy framework

The Technical Paper VI of the Intergovernmental Panel on Climate Change (IPCC) highlights that observational records and climate projections provide abundant evidence that freshwater resources are vulnerable towards climate change, with wide-ranging consequences for human societies and ecosystems in Europe and worldwide. In particular, observed warming over several decades has been linked to changes in the large-scale hydrological cycle (e.g. effects on atmospheric water vapour content, changes of precipitation patterns with consequences on extreme floods and droughts). The consequences of climate change, in particular the increased frequency and severity of extreme hydrometeorological events, may alter the reliability of current water management systems. While quantitative projections of changes in precipitation, river flows and water levels at the river-basin scale remain uncertain, it is very likely that hydro(meteoro)logical characteristics will change in the future. These considerations lead to the development of a complex policy framework designing adaptation and mitigation options to tackle impacts of global warming on water resources and risks to society and assets. These options are closely linked to a range of policies. This chapter gives an outline of some policies relevant to hydrometeorological events, with no pretention of exhaustiveness. Some considerations have been adapted from previous publications of the author (Quevauviller *et al.*, 2011; Quevauviller, 2011).

[1] The views expressed in this chapter are purely those of the author and may not in any circumstances be regarded as stating a formal position of any EU institution

Hydrometeorological Hazards: Interfacing Science and Policy, First Edition. Edited by Philippe Quevauviller.

2.3.2 Climate change impacts on water

There is far-reaching consensus among scientists that climate change is, at least to a certain extent, caused by human activities. According to the terminology of the EU Water Framework Directive (2000/60/EC) discussed in Section 2.3.4, direct climate change impacts on water resources should not be classified as an "anthropogenic pressure" in a narrow sense, since they cannot be mitigated by water managers' action (Wilby *et al.*, 2006). However, climate change impacts interact with and potentially aggravate other anthropogenic pressures and *could* therefore be considered as an anthropogenic pressure, for example changes in the frequency and severity of precipitation or drought events have direct impacts on water resources and society as a whole. In particular, with the change of rainfall patterns, seasonality and spatial distribution, impacts of climate change are reflected in influences on the quantity and quality of water resources and impacts on their uses, e.g. abstraction of both surface and groundwaters. Sustainable water resource management is hence closely connected to various drivers, including climate change, land cover and increasing water consumption (Figure 2.3.1).

Long-term threats to groundwater resources could be linked to intense rainfall, resulting in surface flooding rather than infiltration to groundwater. Water quality will also be affected in that run-off takes nutrients and pesticides from agricultural land and transfers them into rivers and lakes, for instance. Less availability of water resources will mean lower quality in some cases, for example droughts can have an impact on the ecology of rivers. Extremes may also have adverse effects on aquatic ecosystems, leading to not only a water shortage problem but also a large environmental problem

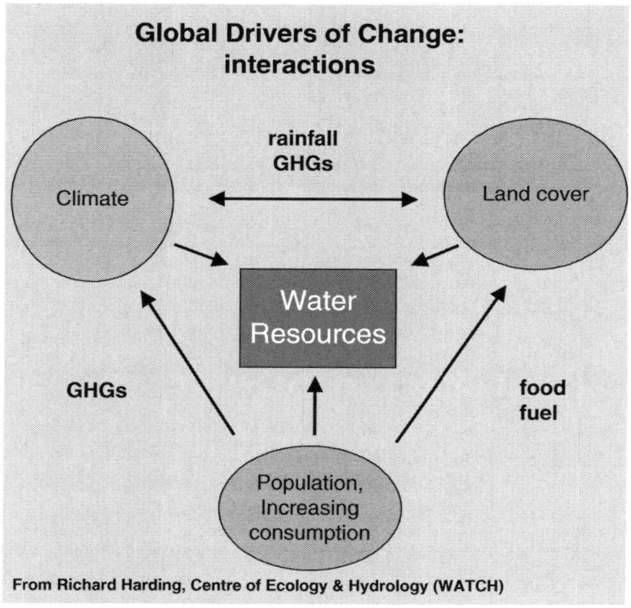

Figure 2.3.1 Sustainable water management in the context of global drivers of change. *Source:* Courtesy of R. Harding, WATCH project 'Water and Global Change'.

(up to desertification in some areas). Possible impacts to be anticipated also concern issues of water demand (increased demand and/or differing patterns), water infrastructures (e.g. in low-lying and coastal areas prone to flood risks, soil movement, effects on water treatment etc.), sewer network operations (e.g. disruption due to extreme rainfall with risks of pollution in industrial areas), threats to economic development of the water sector and so on.

The assessment of impacts of hydrometeorological extreme events on water resources implies a good knowledge of their global/regional distribution. This implies that the most sensitive hydrological systems should be identified at the scale of river basins (this is linked to the WFD (Water Framework Directive) river basin management principles, see Section 2.3.4). It also means that climate-induced changes in hydrological systems and processes should be better understood; in particular, variables such as river flows, groundwater and lake levels, soil moisture, evapotranspiration, snow cover, glacier extent, permafrost and so on, as well as impacts on biodiversity. Besides impacts on natural conditions, climate change impacts (including hydrometeorological extremes) on sectoral water uses should also be evaluated, for example on agriculture (rain-fed and irrigated), forestry (including forest fires and deforestation), hydropower, navigation and water supply (domestic, agricultural and industrial). Furthermore, water-related impacts on infrastructure, health, transport, financial services (e.g. insurance sector), energy and tourism should also be reviewed.

While research on climate change in general, and hydrometeorological events in particular, is progressing fast, there is still much uncertainty, particularly with regard to water-related changes. There is also large uncertainty about future projections of climate change impacts on waters during the forthcoming decades. Over this time frame, mean temperatures are expected to continue to rise but large year-to-year variations in precipitation will probably mask underlying regional trends for several decades. This implies that temperature-dependent processes (such as seasonal snowmelt, species' distribution and phenology etc) will probably manifest change in the first instance. An increase in extreme events (floods and droughts) is also likely to occur. Uncertainties stem from different sources, for example difficulties in predicting future socio-economic development (scenario uncertainty), unsatisfactory model resolution and insufficient mathematical description of all global circulation processes (model uncertainty, especially for precipitation), lack of local hydrological localised models and so on. Attributing these hydrometeorological extremes to climate change is still uncertain because of a lack of accurate data and full scientific understanding of the functioning of the climate system.

2.3.3 Policy background

The need for policy responses to tackle climate change impacts on water, including hydrometeorological extreme events, is recognised worldwide. This is extensively expressed in the IPCC Technical Paper on Water (Bates *et al.*, 2008), which is addressed primarily to policy-makers engaged in all areas related to freshwater resource management, climate change, strategic studies, spatial planning and socio-economic development. This Technical Paper evaluates the impacts of climate change on hydrological processes and regimes, and of freshwater resources (availability, quality, uses and management), at a worldwide scale, and highlights their implications for policy, looking at different sectors. In particular, it provides recommendations regarding adaptation measures in regions prone to climate-change-related extremes about water resource management, ecosystems, agriculture and forestry, coastal systems, sanitation and human health. Some statements issued from the IPCC Technical Paper are summarised in Table 2.3.1. This short introduction only serves to highlight that awareness for policy actions is growing worldwide but that, to date, no legal framework is in place to tackle climate change impacts on water at a global scale.

2.3.4 International policies

While climate change impacts on hydrometeorological events and proactive management efforts are recognised worldwide, and recommendations are issued in many expert groups, there is not yet a 'global policy' dealing with specific measures related to extreme events. This need is, however, clearly expressed by the UN Economic Commission for Europe (UNECE) in a guidance document published in 2009 (United Nations, 2009a). The key messages go on the same paths as recommendations expressed in the IPCC Technical Paper on water (Bates *et al.*, 2008), in particular concerning the negative impacts on nearly all UNECE countries, among others those linked to increased frequency and intensity of floods and droughts and their impacts on ecosystems and human health. Regarding policy, UNECE recommends that any policy needs to consider climate change (including extreme hydrometeorological events) as one of the many pressures affecting resources and economic developments. Effective adaptation in this respect will require a cross-sectoral approach, including at the trans-boundary level in order to present possible conflicts between different sectors and consider trade-offs and synergies between adaptation and mitigation pressures. The guidance insists on the fact that legislation should not present barriers for adaptation and should be flexible enough to accommodate continuing environmental and socio-economic changes. Besides national legislation, a number of international agreements include provisions that can support the development of adaptation strategies (this includes the WFD, see Section 2.3.5), and trans-boundary cooperation and policy might certainly build upon this basis (UNECE, 2009). Finally, the UNECE guidance stresses that uncertainty should not be a reason for inaction and highlights that action, knowledge and experience sharing, and research on adaptation should be pursued simultaneously and in a flexible way.

One of the key policy trends at international level is represented by the Hyogo Framework for Action 2005–2015 (HFA). This programme was adopted by the United Nations in January 2005, with 168 nations committed to substantially reducing the loss of life and livelihoods from disasters. The scope of the HFA goes clearly beyond the sole water-related disasters (as it includes, for example, earthquakes, tsunamis, volcanic eruptions and storms) which are, however, being considered; in particular, extreme floods and droughts. The implementation of the HFA is under the responsibility of the United Nations International Strategy for Disaster Reduction (UN-ISDR) which is the focal point in the UN system for the coordination of disaster reduction and to ensure synergies among the disaster-reduction activities of the UN and regional

Table 2.3.1 Summary of IPCC statements about climate change and water. *Source:* After Bates *et al.*, 2008.

Key statements	Additional comments
Observed warming over several decades has been linked to changes in the large-scale hydrological cycle	This is reflected in increasing atmospheric water vapour content, changing precipitation patterns, intensity and extremes, reduced snow cover, and changes in soil moisture and run-off.
Climate model simulations for the twentyfirst century are consistent in projecting precipitation increases in high latitudes (*very likely*) and parts of the tropics, and decreases in some sub-tropical and lower mid-latitude regions (*likely*)	Outside these areas, the sign and magnitude of projected changes varies between models, leading to substantial uncertainty in precipitation projections. Thus, projections of future precipitation changes are more robust for some regions than for others. Projections become less consistent between models as spatial scale decreases.
By the middle of the twentyfirst century, annual average river run-off and water availability are projected to increase as a result of climate change at high latitudes and in some wet tropical areas, and decrease over some dry regions at mid-latitudes and in the dry tropics	This statement excludes changes in non-climatic factors, for example irrigation. Many semi-arid and arid areas (e.g. the Mediterranean Basin, Western USA, southern Africa and north-eastern Brazil) are particularly exposed to the impacts of climate change and are projected to suffer a decrease in water resources due to climate change (*high confidence*).
Increased precipitation intensity and variability are projected to increase the risks of flooding and drought in many areas	The frequency of heavy precipitation events (or proportion of total rainfall from heavy falls) will be *very likely* to increase over most areas during the twentyfirst century, with consequences for the risk of rain-generated floods. At the same time, the proportion of land surface in extreme drought is projected to increase (*likely*).
Water supplies stored in glaciers and snow cover are projected to decline in the course of the century	This is linked to a projected reduction of water availability during warm and dry periods (through a seasonal shift in streamflow, an increase in the ratio of winter to annual flows, and reductions in low flows) in regions supplied by melt water from major mountain ranges, where more than one-sixth of the world's population currently live (*high confidence*).
Higher water temperatures and changes in extremes, including floods and droughts, are projected to affect quality and exacerbate many forms of water pollution	Water pollution is projected to increase from sediments, nutrients, dissolved organic carbon, pathogens, pesticides and salt, as well as thermal pollution, with possible impacts on ecosystems, human health, and water system reliability and operating costs (*high confidence*). In addition, sea-level rise is projected to extend areas of salination of groundwater and estuaries.
Globally, the negative impacts of future climate change on freshwater systems are expected to outweigh the benefits (*high confidence*)	By the 2050s, the area of land subject to increasing water stress due to climate change is projected to be more than double that with decreasing water stress. In many regions, the benefit linked to increased water supply is likely to be counterbalanced by the negative effects of increased precipitation variability and seasonal run-off shifts in water supply, water quality and flood risks (*high confidence*).

(*continued overleaf*)

Table 2.3.1 (*continued*)

Key statements	Additional comments
Changes in water quantity and quality due to climate change are expected to affect food availability, stability, access and utilisation	This is expected to lead to decreased food security and increased vulnerability of poor rural farmers, especially in the arid and semi-arid tropics and Asian and African megadeltas.
Climate change affects the function and operation of existing water infrastructure – including hydropower, structural flood defences, drainage and irrigation systems – as well as water management practices	Adverse effects of climate change on freshwater systems aggravate the impacts of other stresses, such as population growth, changing economic activity, land-use change and urbanisation.
Current water management practices may not be robust enough to cope with the impacts of climate change	This may affect water supply reliability, flood risk, health, agriculture, energy and aquatic ecosystems. In many locations, water management cannot cope satisfactorily even with current climate variability, so that large flood and drought damages occur. Climatic and non-climatic factors, such as growth of population and damage potential, would exacerbate problems in the future (*high confidence*).
Climate change challenges the traditional assumption that past hydrological experience provides a good guide to future conditions	The consequences of climate change may alter the reliability of current water management systems and water-related infrastructure. While quantitative projections of changes in precipitations, river flows and water levels at the river basin scale are uncertain, it is *very likely* that hydrological characteristics will change in the future.
Adaptation options designed to ensure water supply during average and drought conditions require integrated demand-side as well as supply-side strategies	The former improve water-use efficiency, for example by recycling water. An expanded use of economic incentives, including metering and pricing, to encourage water conservation and development of water markets and implementation of virtual water trade, holds considerable promise for water savings and the reallocation of water to highly efficient water uses.
Mitigation measures can reduce the magnitude of impacts of global warming on water resources, in turn reducing adaptation needs	However, they can have considerable negative side effects, such as increased water requirements for afforestation/reforestation activities or bio-energy crops. On the other hand, water management policy measures, for example hydrodams, can influence greenhouse gas emissions.
Water resources management clearly impacts on many other policy areas	Other areas concerned are, for example, energy, health, food security and nature conservation. Thus the appraisal of adaptation and mitigation options needs to be conducted across multiple water-dependent sectors.
Several gaps in knowledge exist in terms of observations and research needs related to climate change and water	Observational data and data access are prerequisites for adaptive management, yet many observational networks are shrinking. There is a need to improve understanding and modelling of climate changes related to the hydrological cycle at scales relevant to decision-making.

organisations, and activities in socio-economic and humanitarian fields. More information about the UN-ISDR objectives can be found in the 2010–2011 Biennial Work Programme (United Nations, 2009b). The so-called HF2 (post-Hyogo Framework) was under preparation at the time of writing this chapter, for a continuation after 2015.

2.3.5 EU water policies

2.3.5.1 The water framework directive

The Water Framework Directive (European Commission, 2000) is based on integrated water resources management principles, following a stepwise approach regarding risk characterization (including hydrometeorological risks), monitoring, programmes of measures embedded into river basin management planning (Chave, 2001). Besides "classical" risks of water quality deterioration and overexploitation (which have to be considered in the light of "good status" achievements by 2015), hydrometeorological extreme events (in particular floods and droughts) are increasingly discussed in relation to their implications for water management. The basic feature to be kept in mind in the context of this chapter is that the WFD is built upon the principles of river basin management planning (Figure 2.3.2), considering all types of waters and pressures that may affect them.

Climate change is all the more a highlight with the development of the first river basin management planning under the WFD, which started operationally in 2010. The integration of knowledge about possible climate change impacts on water policy implementation concerns various above-mentioned technical aspects (risk characterization,

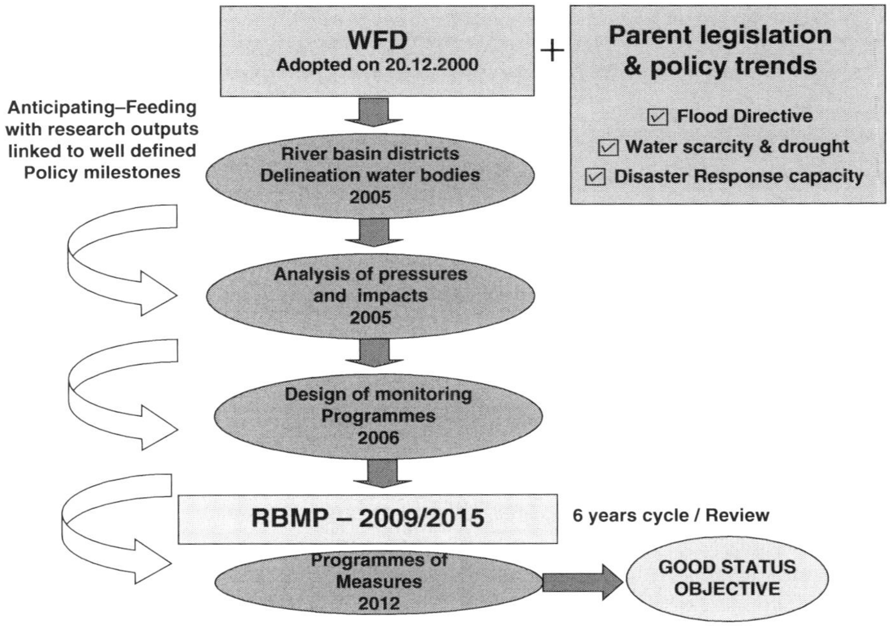

Figure 2.3.2 EU water policy milestones.

monitoring, action programmes) as well as the evaluation of the "good status" objective's achievements (in 2015). These questions have been and are currently discussed with a wide range of experts and stakeholders in the framework of working groups operated under the so-called WFD Common Implementation Strategy (CIS).

2.3.5.2 WFD and climate change

Climate change might affect and interact with all steps of WFD implementation, and thus on the status objectives, and this has been subject to in-depth discussions within the policy and scientific communities over the years 2008–2010, as reflected in the literature (Wilby *et al.*, 2006; Quevauviller *et al.*, 2011) and in a guidance document of the WFD Common Implementation Strategy examining river basin management in a changing climate (European Commission, 2009a).

Through its stepwise approach, the WFD makes IWRM principles operational with the aim of achieving good water status by 2015 (chemical status for all waters, ecological status for surface waters and quantitative status for groundwaters). The different milestones are summarised below:

- **Characterization of water bodies and pressures**. In the context of the WFD, this involves a review of the impact of human activities (and related pressures) on the status of surface and ground waters. Several factors used in this review are based on water system typologies that are themselves variables according to the climate (hence to climate change). This includes, for instance, river flow categories, energy of flow, precipitation patterns, water level fluctuations and so on, with indirect impacts on pollution patterns (affecting both point and diffuse sources through changes of flows, run-offs etc.). This means that some characteristics of water bodies might be modified due to climate change, with effects on their status. This classification had to be carried out by Member States between 2004 and 2005 and the results reported to the European Commission. A report giving a synthesis of Member States' reports was prepared by the European Commission and published in March 2007.
- **Risk assessment**. The characterization of water bodies is an essential part of the WFD as it aims to forecast risks and calculate costs and benefits of the programme of measures. As expressed above, modifications of water bodies' characteristics due to climate change could lead to potential impacts on good status achievements (due to, for example, changes in water temperature, decreased dilution capacity of receiving waters, exceedence of water quality standards, changing metabolic rates of organisms, fish migration patterns, increased eutrophication, changes of river flows etc.; the list is not exhaustive).
- **Prevention of status deterioration**. Changes in the flow regime and physicochemistry of rivers could have significant impacts on key species that could alter ecological status achievements, in particular in protected water bodies, e.g. effects on spawning conditions for salmon, climate-driven shifts in species and community composition and so on.
- **Achievement of good status**. Following from the above, shifts in surface water bodies' characteristics might have effects on WFD status achievements, in particular concerning compliance to environmental water quality standards (chemical status), impacts on fish mortality and biota composition (ecological status) and

so on. At the other end of the spectrum, increased flood frequency might also impact on status objectives through increased sediment loads and mobilisation of contaminated sediments. Groundwater bodies may also be affected, for example through baseline shifting from natural conditions, enhanced downward migration of, for example, agricultural pollutants, saline intrusions in coastal aquifers due to rising sea levels, reduced groundwater recharge (with effects on quantitative status) and so on.

- Establishment of monitoring networks based on the results of characterization and risk assessment (performed in 2004–2005) to provide a comprehensive overview of water status. Monitoring programmes had to be designed by Member States and made operational by the end of 2006. This step is an essential part of the overall management cycle as monitoring data will constitute the backbone of status assessment. Monitoring efficacy to check compliance to WFD objectives might also be affected by shifts in water body characteristics, for example increased river flows with greater dilution making sites "more compliant" towards environmental quality standards. Also, impacts of extreme events may be problematic at low monitoring frequencies. Monitoring strategies in the light of possible impacts of climate change would hence need to be reviewed at regular intervals (this is actually foreseen under the WFD framework).
- Establishment of registers of protected areas within each River Basin District, which has been designated as requiring specific protection of its surface and groundwaters or for the conservation of habitats and species directly dependent on water. These include water used for the extraction of drinking water and all protected areas covered under the following directives: the Bathing Water Directive[2], the vulnerable zones under the Nitrates Directive[3], the sensitive areas under the Urban Wastewater Directive[4], and the areas designated for the protection of habitats and species including relevant Natura 2000 sites designated under the Habitats[5] and Wild Birds[6] Directives.
- Development of a River Basin Management Plan (RBMP) for each river basin district in the European Union (including transboundary river basins), including a summary of pressures and impacts of human activity on water status, a presentation in map form of monitoring results, a summary of the economic analysis of water use, a summary of protection programmes and control and remediation measures. The first RBPM was published by the end of 2009 further to a public consultation which took place in 2008. A review is planned by the end of 2015 (prior to the launching of the second river basin management plan, which will also be subject to public consultation) and every six years thereafter.
- Design of a programme of measures for achieving WFD environmental objectives that should be operational by the end of 2012. Basic measures include, for example controls of groundwater abstraction, controls of point source discharges and diffuse sources liable to cause pollution and so on. Supplementary measures include for example taxation, research and so on (a full list of measures is available in Annex VI

[2] Directive 76/160/EEC, OJ L31 of 5.02.1976
[3] Directive 91/676/EEC, OJ L375 of 31.12.1991
[4] Directive 91/271/EEC, OJ L135 of 30.05.1991
[5] Directive 92/43/EEC, OJ L206 of 22.07.1992
[6] Directive 79/409/EEC, OJ L103 of 25.04.1979

of the WFD). The programme of measures has to be reviewed and if necessary updated by 2015 and every six years thereafter. Programmes of measures necessary to deliver WFD objectives may also be affected directly or indirectly as these depend upon the above operational steps (characterization, analysis of pressures and impacts, in particular). The success of the programme of measures will be closely related to the accurate characterization in the first place, and flexibility to future changes in climate. They should also accommodate possible changes in behaviour ahead of climate change, such as adaptation measures in spatial strategies.

It may, therefore, be considered that the successive steps of the WFD RBMP process provide a convenient structure for incorporating adaptation to climate-related water risks (including extreme events leading to possible disasters) through risk assessment, monitoring, environmental objective setting, economic analysis and action programmes to achieve well-defined environmental objectives (European Commission, 2009a).

Besides the management framework established by the WFD, climate-related disasters (in particular floods and droughts) are considered in the Flood Directive (2007/60/EC) and the Water Scarcity and Drought Communication highlighting current policy developments in this area. It should be noted that climate change is not classified as an anthropogenic pressure in a narrow sense in the terminology of the WFD, even if there is a general consensus among scientists that climate change is at least to a certain extent caused by human activities. Indeed, climate change impacts cannot be mitigated by current WFD programmes of measures (linked to the implementation of various EU directives), which are essentially directed towards anthropogenic pressures (mainly pollution but also overexploitation of water resources).

2.3.5.3 The flood directive

The Floods Directive requires EU Member States to assess and manage flood risks, with the aim of reducing adverse consequences for human health, the environment, cultural heritage and economic activity associated with floods in Europe (European Commission, 2007a). This directive is included in a larger "Flood Action Programme" (Figure 2.3.3) for the assessment and management of flood risks aimed at reducing the adverse consequences for human health, the environment, cultural heritage and economic activity associated with floods in Europe.

Besides the legislative framework, the package covers an optimal use of funding instruments (e.g. capacity-building projects), information exchange and research. This directive is coordinated with the implementation of the WFD from the second RBMP onward. It therefore provides a comprehensive mechanism for assessing and monitoring increased risks of flooding due to climate change and for developing appropriate adaptation approaches. The coordinated approach with the RBMPs will ensure an overall effective adaptation approach.

2.3.5.4 The communication on drought

Alongside the increased risk of flooding, the policy implications of increased frequency of droughts due to climate change have to be considered. In this respect, the European

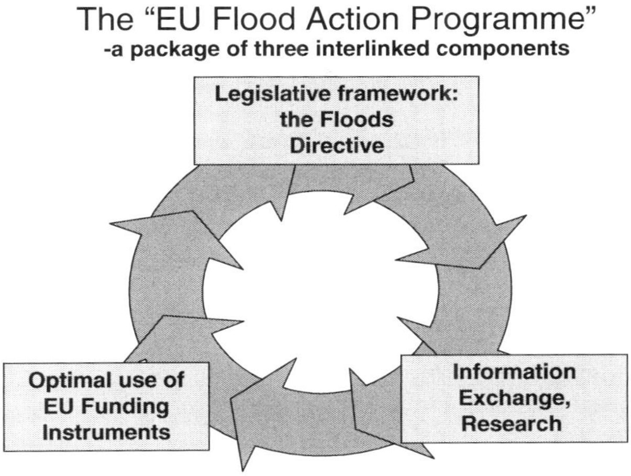

Figure 2.3.3 The EU Flood Action Programme.

Commission has adopted a Communication addressing the challenge of water scarcity and droughts in the EU (European Commission, 2007b), which sets out a number of policy options for addressing the challenge of water scarcity. The Commission is conducting an annual European assessment of water scarcity and droughts, making it possible to monitor changes across Europe and to identify where further action is needed in response to climate change. In addition, a review of the strategy for water scarcity and droughts is planned for 2012.

2.3.6 Climate adaptation strategy

In April 2013 the European Commission adopted an EU strategy on adaptation to climate change which has been welcomed by the EU Member States (European Commission, 2013). The strategy aims to make Europe more climate-resilient. By taking a coherent approach and providing for improved coordination, it will enhance the preparedness and capacity of all governance levels to respond to the impacts of climate change. The strategy focuses on three key objectives:

- **Promoting action by Member States**: The EU encourages all Member States to adopt comprehensive adaptation strategies (currently 15 have strategies) and provides funding to help them build up their adaptation capacities and take action, including actions related to preparedness to extreme climatic events.
- **'Climate-proofing' action at EU level** by further promoting adaptation in key vulnerable sectors, such as agriculture, fisheries and cohesion policy, ensuring that Europe's infrastructure is made more resilient, and promoting the use of insurance against natural and man-made disasters. This climate-proofing goes along the way the second river basin management planning of the WFD is being developed in a "climate-resilient" fashion.

- **Better informed decision-making** by addressing gaps in knowledge about adaptation and further developing the European climate adaptation platform (Climate-ADAPT) as the 'one-stop shop' for adaptation information in Europe.

EU adaptation actions include mainstreaming of climate change (mitigation and adaptation) into EU sector policies and funds, including not only marine and inland water issues, forestry, agriculture, biodiversity, infrastructure and buildings, but also migration and social issues.

The EU also addresses knowledge gaps through research and Climate-ADAPT. This platform, launched in March 2012, provides several useful resources to support adaptation policy and decision making, such as: a toolset for adaptation planning; projects and case studies' database; and information on adaptation action at all levels, from the EU through regional and national to the local level.

Moreover, stakeholders from the local, regional and national level are encouraged to participate in the development of the EU Adaptation Strategy. The EU is providing guidelines on integrating climate into policies and investments and on how to use the instruments and funds provided by the Commission for climate change adaptation. For instance, EU-Cities Adapt is an EU initiative to train and exchange knowledge among stakeholders at city level.

In the hydrometeorological sector, climate adaptation needs are reflected in a guidance document published by the European Commission, which examines river basin management in a changing climate (European Commission, 2009a). In this respect, possible hydrometeorological risks affecting water management (mainly related to floods and droughts) are not specifically addressed in the WFD, which, however, provides a framework to include climate change impacts into the planning process. In particular, the requirement to identify and characterize 'significant pressures' affecting waters could be considered as including climate change (Wilby *et al.*, 2006). The guidance document has actually been built upon the principles of the European Commission's White Paper on 'Adapting to climate change' (European Commission, 2009b), which itself is the precursor of the EU Climate Change Adaptation Strategy.

2.3.7 Conclusions

While hydrometeorological risks are not considered in the framework of one single policy, this chapter shows that various policy instruments are directly or indirectly taking these risks into consideration at different operational levels (international, European, national or even regional). The absence of a tailor-made framework leads to some possible lack of coordination which may be observed either at international or national level, owing to the wide variety of key players and the lack of appropriate interfacing mechanism. Overall, policy tools are in place for an efficient management of hydrometeorological extreme events; operationally, the implementation is, however, prone to many drawbacks which lower the efficiency of concrete actions. Considerable progress has been made from 2010 onward with an improved strategic thinking; the challenge ahead is now to implement identified policy actions.

References

Bates B.C., Kundzewicz Z.W., Wu S. and Palutikof J.P. (eds) (2008) *Climate Change and Water*, Technical Paper of the Intergovernmental Panel on Climate Change, IPCC Secretariat, Geneva, 210 pp.

Chave P. (2001) *The EU Water Framework Directive*, IWA Publishing, 208 pp.

European Commission (2000) Directive 2000/60/EC of the European Parliament and of the Council of 23 October 2000 establishing a framework for Community action in the field of water policy, *Official Journal of the European Communities*, L 327, 22.12.2000, 1.

European Commission (2007a) Directive 2007/60/EC of the European Parliament and of the Council of 23 October 2007 on the assessment and management of flood risks, *Official Journal of the European Communities*, L 288, 6.11.2007, 27.

European Commission (2007b) Communication to the European Parliament and the Council – Addressing the challenge of water scarcity and droughts in the European Union, COM/2007/04141 final.

European Commission (2009a) *River Basin Management in a Changing Climate*, Common Implementation Strategy for the Water Framework Directive, Guidance document No. 24, ISBN 978-92-79-14298-7.

European Commission (2009b) White Paper Adapting to climate change: Towards a European framework for action, COM(2009) 147 final.

European Commission (2013) An EU Strategy on Adaptation to Climate Change, COM(2013) 216 final.

Quevauviller Ph. (2011) Adapting to climate change: Reducing water-related risks in Europe – EU policy and research considerations, *Environmental Science and Policy*, **14**(7), 722.

Quevauviller Ph., Borchers U., Thompson K.C., Simonart T. (eds) (2011) *The Water Framework Directive – Action Programmes and Adaptation to Climate Change*, RSC Publishing, Cambridge, ISBN: 978-1-84973-053-2, 214 pp.

United Nations (2009a) *Guidance on Water and Adaptation to Climate Change*, United Nations, New York and Geneva, ISBN: 978-92-1-117010-8.

United Nations (2009b) *Reducing Disaster Risks through Science: Issues and Actions*, the full report of the ISDR Scientific and Technical Committee, UN-ISDR, Geneva, Switzerland, 213 pp.

Wilby R.L., Orr H.G., Hedger M., Forrow D., Blackmore M. (2006) *Environment International*, **32**, 1043.

Part Three

Outline of Scientific Features

3.1
Hydroinformatics and Its Role in Flood Management

Philippe Gourbesville

Université Nice Sophia Antipolis, Polytech Nice Sophia, I-CiTy Lab, Nice, France

3.1.1 Background

In 2013, floods were the leading natural hazard. Over the last five years, major events have affected more than 600 million inhabitants in Asian countries and constitute one of the main obstacles to economic and social development. The recent events in Pakistan (2010), in Thailand (2011) and in China (Beijing, 2012) have demonstrated the magnitude of the hydrological processes and the consequences for the impacted communities. Many factors contribute to increasing the vulnerability of human societies regarding flooding processes: growing urbanization, changes in land use, climate change, extreme rainfall events and so on. In order to address part of the flooding challenge, the use of hydroinformatics methods, combining information and communication technology (ICT), water sciences and decision-making methods, can be promoted efficiently (Gourbesville, 2009, 2011). This approach, closely related to the technological evolution, could be able to provide a real support in order to cope with the risk and to promote more resilient societies.

3.1.2 Flood management in water-related activities

Flood management involves engineering activities which are dedicated to minimizing the impact of waters on human societies. The definition of this set of activities has to be connected to the actions taking place in the global domain of water management.

The water cycle is frequently defined as the hydrologic cycle which describes the continuous movement of water on, above and below the surface of the Earth. The hydrologic cycle involves the exchange of heat energy, which leads to temperature changes and drives states of water. The water cycle figures significantly in the maintenance of life and ecosystems. In order to preserve this essential resource, the concept of integrated water resources management (IWRM) has been developed (Jønch-Clausen, 2004). The purpose of the approach is to "promote the coordinated development and management

Hydrometeorological Hazards: Interfacing Science and Policy, First Edition. Edited by Philippe Quevauviller.

of water, land and related resources, in order to maximize the resultant economic and social welfare in an equitable manner without compromising the sustainability of vital ecosystems." Operationally, IWRM approaches involve applying knowledge from various disciplines as well as the insights from diverse stakeholders to devise and implement efficient, equitable and sustainable solutions to water and development problems. As such, IWRM is a comprehensive, participatory planning and implementation tool for managing and developing water resources in a way that balances social and economic needs, and that ensures the protection of ecosystems for future generations.

In such an approach, the flood management does not appear as a key issue and is viewed as an action among many others. In order to identify which and how technical solutions – hydroinformatics methods and tools – can be implemented, it is necessary to look at the water cycle through an approach based on functional domains and business processes. This methodology allows one to consider each action involved in the resource management and identify the potential needs.

The water cycle can be divided into three domains which are associated with specific activities and business processes (Figure 3.1.1):

- Water uses
- Protection of natural environment and ecosystems
- Natural hazards mitigation and disaster prevention.

The first domain covers the added influence of human activity on the water cycle. Generally, the water uses refer to use of water by agriculture, industry, energy production and households, including fishing, recreation, transportation and waste disposal. All of these uses are directly linked to specific activities and processes which are potential targets for deployment of ICT solutions. In order to stick to the reality of the water management operated by entities in charge of water services, the traditional classification can be reviewed. The main water uses appear then as: agriculture, aquaculture, industry, recreation, transport/navigation, and urban.

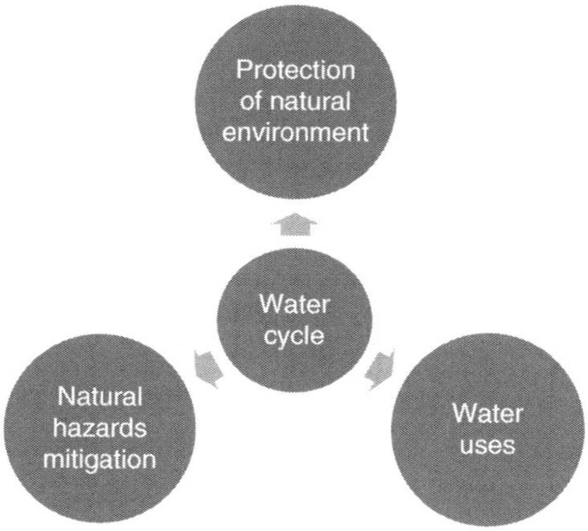

Figure 3.1.1 Domains of the water cycle.

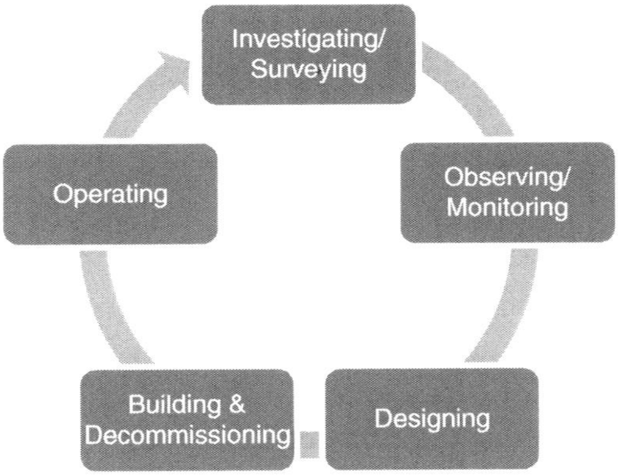

Figure 3.1.2 Invariant activities taking place in the various domains and water uses.

The second domain considers all actions needed to assess and advise on the environmental impacts of development proposals and projects related to specific water uses. The results are used by regulatory services. The domain also covers all conservation actions of water-related ecosystems.

The third domain is focused on water-related natural hazards mitigation actions. Floods, water-borne and vector disease outbreaks, droughts, landslide and avalanche events and famine are the processes covered by this domain. Every year, disasters related to meteorological, hydrological and climate hazards cause significant loss of life, and set back economic and social development by years. The disaster is defined as a serious disruption of the functioning of a community or a society causing widespread human, material, economic and/or environmental losses.

In the three domains, five major activities are taking place and appear as invariants (Figure 3.1.2). These key activities are: investigating/surveying, observing/monitoring, designing, building and decommissioning, operating. Each activity can be defined.

- Investigating/surveying: consists in the gathering of information of the previous and actual state and/or working of the domain under study. This assembly of information can be done either by a systematic collection of field data (survey) or a collection of information or data from a methodical research of available documents and/or the production of new ones in order to understand or to improve the actual state of the domain.
- Observing/monitoring: from a general point of view, this activity refers to being aware of the state of a system. It describes the processes and activities that need to take place to characterize and monitor the quality and/or state of the domain under study. All monitoring strategies and programmers have reasons and justifications which are often designed to establish the current status of the domain or to establish trends in its parameters. In all cases the results of monitoring will be reviewed and analysed. The designer of a monitoring programme must therefore have regard to the final use of the data before monitoring starts.
- Designing (including risk assessment): refers to the process of devising a system, component, or process to meet desired needs. It is a decision-making process

(often iterative) in which the basic sciences, risk assessment and engineering sciences are applied to convert resources optimally to meet a stated objective. Among the fundamental elements of the design process are the establishment of objectives and criteria, synthesis, analysis, construction, testing and evaluation. In order to obtain a design that achieves the desired needs for the domain under study, the two previous steps should have been accomplished and taken into account.

- Building and decommissioning: consists in carrying out the proposed solution (design) for the domain. In order to execute this design, construction and/or decommission activities may be executed. It is essential to have a minimal environmental impact when accomplishing these activities. The tolerable environmental impact will be obtained from the risk assessment of the designing step.
- Operating: refers to the action of manoeuvering a system. It may include the combination of all technical and corresponding administrative, managerial, and supervision actions. Operation may also include performing routine actions which keep the system in working order. These latest actions might turn out to be in response to problems detected during monitoring.

According to the analysis, the flood management activities appear mainly related to the natural hazards mitigation domain. Obviously, to a lesser extent, such activities are also related to the water uses domain and represent a constraint which has to be taken into account in the operating activities. Due to the specificity of the hazards mitigation domain, the predominant invariant activities are investigating and observing. These activities are strongly supported by technological devices and simulation tools which allow one to analyse and anticipate the flood phenomena. The modeling methods and simulation results appear today as an essential component which may efficiently assist the decision makers. Velocities, flood maps and real time forecasts represent strategic elements that could be used at different stages of the flood management, such as public awareness or urban planning development. The deep integration of the modelling environments – due to their performance – within the decision procedures promotes the development of hydroinformatics methods.

3.1.3 Why hydroinformatics?

The evolution of human activities, in the foreground of climate evolution and growing world population, induces situations which are more and more complex to manage. The sustainable development and, above all, management of water resources within the context of the aquatic environment represents today, and will represent within the foreseeable future, a major challenge for the post-industrial economy and social organization. The essential aim of such management is to avoid, if possible, or at least minimize, the risks of crises in water supply and waste water treatment for the population, in water scarcity for irrigation, in management of the consequences of floods, and so on.

Initiated in Europe more than 30 years ago, hydroinformatics emerges as the central element for the progress of modelling activities and management of capacities on the theoretical side as well as in the operational field. ICT has extended greatly the potential of the tools and, at the same time, modified the engineering activities in the water field.

The development of ICT allows synergetic use of simulation tools and communication technologies within a single methodological approach dealing with physical, social and economic aspects. The only possible way to manage successfully the problems evoked above implies consensual views and actions of decision-makers and users (population, governments, administration, elected bodies, NGOs), on the one hand, and the executive body: engineers, on the other hand. The idea of hydroinformatics is precisely this: to make available and intelligible the results of engineering thinking to decisions-makers and to make clear for engineers what are the social requirements that they are to satisfy. Theoretically, this can be done thanks to ICT but, in practice, it means that there is a need, within administrations, consulting institutions and contracting enterprises, for professionals who understand this and have the knowledge and ability necessary to implement such a, still new, vision. In short there is a need for "ferment", for a catalyst, for engineering professionals who not only know how to build a dam or water supply system but who are also able to talk to the outside world and modify by their ideas their own surrounding professional environment.

The term hydroinformatics was introduced in 1991 by Professor M.B. Abbott (Abbott, 1991) who defined it as the integration of computational hydraulics and artificial intelligence. Computational hydraulics is the fusion of numerical methods of applied mathematics, hydrodynamics and hydraulics, the latter being originally very much an experimental science. This means that numerical simulation and modelling are integrated with artificial intelligence in single tools. The meaning has evolved since. Probably the most consensual definition today would be that hydroinformatics is about making the best use of information technologies to manage water in the environment. The term "manage" is used here in a very wide sense: certainly not limited to the ordinary understanding of "water management practice". Indeed, management here means dealing with the whole gamut of information. On the one hand are the very engineering-based subjects such as: data collection, measurement, interpretation, design of river basin management strategies including civil engineering structures, modelling (simulation) of river, groundwater and coastal flows and water quality to forecast natural events (floods) as well as the impacts of human activities. On the other hand it is concerned with information management: the time when Society left the engineer with the decisions concerning water problems is over. Decision makers today are numerous, outside the engineering profession and they cannot be asked to understand engineering language or argumentation. Inversely, the engineering profession is still very far from understanding what is desired by the decision makers (citizens, elected bodies, NGOs, etc.).

Hydroinformatics gathers a range of tools that, based on information technology, may be used to explain exactly what these various groups wish and how engineers can realize these wishes and, especially, what the various impacts of the realization could be. Moreover, these tools allow participation of stakeholders in the process of execution of the projects and there are examples of such participation in recent years. Thus hydroinformatics is also about movement, flow of information, not only about its content; it is at the heart of the process of communicating information. Here we come to the meaning of another term of the above definition, that is " … the best use of information technologies to manage … ". What is meant by the word best? This concept is directly connected to the domain of social rightness, and also ethics, in the general sense as well as of engineering ethics. Thus hydroinformatics became a technology in itself that asks for a specific state of mind (or culture) and for specific knowledge.

3.1.4 Towards integrated flood management

Flood control and protection measures in the past have played an important role in protecting people and socio-economic development from flooding. Until recently, they have been engineering-centred, with little or no consideration of the social, cultural and environmental effects of the chosen strategy or of itslong-term economic sustainability. They have largely relied on structural solutions, such as embankments, bypass channels, dams and reservoirs. Although structural flood control measures over the last 50 years have been complemented with nonstructural measures, such as flood forecasting and land use regulations, the need for a paradigm shift from flood control to flood management has been recognized only during the past decade. This shift is enshrined in the integrated flood management (IFM) approach promoted by WMO (2009) which aims at:

- maximizing the net benefits from flood plains,
- reducing loss of life as a result of flooding,
- reducing flood vulnerability and risks, and
- preserving the ecosystem and its associated biodiversity.

IFM addresses the interplay between the beneficial uses of floods, on the one hand, and risks posed by extreme events to the sustainable development in flood-prone areas, on the other. The attempt is, therefore, to try to improve the functioning of the river basin as a whole, while recognizing that gains and losses arise from changes in interactions between the water and land and that there is a need to balance development requirements and flood losses. It has to be recognized that the objective in IFM is not only to reduce the losses from floods but also to maximize the efficient use of flood plains, particularly where land resources are limited. However, while reducing loss of life should remain the top priority, the objective of flood loss reduction should be secondary to the overall goal of optimum use of flood plains. In turn, increases in flood losses can be consistent with an increase in the efficient use of flood plains in particular and the basin in general.

IFM aims at a fundamental re-orientation of the social perception of floods from the 'need to control' to the 'need to manage' and shifting the focus from a reactive to a more proactive response. Living harmoniously with floods is an important strategic option in IFM. Over the centuries people have been living with floods. Unfortunately, such areas have not seen much economic prosperity and improvement in the quality of life compared to those that have adopted flood protection measures. IFM, by reducing the vulnerability of the people and activities in the flood plains, addresses these issues.

Flood issues are influenced not only by the physical causes of flooding but the overall social, economic and political setting of the area concerned. Therefore, those issues are addressed within the framework of social and economic development planning, particularly water resources development, within the framework of integrated water resources management (IWRM). IFM is based on the following principles:

- adopting a basin approach to flood management
- bringing a multi-disciplinary approach to flood management
- reducing vulnerability and risks due to flooding
- addressing climate variability and change
- enabling community participation.

IFM recognizes the river basin as a dynamic system in which there are many interactions and fluxes between land and water bodies. Therefore, IFM, like IWRM, requires the adoption of a river basin approach to planning through multidisciplinary inputs. A number of flood risk reduction options tend to simply transfer the flood risks to elsewhere in the basin. At the same time, certain types of infrastructure, like flood diversion works, bridges, railways and highways, if not provided with appropriate waterways, tend to increase the flood risks upstream, thereby increasing the overall flood risks. Instead of fixing flood problems at local level in an ad hoc manner, it lays emphasis on the development of flood management plans for the river basin as a whole. As such the starting point is a vision of what the river basin should be. Incorporating a sustainable livelihood perspective means looking for ways of working towards identifying opportunities to enhance the performance of the system as a whole.

With the need to address the vulnerability and risks, the IFM requires a close collaboration and coordination between various development ministries, sectors and institutions from various level of administration, through a multi-disciplinary approach. It requires the involvement of all stakeholders, including the civil society and communities that are directly affected.

IFM enables the adoption of a best mix of strategies, both structural and nonstructural through short- and long-term measures. It attempts to manage the land phase of the water cycle as a whole while considering all floods: small, medium and extremes. The concept recognizes the benefits of smaller and more frequent floods, the importance of flood plains and the increasing development demands they face, while at the same time recognizing the disruptive nature of floods, the influence of floods on the ground water recharge, which forms an important source of water during the dry periods and the other extreme of the hydrologic cycle, the drought.

The measures adopted in IFM take full account of the impact of land use and spatial planning on the flood generation processes. Certain other hazards such as landslides within the basin have a potential to modify the flood risks downstream and, combined with floods, can generate mudflows. Such mutually impacting hazards are accounted for and addressed by adopting a multi-hazard approach, wherever they coexist and influence each other.

In order to achieve the required objectives of IFM, there is need for an appropriate legal and institutional framework for collaboration and coordination between different levels of government entities and sectoral planning. The basic data and information should be available to all stakeholders, thus creating an enabling environment or participation in decision-making. Such a legal framework has to clearly define the roles and obligations of various stakeholders. The process of making choices among various options, particularly where a compromise has to be made between development and ecosystem preservation needs, should be based on an environmentally sensitive framework. Such a framework should be based on scientific understanding and analysis of various processes, environmental assessment and adaptive management through monitoring and evaluation. Where an appropriate level of certainty in the knowledge is missing, such as in the case of climate change impacts, the adaptive approach is necessary. This would help in developing a three-fold approach of avoiding, reducing and mitigating adverse environmental impacts. Stakeholders' involvement is essential in all stages of the decision-making process, policy formulation, development of basin flood management plans, making choices among available options, implementation of the activities and, above all, building resilience in the society, thereby developing ownership. Such an approach allows all stakeholders to share their views, raise concerns, build consensus and commitment, thereby ensuring sustainability of the measures taken. An institutional mechanism providing an appropriate platform for various stakeholders, such as a river basin organization, smooths the process. Such a process requires building the capacity of all the stakeholders, by building awareness about the issues and sharing relevant data and information.

3.1.5 Hydroinformatics and floods

As mentioned in the previous sections, hydroinformatics methods and tools may contribute significantly to the flood management activities, especially for the investigating and observing activities. The main contributions are based on numerical models which can represent accurately the hydrological and hydraulic processes in order to qualify the flooding process. The results are used according two different protocols:

- Supporting the design actions (structural and nonstructural measures) which are implemented in order to reduce the impacts of flooding (protection, awareness, response, resilience). The most emblematic products are the various "flood maps" which cover extent of flooding, water depth, velocities, duration, vulnerability and damages.
- Providing information on the on-going flooding process through a real time system.

Before addressing the details of the hydroinformatics methods, it is essential to underline the diversity of the flooding processes which have to be covered by the modelling and simulation tools.

Riverine floods. Riverine floods are primarily caused by prolonged, heavy precipitation, snowmelt or combinations of both in the catchment area. This leads to high flows and water-stages in the receiving streams and rivers. These floods are often large-scale events in the sense of accumulating rainfall from a large drainage area with flood waves passing through the river corridor for extended periods of time (from several days to several weeks). When the bank level or design level of levees is exceeded, or embankments fail, flooding can occur over large areas of the adjacent land. It is important to note that floods are natural events that can never be fully controlled and that they have both positive and negative impacts, as perhaps the only natural hazard. Flood extent is expressed in return periods (e.g. 10-year, 100-year flood) or annual exceedance probabilities (having a chance of being equalled or exceeded by $x\%$ in any given year).

Pluvial floods. Pluvial flooding – or surface water flooding – concerns mainly urban areas and is defined as flooding that results from rainfall-generated overland flow before the run-off enters any watercourse or sewer system. It is usually associated with high intensity rainfall cells within longer duration events or convective thunderstorms. The intensity of rainfall can be such that the run-off totally overwhelms surface water and underground drainage systems. In theory, no area is entirely free from the risk of pluvial flooding although depressions in the topography and surface flow paths may be more susceptible.

Flash floods. A flash flood is generated by a causative event – intense rainfall – in a short period of time and is characterized by a sudden increase in level and velocity of the water body. The term "flash" reflects a rapid response to the rainfall event, with rising water levels in the drainage network reaching a crest within minutes to a few hours of the onset of the flood event, leaving extremely little time for warning. Thus, flash floods are localized phenomena that occur in catchments with maximum response times of a few hours. At spatial scale and according to the catchment characteristic, the risk corresponds to surfaces below 3000 to 4000 km^2. The two key elements are rainfall intensity and duration. Topography, soil conditions, and ground cover also play an important role. Dam failure and similar phenomena could be described as flash floods.

Coastal flooding. Coastal flooding, the inundation of land areas along the oceanic coast by sea waters over and above normal tidal action, is caused by a combination of high astronomical tides and waves. It occurs when intense, offshore low-pressure systems or tsunami drive ocean water inland. Storm winds cause the sea water to pile up on the coast, leading to sudden inundation and flooding of coastal regions. The shallowness of water may considerably modify the surge heights in the region. The storm surge due to a tropical storm is confined to a couple of hundred kilometres. Due to the immense volumes of water and energy involved, the effects of tsunami can be devastating, as during the event recorded in Japan in 2011.

Flooding due to groundwater rise. A groundwater flood results from a rise in groundwater level sufficient for the water table to intersect the ground surface and inundate low lying areas. The phenomenon is rather sporadic in time and location, but when it occurs inundation usually lasts longer than flooding inflicted by riverine floods.

Flooding due to lake overflows. When lake inflow exceeds the outflow the lake levels rise. In contrast to riverine flooding not the peak discharge, but the volume of inflow in excess to outflow capacity is decisive. Total rise and rising velocity are determined by the size of the lake surface in relation to the excess inflow.

Flooding due to ice-jams. Ice conditions impact many rivers throughout the world, particularly in the Northern Hemisphere. Some rivers occasionally develop partial ice covers while other rivers are completely ice covered for more than six months of the year. The uncertainties associated with climate variability create uncertainties in the duration, extent and severity of future river ice covers and their associated flood potential. River ice processes associated with river ice jam flooding occur during the ice formation process, with the occurrence of periods of warm weather in mid-winter and with the spring warming trend that transitions the river from an ice covered to an open water state. An ice jam (or ice dam) occurs when sufficient ice has been accumulated in the river channel to impede the river flow.

3.1.6 Flood maps production

For decades flood mapping has been based on observed and recorded events. Water levels and direct observations have been converted into maps based on topographic or cadastral layers. The increase in vulnerability in many places has driven the development of hydrological knowledge and the request for the possibility to predict the consequences of extreme events. To achieve this objective, mathematical models have been gradually developed and introduced as essential tools to support the flood mapping exercise. The models have reached a level of refinement and sophistication which opens the possibility to investigate complex flow situations, such as pluvial flooding in highly urbanized environments, and to provide an accurate description of the flow dynamics in space and time. In order to provide suitable data to produce the flood maps, the modelling approach is based on two sequential steps:

- the flood hydrograph analysis
- the routing of the flood hydrograph through the domain representing the river reach exposed to flooding: river networks, floodplains, sewage system, urban environment and so on.

The flood-hydrograph analysis is required to determine the flood characteristics – mainly discharges according to time. It will result in hypothetical hydrographs associated with the probability of occurrence (or return period) which are entering the flow domain at the upstream boundary. This task could be achieved through different hydrological approaches which are based on recorded events – gauged catchments – or/and on simulations produced by different kind of models focused on rainfall/run-off transformation for ungauged catchments.

The recorded flood events when available and with a sufficient chronicle – several decades – could be assimilated to random events. The theory of probability is used to estimate the probabilities of occurrence of these events. The emphasis in statistical analysis is on events rather than on the physical processes that generate them. In other words, the analysis will not be able to adjust the results if some modifications took place within the catchment and strongly affected the shape and the magnitude of the hydrograph. For these reasons and as for the ungauged catchments, evaluation of the hydrograph must be based on deterministic hydrological models (Guinot and Gourbesville, 2003). Today several hydrological distributed modelling systems are available and could provide an efficient answer for the flood hydrograph definition.

For modelling the flood propagation along the river network (flood routing), priority must be given to deterministic approaches resolving the physics of the processes. It is important to remember that no matter what level of refinement the model has accomplished it still remains a mathematical model which includes simplifications of reality and thus is restricted to certain aspects of reality. In this sense a deterministic modelling tool is based on a number of physical laws that have been formulated in suitable mathematical form – usually in the form of equations – the latter being solved numerically through the application of appropriate algorithms. The laws that provide the foundations of the tool define its domain of application and the user may choose one or another tool as a function of the adequacy of the laws upon which the tool is built, together with the user's requirements and needs. For the deterministic approach, each such formulation is related to a physical reality. A modelling tool is an informatics code for the numerical computerized solution of mathematical formulations.

For the flood map production, the models which are able to resolve the dynamics of flow in space – streams and adjacent floodplains – are obviously the most suitable modelling systems. Hydrodynamic methods of flood routing seem to fulfill this task best. Both one- and two-dimensional approaches can lead to good results but more and more priority is given to two-dimensional models, for they accomplish a better geometric discretization of the irregular topography in natural rivers and urban areas and computing capacities have increased rapidly in the last decade so that two-dimensional flow studies on are now state of the art.

3.1.6.1 Producing the hydrograph

Since the 1930s the science of hydrology has concentrated on the development of mathematical models to predict a flood hydrograph in catchments (deterministic hydrology). It started in 1932 with the unit hydrograph model of Sherman (Sherman, 1932) and Horton's simplistic infiltration approach in 1933. Although further refined these two concepts were the only modelling approach for more than 30 years. They are still today the theoretical basis of the conceptual models in which the catchment is regarded as a system in which the input (precipitation) is transferred by a system function (unit hydrograph) to a run-off hydrograph at the outlet of the catchment. These rainfall-run-off models are also called "lumped models". In the 1970s the science of hydraulics joined the research field of run-off modelling. In the beginning the models concentrate on the prediction of the changing magnitude, speed and shape of a flood wave as it propagates through rivers and channels (flood routing). The approach resolves the flow path by dividing the stream channel into profiles and applying the equations of motion (Cunge, 1975). Also at the end of the 1970s hillslope hydrology (Kirkby, 1978) provided new knowledge on the relationship between precipitation input and hillslope discharge output in terms of the spatial soil distribution and terrain contour. With the availability of powerful computers these physically-based theoretical concepts (physical hydrology) have stimulated the development of a new generation of rainfall-run-off models, the physically-based distributed models (Beven, 1985; Beven and Kirkby, 1979; Abbot *et al.*, 1986). The method suggests subdividing the catchment into a regular grid of cells. For each cell they solve the equations of motion to get the run-off pattern of overland and subsurface storm flow. At first these models were restricted to one-dimensional vertical simulations. Later they were extended to two- and even three-dimensional approaches to give the spatial pattern of overland flow, interflow and groundwater flow. The most sophisticated and applied distributed models are the SHE-model (Abbot *et al.* 1986) and the hillflow-3d model (Bronstert and Plate, 1996). All use a one-dimensional approach to simulate the flow in the unsaturated soil layer. While the SHE-model and hillflow-3D models solve the one-dimensional Richards equation, the Topmodel (Beven and Freer, 2001) applies a simple exponential function of water content in the saturated zone. The latest version of the SHE-model, MIKE SHE (DHI, 1998a,b) provides a three-dimensional groundwater model and calculates the overland flow on the basis of a two-dimensional implicit finite difference model based on the kinematic wave theory. The channel flow is described by using river modelling systems (MIKE 11 – DHI, 1998a or MIKE 21–DHI, 1998b) which solve the full one-dimensional Saint-Venant equations or the two-dimensional shallow water wave equations (dynamic wave). This system of models has reached the highest grade of hydraulic components to simulate the rainfall-run-off process. It comes closest to the target to describe fully the flow path of precipitation from the moment it reaches the earth's surface to the outlet of the catchment at the end of the river.

Parallel to the development of the distributed models the lumped models have been further refined and improved to more physically-based models which fully cover the whole hydrological cycle of land-bound water movement and refine its spatial resolution by sub-catchments, areas of equal hydrological characteristics with respect to the vertical processes of interception, infiltration, evapotranspiration and groundwater recharge. Some of the lumped models use the same mathematical approaches as the distributed models to simulate these vertical processes, such as the Richards equation or exponential functions. Most of these models, however, use simpler methods such as the SCS method to determine the infiltration rate of precipitation into the ground. This method estimates precipitation excess as a function of cumulative precipitation, soil cover, land use and antecedent moisture. The explicit consideration of the various factors that are thought to affect flood run-off makes the method attractive.

The overall trend in rainfall-run-off modelling is directed towards more refined physically-based modeling. These models have the capacity to provide the needed information for the different stages of the flood management activities, such as flood map production and implementation of real time systems.

3.1.6.2 Hydraulic models

Many questions concerning fluvial or pluvial flooding cannot be answered in a simple way since flow in rivers is dominated by structural variability and complex geometries, which lead to multidimensional flow systems depending on a number of parameters. Therefore a hydraulic engineer needs powerful and reliable modelling instruments. Numerous promising scientific examinations in the past that decisively improved the understanding of fluvial flow mechanics have responded to these demands in engineering practice (Rouvé, 1987) and have led to a variety of mathematical methods, from simple, one-dimensional analytical models up to multi-dimensional numeric models with sophisticated turbulence modelling (Giammarco, Todini and Lamberti, 1995; Pasche, 1984; Wenka, 1992; Malcherek 2001).

Although, in the 1980s, engineering practice still preferred simple hydraulic methods that could be solved with only the help of a calculator, and complicated questions demanding high precision were answered with physical models, computer-aided numerical flow models dominate at present. The array of models goes from one-dimensional profile line models over two-dimensional shallow water models up to complete three-dimensional mathematical modelling. The practitioner is thus confronted with the problem of choosing the right model for his solution. This choice should depend mainly on professional criteria of course. However, the question of adequacy and the available database are crucial in practical applications, since the motto in business is "not more accurate than necessary and as inexpensive as possible". This point of view explains why one-dimensional profile line models are most common in engineering practice today. They compose a good trade-off between the technically needed and the economically feasible. The considerable increase in performance of computer hardware and of numerical methods and software ergonomics let us look forward to a more widespread application of two-dimensional models in practice. Three-dimensional models are gradually entering engineering practice with the improvement of computing capacities and availability of high performance resources. Consequently, the following presentation of theoretical foundations is limited to one- and two-dimensional mathematical models.

Good consensus exists between engineers in the formulation of the basic flow equations for rivers. For backwater surface calculation under steady non-uniform flow conditions the energy equation is well established (BWK 1/1999) (BWK, 1999) for the case of inflow and outflow into the river:

$$\frac{1}{2g}\left(\alpha'_{i+1}\frac{Q^2_{i+1}}{A^2_{i+1}} - \alpha'_i\frac{Q^2_i}{A^2_i}\right) + \cos\Theta\,(z_{\mathrm{WSP},i+1} - z_{\mathrm{WSP},i}) + \int_{x_i}^{x_{i+1}} I_{\mathrm{R}}\ \mathrm{d}x + \int_{x_i}^{x_{i+1}}\left(\frac{2\alpha' Q}{A} - \cos\phi\ v_{\mathrm{e}}\right)\frac{q_{\mathrm{e}}}{gA}\ \mathrm{d}x = 0 \qquad (3.1.1)$$

with Q = total discharge [$\mathrm{m}^3\ \mathrm{s}^{-1}$], A = cross-section area [m^2], q_{e} = inflow/outflow [m^3/sm], z_{wsp} = water surface elevation [m], I_{R} = friction slope [-], α' = energy coefficient [-], Θ = longitudinal slope of river bottom [°], φ = angle between inflow/outflow and main flow [°], $i,i+1$ = downstream, upstream profile [-], x = length of river along thalweg [m].

The unsteady flow in rivers can be described by a one-dimensional momentum balance which leads to the well-known Barré de Saint Venant equation (Pasche *et al.*, 2005)

$$\frac{1}{gA}\frac{\partial Q}{\partial t} + \frac{1}{gA}\frac{\partial(\alpha' Q^2/A)}{\partial x} + \cos\Theta\,\frac{\partial z_{\mathrm{WSP}}}{\mathrm{d}x} + I_R - \cos\phi\,\frac{v_{\mathrm{e}} q_{\mathrm{e}}}{gA} = 0 \qquad (3.1.2)$$

The assumption of hydrostatic water pressure and the integration of the Reynolds stress equations over the depth gives the two-dimensional shallow water equations. In rivers the Coriolis force and the shear stress at the water surface are of minor relevance so that the equations can be further simplified to the following form:

$$\frac{\partial u_i}{\partial t} + u_j\frac{\partial u_i}{\partial x_j} = -g\frac{\partial}{\partial x_i}(z_{\mathrm{Wsp}}) + \frac{1}{h}\frac{\partial}{\partial x_j}\left[h\left(\frac{1}{\rho}\tau_{\mathrm{L},ij} + \frac{1}{\rho}\tau_{\mathrm{t},ij} - \widetilde{u}_i\widetilde{u}_j\right)\right] - \frac{1}{h}\frac{\tau_{\mathrm{so},i}}{\rho} \qquad (3.1.3)$$

with u = the local velocity component [$\mathrm{m\ s}^{-1}$], h = local water depth [m], τ_{L} = viscous shear stress [$\mathrm{N\ m}^{-2}$], τ_{t} = turbulent shear stress component [$\mathrm{N\ m}^{-2}$], $\widetilde{u}_i\widetilde{u}_j$ = dispersion terms [$\mathrm{m}^2\ \mathrm{s}^{-2}$], τ_{So} = bed shear stress [$\mathrm{N\ m}^{-2}$], ρ = density of water [$\mathrm{kg\ m}^{-3}$], i,j = 1,2 (x-, y-component and index of the Einstein summation convention).

All three equations need empirical parameters. The most relevant parameters for the one- dimensional models are the friction slope I_{R} and the bed shear stress τ_{So}. In the shallow water equation further parameters are needed to quantify the turbulent shear stresses and the dispersion terms, which consider the effect of secondary current.

3.1.6.3 Parameter estimation in one-dimensional flow models

The simplest approaches use only one parameter to quantify the friction slope and bed shear stress in the one-dimensional equations. It includes all flow losses and represents the variable roughness along the wetted parameter by a mean value. Depending on which flow formula is used, this parameter is named Manning's n (GMS-formula) or the Darcy Weisbach coefficient λ (DW-formula).

$$I_{\mathrm{R}} = \frac{1}{8g}\frac{\lambda}{r_{\mathrm{hy}}}\frac{Q^2}{A^2} \quad (3.1.4)$$

$$\tau_{o,i} = \frac{\lambda}{8}\rho\ u_i\ \sqrt{u_i^2 + u_j^2} \quad (3.1.5)$$

Darcy-Weisbach (DW) formula

$$I_{\mathrm{R}} = \frac{n^2}{r_{\mathrm{hy}}^{4/3}}\frac{Q^2}{A^2} \quad (3.1.6)$$

$$\tau_{o,i} = \frac{\rho g n^2}{h^{1/3}}\ u_i\ \sqrt{u_i^2 + u_j^2} \quad (3.1.7)$$

Manning (GMS) formula

Only the DW-formula is physically-based, which can be seen by the units of the empirical parameters. While the DW-coefficient λ is dimensionless the Manning's n parameter has the physically senseless units of [$s^2\ m^{-1/3}$]. Making use of the boundary layer theory and the theory of flow around bodies physically-based equations can be derived in which the DW-coefficient λ is expressed in terms of directly determinable geometric parameters or equivalent parameters, like the equivalent sand roughness k_s in the Colebrook–White formula. Pasche, 1984 has shown that, especially for natural rivers, the DW-formula has advantages as the flow losses caused by non-submerged wooden vegetation and by momentum transfer at the interface between river and flood plain can be well described by the following formulas.

$$\lambda_{\mathrm{P}} = \frac{4h_{\mathrm{P}}d_{\mathrm{P}}}{a_x a_y} c_{\mathrm{WR}} \cos(\alpha_{\mathrm{lat}}) \quad (3.1.8)$$

$$\frac{1}{\sqrt{\lambda_{\mathrm{T}}}} = -2 \times \log\left[0.07\left(\frac{cb_{\mathrm{m}}}{b_{\mathrm{III}}}\right)^{1.07} \Omega\right] \quad (3.1.9)$$

$$\Omega = \left[0.07 \times \frac{a_{\mathrm{NL}}}{a_x}\right]^{3.3} + \left[\frac{a_{\mathrm{NB}}}{a_y}\right]^{0.95} \quad (3.1.10)$$

with a_x, a_y = distance of vegetation elements in both horizontal directions [m]; h_{P} = water depth in front of the vegetation element [m]; d_{P} = diameter of vegetation element [m]; α_{lat} = lateral inclination of bottom [°]; c_{WR} = drag coefficient of wooden vegetation [-], according to Pasche, 1984; a_{NL}, a_{NB} = wake length and wake width [m] according to BWK 1/1999, c = form factor [-], b_{m}, b_{III} = contributing width of vegetation zone and river [m] according to BWK 1/1999.

The vegetation parameters and the contributing vegetation width represent real parameters, which can be directly determined in nature or derived on this basis. In contrast the GMS-formula uses a parameter which can be only quantified on the basis of tables, own experience or by calibration.

3.1.6.4 Parameter estimation in two-dimensional flow models

For evaluating the turbulent shear stresses τ_t in two-dimensional hydrodynamic models only the Boussinesq approach is applied in the engineering world:

$$\tau_{t,ij} = \rho \left(\nu_T \left(\frac{\partial u_i}{\partial x_j} + \frac{\partial u_j}{\partial x_i} \right) - \frac{2}{3} k \, \delta_{ij} \right) \tag{3.1.11}$$

with k = turbulent kinetic energy [m^2 s^{-2}], ν_T = turbulent viscosity [m^2 s^{-1}], δ_{ij} = Kronecker delta [-].

The simplest models assume that the turbulent viscosity ν_T is a constant for a given flow domain (constant-eddy-viscosity-approach). Indeed it could be shown that this assumption leads to good results but only if the parameter could be calibrated. Theoretical and experimental analysis of turbulence, however, demonstrates that the turbulent viscosity is not a constant but varies in dependence on the flow situation. These characteristics of the turbulent viscosity are well considered by the mixing length approach (term 1 in equation (3.1.12)) in combination with the bed shear stress approach (term 2 in equation (3.1.12)) (Lippert, 2005):

$$\nu_{t,ij} = l_m{}^2 \left[2\left(\frac{\partial u_i}{\partial x}\right)^2 + 2\left(\frac{\partial u_j}{\partial y}\right)^2 + \left(\frac{\partial u_i}{\partial y} + \frac{\partial u_j}{\partial x}\right)^2 \right]^{1/2} + e^* \sigma_t h \sqrt{\frac{\lambda}{8}} \sqrt{u_i^2 + u_j^2} \tag{3.1.12}$$

with $e^* \sigma_t = 0.15$. The mixing length l_m can be quantified on the basis of a free shear layer approach from Rodi (1983):

$$l_m = 0.09 b_S \tag{3.1.13}$$

with b_S = mean width of the free shear layer (free-shear-layer approach). In an approach of Smagorinsky, the mixing length is correlated to the geometry of the numerical grid to compensate the numerical diffusion which occurs in most numerical grids (Malcherek, 2001)

$$l_m = 0.197\Delta \tag{3.1.14}$$

with Δ = equivalent length of one element side of the numerical grid.

More refined turbulence models such as the k-ϵ-model have no relevance in the engineering world, as the numerical effort is considerably more than in analytical approaches without improving the quality of the result substantially (Pasche, 1984).

As the shallow-water equations cannot simulate secondary currents directly, they need to be included by analytical approaches. Most often this effect is included in the turbulent viscosity, which is without physical meaning. A more realistic and physically based approach is the quantification of the dispersion terms $\overline{\tilde{u}_i \tilde{u}_j}$ by analytical equations (Lippert, 2005).

Figure 3.1.3 shows an example of 2D hydraulic modelling in an urban environment.

Figure 3.1.3 Two-dimensional hydraulic modelling in an urban environment – velocity vectors and water depth. See plate section for colour representation of this figure.

3.1.6.5 Validation of results

The data quality of topography is an essential requirement for obtaining accurate and reliable results. The different flood conveying elements, such as streams, river, sewage system, streets and so on. have to be represented in an accurate manner in order to get a realistic representation of the flooding dynamic and a correct extent of the inundated areas. In addition to the work focused on the hydraulic parameters (Guinot and Gourbesville, 2003), a specific effort has to be made to collect, in the most efficient way, the data to describe the topography. During the last 10 years, some new technologies like LIDAR (light detection and ranging), push broom cameras or remote sensing have greatly improved the data acquisition phase and now allow production of accurate topography which can be used with hydraulic models.

For the deterministic approach which is strongly recommended for both stages, the following paradigm, defined as "good practice", is proposed with the different stages in the modeling process (Cunge, 2003):

- Instantiation or set-up or "construction" of the model; definition of the methodology necessary to define the range of uncertainty in the results of the computations.
- Validation, which consists of executing a number of simulations of past-observed events with the model, to find the range of uncertainty in the results and physically-based reasons for the differences between the simulated and observed results. After this, the impact of these deviations and uncertainties needs to be explored with respect to the exploitation results.
- Exploitation runs (studies): supplying the results and impacts and their range of uncertainty to the end-user in a comprehensible form.

The proposed paradigm eliminates the well-known "calibration" stage as such. Calibration usually consists in executing a number of simulations of past observed events while the parameters of the model are varied until an acceptable level of error between observation and computation is obtained.

For the inundated surfaces, indeed the only past-observed data concerning the unsteady evolution of water stages that can be found on inundated plains are those rare marks of the highest elevations attained during historical floods. In most cases, the environment has changed drastically: river bed, hydraulic works, land use, extension of urban areas and so on. At the same time the quality of the observations is uncertain and the global coherence of the data has to be validated. In very exceptional situations like for the Mekong delta in 1963, 1964 and 1965 (Zanobetti *et al.*, 1968; Cunge, 1975) measurement campaigns are organised but this approach represent a financial effort which is rarely engaged.

An obvious principle is that the calibration must be limited to the model parameters that are invariant between the instantiation and exploitation stages, unless the purpose is to study the sensitivity of the model to modifications in its parameters. To calibrate parameters that will subsequently be modified during the exploitation runs used for simulating the flood extension is most often a useless, as well as costly, exercise. It is better by far to let a model be truly deterministic, that is a model without 'inner black boxes' describing physical processes. But still these models need empirical input in the form of parameters, for example the roughness coefficient, head-loss coefficients and discharge coefficients of structures (weirs, gates, culverts, etc.). This quality is not given for the Mannings/Strickler coefficient. It is not dimensionless which is a requirement for physically-based parameters. Additionally the parameter is not only dependent on the roughness conditions but also on the water depth (Pasche, 2007), which makes a reliable evaluation of this parameter more difficult. Based on the boundary layer theory and the theory of flow around bodies the degree of physical soundness can be substantially increased leading to the DW-coefficient. It can be quantified in terms of directly determinable geometric parameters or equivalent parameters, like the equivalent sand roughness k_s in the Colebrook–White formula or the diameter and distance of wooden vegetation in various resistance formulas for non-submerged vegetation (Pasche, 2007).

In the proposed paradigm, the validation stage is reinforced. In a way it incorporates the calibration stage. The past measured data will, of course, be as useful and as necessary as for calibration under the currently admitted paradigm but they will be used for a validation analysis of the computed results. It is claimed that a deterministic model, with values of parameters defined by inspection on the basis of engineering practice, should simulate reality correctly and its results should be close to past observed results.

Making computed results nearer and nearer the observed ones must not be carried out through a calibration process as it is currently understood and applied. Indeed:

- If the differences between the computed and the observed lie within an acceptable interval of uncertainty, or can be explained by physical reasons, and if the consequences of differences upon exploiting the model as it is are analysed and acceptable, then there is no reason to go any further with the modification of parameters.
- If the differences are greater than the uncertainty interval, then they must be explained. The reasons must be found and analysed, taking into account, once more, the consequence of using the model as it is or amending it. Most often the findings lead to modifications of originally erroneous data, such as topography, hydraulics characteristics or boundary conditions and have not much to do with parameters. Sometimes there are factually important errors in the values of parameters assessed during a visual inspection. However, sometimes, one may find that the modelling tool is not adequate: such often occurs when using one-dimensional models where only two-dimensional ones can simulate the real flows.

The paradigm insists on the fact that this approach is not a calibration under a new name. Good practice requires the collection and analysis of data for the purpose of validation, and validation is not just a check that computed values are not very far from observed ones, it is a study of the reasons why there is a difference between the two. Also it must, of course, be substantiated by a report leading to an understanding of how such an analysis was carried out and how the conclusions were reached.

3.1.7 Real-time systems for decisions support

The need for efficient warning systems related to flood is now clearly expressed, especially in the growing complexity of urban environments like those of the megacities. The combined development of new generations of sensors and new communication protocols allows one to revisit the concepts used until now in the flood warning systems. The efficiency of these systems has been seriously questioned during recent events and improvements are seriously expected in order to provide a real support to the exposed populations.

Most of the flood warning systems already implemented are based on the concepts used in the Supervisory Control and Data Acquisition (SCADA) systems. Originally, these systems were designed and implemented for monitoring and controlling a plant or industrial equipment, such as telecommunications, water and waste water control, energy production, oil and gas refining,transport and so on. A SCADA system (Figure 3.1.4) usually consists of several subsystems:

- A human machine interface (HMI) is the apparatus which presents process data to a human operator from which the human operator monitors and controls the process.
- A supervisory system, acquiring and gathering data on the process and sending commands to the process.
- Several remote terminal units (RTUs) connecting to sensors in the process, converting sensor signals to digital data and sending digital data to the supervisory system.
- Several programmable logic controllers (PLCs) used as field devices because they are more economical, versatile, flexible, and configurable than special-purpose RTUs.
- Communication infrastructure connecting the supervisory system to the RTUs.
- Various process and analytical instrumentation.

SCADA systems are clearly data acquisition oriented and they could be associated with distributed control systems (DCSs) that are process oriented and driven. SCADA refers to centralised systems which monitor and control entire sites, or complexes of systems spread out over large areas (anything from an industrial plant to a nation). Most control actions are performed automatically by RTUs or by PLCs. Host control functions are usually restricted to basic overriding or supervisory level intervention. For example, a PLC may control the flow of cooling water through part of an industrial process, but the SCADA system may allow operators to change the set points for the flow and enable alarm conditions, such as loss of flow and high temperature, to be displayed and recorded. The feedback control loop passes through the RTU or PLC, while the SCADA system monitors the overall performance of the loop.

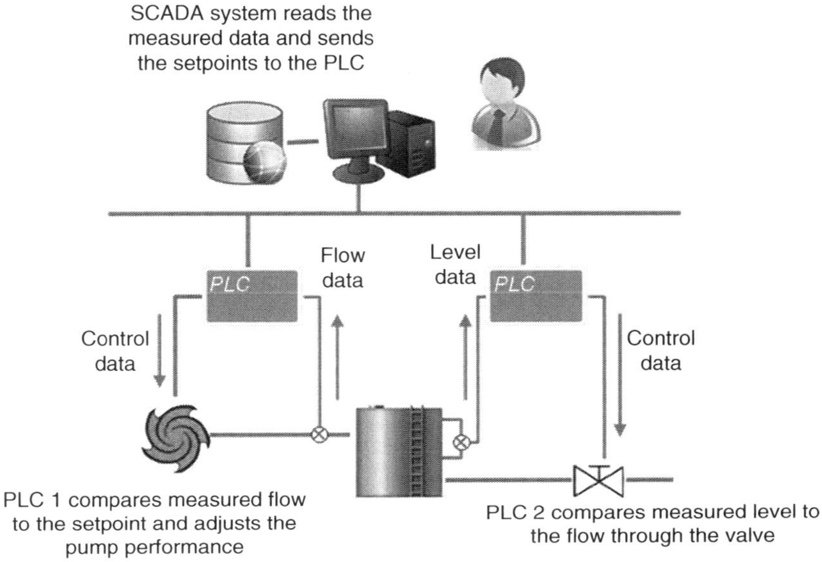

Figure 3.1.4 Principle of SCADA applied to a water system.

In the flood warning systems, the SCADA are frequently associated with hydrological/hydraulic modelling tools which represent the support to the decision and the link to the DCS. However, this strategy considers that the concepts implemented in SCADAs may operate in the specific domain of flood warning.

Flood forecast systems can be classified in several ways as there are many different approaches and features involved. Such systems always need a certain degree of customisation, of course, because each city has a different flood context and also users need help to get hold of the system. Classification can be done based on items that constitute standard architecture of such systems, such as the type of source for weather forecast, the level of pre-treatments and technology involved in data collection and so on. The systems can be organised according to a three-type classification (Henonin *et al.*, 2013):

- **Empirical scenarios-based system**: real-time flood forecast system with no hydraulic model involved in any part of the process. In this approach, a rainfall forecast is used as input for empirical scenarios selection. Scenarios are based on a historical events review and key people's knowledge (drainage network managers and technicians, emergency services and so on), optionally with hydrological model(s). Such systems can be very simple with regard to the technology involved, for example a single rainfall forecast or rain gauge record with warning levels based on previous experience. It can also involve more advanced technology such as a real-time database that stores the rain data for further use and the setting uo of an automatic warning through e-mail/SMS/website.
- **Pre-simulated scenarios-based system**: real-time flood forecast system with scenario and results catalogue built from previous hydraulic simulations. In the approach, once again, a rainfall forecast is used as input for scenario selection, but the scenario

catalogue has been built from a pre-study project that involves hydraulic simulations. Various modelling strategies can be used depending on the available tools and data. The accuracy of simulated scenarios, hence warning levels, depends on both the model(s) and the input data quality (including calibration data). Proper maintenance is of course a main issue, as any major change in the actual drainage system should involve recompilation of the scenario catalogue. Furthermore, the scenario catalogue has to be as exhaustive as possible to cover the whole range of flood events. Rules are used to pick up the most probable scenario depending on the forecast, and to set the proper warning level. Hence, these rules are to be developed with great care and should be reviewed each time a major flood event occurs and within a scenario update schedule. In spite of this, climate change effects are difficult to take into account in such a system.

- **Real-time simulations-based system**: real-time flood forecast system with online and real-time hydraulic model(s). Rainfall forecast is used as input for an online hydraulic modelling system. The hydraulic model, called the "real-time model", has to comply with real-time forecast standards, such as short computation time and fast transfer/display of the results. As the flood simulations are computed as an 'on the go' and continuous process, the necessity to take climate change effects into account is no longer an issue. One main issue for such a system would be the accuracy of the flood simulation results, influenced by the proper update of the hydraulic model (major changes in drainage system, real-time controlled devices status, calibration process and so on). Of course, the more detailed the model, the more difficult to update, hence a proper balance between flood analysis quality and model complexity has to be found. For any real-time forecast system, maintenance is a critical issue. Therefore it may be better to keep the real-time model simple and consequently easier to update and maintain, rather than to use a very detailed model that would face a bigger risk of being rapidly outdated due to lack of maintenance.

Even the best forecast quality is useless if the system does not communicate the right information at the right time to the right people. Forecast and alert information have to be understandable at every level of use, from operational staff and decision makers to a wider public audience (Price and Vojinovic, 2008). The alert system has to rely on adequate media and the content has to fit with the needs of each user. Communication, information level and interactivity are some of the main challenges for existing and future forecast systems. The educational potential of a flood forecast system should also be taken into account in the communication part. Such a system can be a useful tool to help people to accept a reasonable flood risk and thus live with floods. The set-up of proper warning levels is also an important process to ensure the sustainability of any warning system. If the alert thresholds are defined too low, a warning could often be issued, but actually no flood would occur. On the other hand, too high a threshold may give an alert that is too late. Of course, too many false warnings entail loss of trust from the users of the system. Error propagation and uncertainties should be estimated for every element of the system to evaluate the reliability of the forecast and thus take informed decisions. Besides the error and uncertainty calculations, the use of such information is a challenge, particularly when it comes to public information. What kind of uncertainty information can be delivered, and for what kind of user, are questions that are still to be addressed in further research and applications.

3.1.8 Emerging trends for higher efficiency

The growing complexity of the situations regarding flooding requires a higher accuracy for the models in order to provide the relevant support to the decision makers. The necessary improvements could be supported by different actions which are highly related to technological evolution. The availability of data allows one to move forward to high resolution hydraulic models which can provide an accurate representation of flooding processes. In parallel, the development of low cost sensors introduces the paradigm shift from "data scarcity" to "data abundance". This situation allows one to review the concepts and to develop completely new approaches based on a decentralized paradigm.

3.1.9 High resolution data and high resolution hydraulic modelling

To assess run-off hazard generated by local intense rainfall events in urban environments, various methodologies can be applied, including approaches relying on 2D shallow water equations (SWEs) based on numerical modelling tools (Abily *et al.*, 2012). This category of modelling tools is now intensively used for practical studies for overland flow calculation in fluvial and coastal flood events contexts over an urban environment. Recently, for run-off hazard modelling at high resolution, use of standard numerical modelling tools based on 2D SWEs became a 'press forward' practice (Ciliberti *et al.*, 2008; Gomez, Macchione and Russo, 2011). However, the application of such tools represents a real challenge. Industrial and urban areas are complex environments because above-ground surface features that might influence the surface run-off flow path are numerous and densely present and have a high level of diversity, from a few metres high (buildings, sidewalks, roundabouts, crossroads, etc.) to a few centimetres wide (walls, road gutters, etc.). Nevertheless, for run-off modeling over a complex environment, the level of detail of the digital surface model (DSM) has to include features influencing run-off flow paths. Notwithstanding the fact that the framework for such application goes straight from some of the 2D SWE underlying assumptions, questions regarding feasibility of such an approach at a site scale as well as questions about possibilities, added values and limits of high resolution topographic data integration in standard 2D numerical tools remain open.

For decades airborne topographic data-gathering technologies, such as imagery and LIDAR scans were intensively used, mainly by geomatics and remote sensing communities for urban reconstruction (Musialski *et al.*, 2012). These technologies produce highly accurate data for DSM construction using LIDAR data, photogrammetry data or combining the two technologies (Lafarge *et al.*, 2010; Lafarge and Mallet, 2011; Mastin, Kepner and Fisher, 2009). During the last decade, LIDAR and photogrammetry produced topographic data has become widely used by other communities, such as urban planners (for a 3D reconstruction approach) and engineering companies for various applied study purposes including flood risk studies. The availability of unmanned aerial vehicles (UVAs) (Remondino *et al.*, 2011) makes the intensive use of a high resolution LIDAR or imagery produced data easily available for a specific study purpose. Consequently, the use of DSM information from airborne technologies for complex environments, such as dense urban areas or industrial sites, (Tsubaki and Fujita, 2010) appears as a common solution for the modelling community.

Run-off flow paths affecting above ground features are not equally represented in DSM generated based on LiDAR and photogrammetric data. Moreover, optimal use of high definition DSM in standard 2D numerical modelling tools might be challenging in terms of feasibility of data integration within the modelling tools. The possibilities and challenges of including these surface features in highly detailed run-off 2D models for run-off flood hazard assessment deserve specific consideration.

The problem of high density topographic information inclusion in standard 2D modelling tools is a recurrent issue which has been analysed with real sets of data in order to assess the possibilities and impact of the inclusion of small features.

Dissimilar sets of DSM were created based on two different sets of high density topographic data gathered from a LIDAR and a photogrammetric campaign. For the analysis, a site with typical characteristics of an industrial environment has been selected based on the following criteria: a spatial extent compatible with an industrial platform/urban suburbs size, the presence of a wide variety of above-ground structures creating a complex environment having typical characteristics of an industrial site and availability of high resolution topographic data (Figure 3.1.5).

Thetandard 2D numerical modelling tools used were based on 2D SWEs resolution. This category of modeling tool has different numerical strategies to solve 2D SWEs and discretise the spatial information in different ways. Interest in assessing the possibilities and limits of strategies for spatial discretisation used by modelling tools has been keen, exploring high resolution DSM use with regular grid meshing and flexible meshing approaches.

The feasibility of high resolution topographic data use is confirmed for run-off scenario simulations at an industrial site scale with different categories of standard numerical modelling tools (Mike 21 and TELEMAC-2D) (Figure 3.1.6). The results are highly

Figure 3.1.5 Example of two-dimensional hydraulic modelling based on a LIDAR data set (DSM). See plate section for colour representation of this figure.

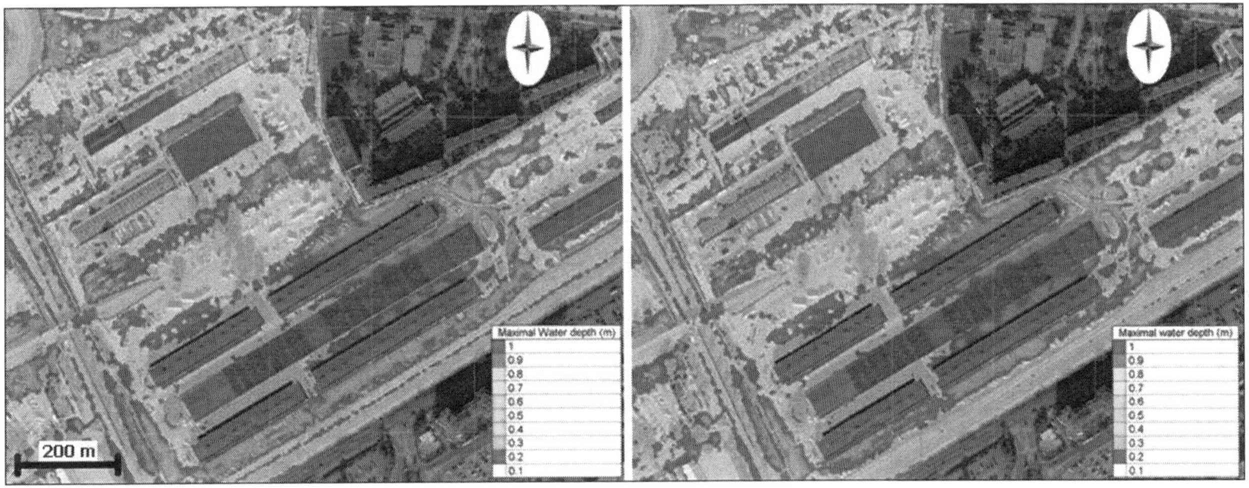

(a) (b)

Figure 3.1.6 Maximum water depth using photogrammetry combined with LIDAR DSM with (a) a structured (Mike 21) and (b) a flexible mesh (TELEMAC-2D). See plate section for colour representation of this figure.

influenced by the quality of the high resolution DSMs. Therefore, when using high resolution topographic data for practical hydraulic application, it is important to have hindsight concerning the quality of the raw data and the quality of the created DSM.

The possibilities of run-off influencing above-ground features inclusion in DSM is revealed to be unequal, depending on the high resolution topographic data gathering technology. In the present case, the LiDAR data resolution (2m per 2m) is not fine enough to catch thin small-scale above-ground features (i.e. concrete walls) which can highly influence the run-off drainage path. Inclusion of small-scale above-ground structures information in DSMs was achieved using 3D classified data from photogrammetry technology combined with the LIDAR data. The added value of this approach for thin run-off influencing features inclusion in DSM is clearly observed regardless of the resolution used for their inclusion (either 0.3 m or 1m).Thin above-ground features inclusion can lead locally to up to a 0.5 m difference in maximal water depth estimations.

Integration of high definition DSM information within 2D SWEs based on standard hydraulic modelling tools depends on a spatial discretisation strategy. Regular and flexible mesh use gives comparable results but the workload for an optimal flexible mesh creation is much more important as required refinement lines will tend to over constrain the meshing algorithm due to their increased number and density.

Most topographic data used in practical engineering application for flood risk assessment have not been acquired specifically for hydrological application but have been acquired for multipurpose applications. Consequently for LIDAR data, an aspect such as removal of flow blocking macrostructures (bridges, elevated roads and so on) has to be properly handled by the modelling practitioner when creating DSM. With photogrammetry produced 3D classified data, the choices of the practitioner regarding which categories of above-ground features have to be included in the DSM will highly impact the results. Feedback to photointerpretation operators, such as differentiating vertical structures classes or subclasses depending on their permeability nature (e.g. fences or concrete walls) is technically accessible and might improve the ease of data use for hydraulic orientated application. Unavoidable errors of classification in photointerpretation might lead to important consequences for simulations and have to be carefully controlled. This might lead to restriction of 3D classified data use at a larger scale (city scale), independently of computational cost concern, as it would imply a large workload of data control.

The use of standard numerical modelling tools for high resolution run-off modelling spotlights disparities among tools in terms of practical aspects. High resolution topographical data integration can be properly done but modelling tools cannot equally fulfill the requirement for establishment of an adapted discretisation. Here, this is the case in terms of the ease of creating an adapted mesh and in terms of temporal discretisation limitations. Modelling tools cannot possibly represent all categories equally and variations of scenarios (rainfall/initial water depth). Modelling tools stability to handle high gradient occurrences, moving boundary treatment and spurious oscillations induce different limits related to discretisation and the numerical scheme used. In terms of relevance, the numerical modelling approach for run-off flood risk assessment can give a valuable insight into evaluation of the exposure to run-off and flooding risk. Moreover, the approach may highlight comprehension of dynamical aspects of the phenomena at a specific location.

3.1.10 From centralised to distributed and ubiquitous architecture

If the general principles of the classical warning systems may be accepted without difficulty, the specificity of some flood events – extremely short concentration time (below 10 minutes) and fast propagation in urban environments – require an immediate answer and information for the public who are exposed to the flooding process. In such situations, obviously, the centralised approach with supervision and monitoring cannot match the expectations and provide an efficient answer to the exposed population. The time for analysis, mainly based on human expert analysis, demands a duration which is not compatible with the speed of the process. Many examples can be provided to illustrate such a situation, like the flood events recorded in November 2011 in Genoa, Italy.

In order to improve the performance of flood warning systems, especially in the case of fast processes like urban flash floods, a shift of paradigm is needed to provide instantaneous information to the exposed populations and with the right message. The development of new generations of sensors allows the monitoring of urban streams with a high level of accuracy and for a limited cost. These new sensors now have communicating capacities which are embedded in a similar way to all modern communicating devices. The development of such technology allows introduction of the concept of ubiquitous computing which represents an alternative to the widely used centralised approach.

Ubiquitous computing refers to omnipresent computers that serve people in their everyday lives at home, at work, in transport and so on. Ubiquitous computing was introduced by Mark Weiser in 1991 and raised many challenges across computer science: in systems design and engineering, in systems modelling and in user interface design. However, initially the effective integration and interaction with the physical world sufficed to promote interest in ubiquitous computing due to significantly increased real world visibility as well as real world control, towards ambient intelligence. These first ubiquitous systems were often proofs of concept with a single static configuration with a priori known devices. Ubiquitous computing has now become a reality for applications evolving in the real environment thanks to the evolution of ICT technologies: consumer electronics, telecommunications and software.

With the new devices and their functionalities, ubiquitous computing can now revolutionize interactions between users, their environment and information systems. It offers a post-desktop model for human/machine interfaces to access information systems, not only through laptop, PDA and mobile phone, but also through each equipment used in daily life: alarms within buildings, display networks in cities and so on. As devices become smaller, more connected and more integrated into a global environment which is developing continuously, the technology disappears into our surroundings in everyday life objects and devices. Thus ubiquitous systems can open up new areas of application: smart buildings, industrial automation, healthcare, power distribution and a host of others. This interconnection of devices and objects is expected to amplify the profound effects that networked systems are having on our modern society. The emergence of intelligent cities is one of the main societal impacts of ubiquitous computing. Indeed, the city now has the technology to improve public safety through a better management of natural risks. Especially, the efficiency of warnings to the population can be improved tremendously, by increasing the number of informed people, by improving the relevance of broadcast messages and by reducing the time required for communication with the public. With networked devices in the smart cities, warning systems can become, opportunistically, richer than ever, mainly in data collection from the field and for communications with the public.

The new possibilities offered by the ubiquitous approach have been investigated within a research project called UbiFLOOD [http://www.ubiflood.eu/]. The project has been launched with the objective to develop some ubiquitous approach in different Asian cities which are exposed to regular flooding events. Case study areas within the project are Bangkok, Mumbai and Singapore. The common characteristics for the selected cities are that they are in monsoon regions and that they have a history of flooding problems. However, the characteristics of flood events are very different (from slow to extremely fast) and require specific answers.

Following the analysis made in Mumbai and in Singapore, it appears that one of the most severe difficulties is generated by flash flood events that may appear in less than 10 minutes after the start of intense rainfall events. Singapore authorities are deploying a complete data acquisition chain including radar data treatment in real time but, despite this action, the city centre was exposed to flooding in 2010 and 2011 (Recorded videos are available on the UbiFLOOD website at: http://www.ubiflood.eu/). In order to cope with such situations, a potential solution is to deploy a set of sensors – autonomous and low cost – able to collect information about water levels and to communicate directly to the exposed population by SMS, email and voice messages. This approach allows faster reaction and avoids the delay due to the centralised concept of the traditional warning system. The principle of the new systems is presented in Figure 3.1.7.

In order to validate the concept, a prototype based on an Ijinus communicating sensors (http://www.ijinus.com/) wireless ultrasonic numerical level probe, has been developed and tested in the laboratory in Singapore. The ultrasonic sensors used – with the following characteristics: nominal frequency 120 KHz with a range of 0.2 cm to 3 m – provide an accurate measure of the water level with a minimum resolution of 2 mm. The autonomy of the sensor can reach up to one year with the internal battery. With these characteristics, the installation and the maintenance of the sensors represents a limited investment which allows redundancy in order to optimise the functionalities of the network. The sensors are connected to the communication network and can send information directly to the surrounding devices according to the defined protocol.

Some videos of the conducted experiments are presented on the UbiFLOOD website [http://www.ubiflood.eu/] and describe how the data acquisition process is able to send an alert message to a mobile phone according to the defined rules.

Service-oriented architecture (SOA) (Durairaju Kumaran *et al.*, 2010, Papazoglou and Houvel, 2007) is a well-known set of principles and methodologies for designing and developing software in the form of interoperable services. SOA defines how to integrate widely disparate applications for a Web-based environment, over the Internet, and uses multiple implementation platforms. Rather than defining an API, SOA defines the interface in terms of protocols and functionality. Service orientation requires loose coupling of services with operating systems and other technologies that underlie the applications.

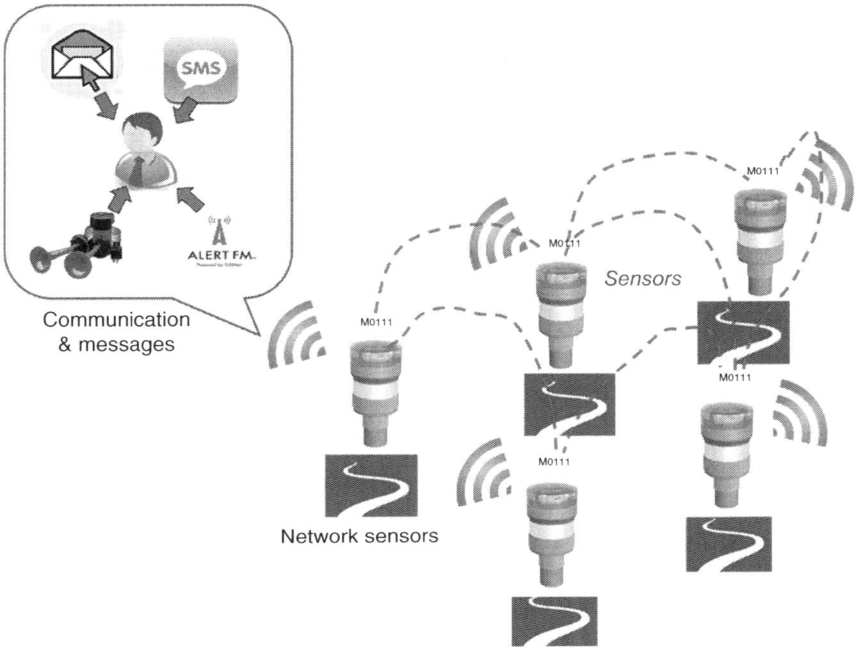

Figure 3.1.7 Architecture of the UbiFLOOD solution based on Ijinus sensors. Each sensor collecting the water level is able to communicate with the other devices and may broadcast a message to experts and exposed persons in case of flooding.

Under pressure from ubiquitous computing, many standards in the business are appearing to promote new concepts like machine to machine (M2M) (Lawton, 2004), Internet of things (IoT) (Mattern and Floerkemeier, 2010) and, more recently, the Web of things (WoT), to federate service-oriented approaches and software development using networked physical devices. The approach used is based on these recent advances. These standards and/or well-known technologies are used to implement services for devices oriented architectures. For example UPnP (universal plug and play from the UPnP consortium (http://www.upng.org) and DPWS (device for Web services from OASIS (http://docs.oasis-open.org/ws-dd/ns/dpws/2009/01) allow devices to connect easily and simplify the implementation of networks in a restricted area like smart buildings (file sharing, communications, and entertainment) or in companies. It allows defining and publishing UPnP and DPWS control protocols on top of standard communication of the Internet.

Both use XML-based protocols above Web protocols to manage services provided by networked physical devices. These protocols allow two kinds of communications between services of devices and software clients: classical remote methods invocations but also events-based communications much more adapted to the device that must be able to collect spontaneous information in the physical environment. Moreover, they introduce protocol to manage appearance, discovery and disappearance of services attached to the device, at runtime. Every device is then usable without knowing beforehand what the services it provides are and their interface. Thus, ubiquitous applications can be deployed on a partly unknown set of devices that can be change at runtime.

The implemented solution is founded on an original approach based on services for devices software infrastructure to self-adapt at runtime any ubiquitous application to the change and the availability of surrounding devices (Durairaju Kumaran *et al.*, 2010; Papazoglou *et al.*, 2007). This approach is based on a set of independent rules to describe how the system must react and evolve when the software infrastructure is changing. Then the main theoretical challenges are the interferences and even conflicts that may appear between rules-based modifications on the shared software application. Indeed the merge algorithm between the modifications to finally modify the common application must maintain the consistency, that is verify proved properties during the merging process, like mastered response time for the adaptation, and some logical and semantic properties described as requirements for the consistency of the application.

In the case of a flood warning and management system, the software infrastructure consists of networked sensors, simple pre-computed rules to define in real time in which geographical area warning messages have to be sent to the population. In this scenario, the sensors provide real time water level data which are more reliable than forecast data. The flooding area/zone is calculated from the pre-created scenario. In the validation case developed in Singapore, two kinds of services and devices are used to send information about flooding. The first set is dedicated to expert users in order to help them to analyse and assess the situation. The second set is composed of various communication devices and services and is dedicated to sending a warning to the public. The population can be informed through a web page, a SMS/MMS and/or an e-mail. The map displayed on the web page can be modified with real-time information on the water levels and discharges collected from the monitoring network. The system sends an SMS/MMS alert in the following cases:

- SMS/MMS alert will be sent to all government officials /experts in the flooded zone based on their phone model. The information to those officials /experts will be stored in the database.
- SMS/MMS alert will be sent to registered public. The public can register to get an alert in their zone of interest. While registering, they need to provide their phone model. This alert is not location based. For example, if heshe registers for zone A, he/she will get information only for zone A even though he/she may not be there while the flood may affect that zone.
- SMS/MMS alert will be sent to registered users. He/she does not want to register for a particular zone/area. This alert is a location-based service. When registered with the system, he/she will get a flood alert if he/she is in the potential flooded zone.

Obviously, the type of alert can be adjusted according to the situation and the availability of data: if upstream details are available, the system may provide a forecast of the potential extent of the flooded area and allows a reaction from the concerned population. If the system is implemented close to the flooded area, the forecast will be based on the duration and on the magnitude of the event. In all cases, the devices monitoring the water levels can almost instantly broadcast the predefined messages according to the established rules.

The ubiquitous approach supported by new sensors and interfaced with communicating devices allows potentially a reshaping of the flood warning systems and could provide an efficient answer for flood events such as thosein Singapore where the concentration time does not exceed 10 minutes. The communicating capacity of the individual devices allows immediate communication to the exposed population and information about on-going risks. However, this progress needs to be associated with a very careful definition of the messages to send to the population in order to generate the proper reaction. At the same time, the data collection has to be carried out in order to nourish a knowledge base which is the key reference for establishing the rules for issuing warning messages. In such context, the future of the warning system has to be seen as a combination of the traditional SCADA approach combined with the ubiquitous devices which represent a great potential for improvement. The availability of both sensors and communication devices like smartphones allow in depth review of the concept of a warning system with the associated workflows.

3.1.11 Perspectives in conclusion

Flood management represents a major issue for modern societies. The growing sophistication of human environments – urban and agricultural – has induced a higher vulnerability which has to be managed in order to minimize impacts on both the population and economic activities. In such a situation, hydroinformatics methods which integrate both models focused on water sciences and on a decision process have a major role to play. In addition to the actions initiated over the last three decades, hydroinformatics has to face the challenge of resilience. The introduction of the resilience concept is following the path initiated by the Integrated Food Management and leads one to accept that the protection measures will not be able to ensure a 100% safe environment. The integration of the risk of flooding in the very first stage of design for urban and rural environments will contribute to improving the capacity of modern societies to adapt and to restore their functionalities. This shift of paradigm represents the major challenge for the hydroinformatics methods which now have a key position within the spreading and the practical implementation of the resilience concept.

Acknowledgement

LIDAR and 3D classified photogrammetric data sets have been kindly provided by Nice Municipality GIS Department (DIGNCA) for research purposes.

References

Abily M., Duluc M.C., Faes J.B. and Gourbesville P. (2012) Performance assessment of modelling tools for high resolution runoff simulation over an industrial site, *Journal of Hydroinformatics*, **15**(4), 1296–1311.

Abbott M.B. (1991) *Hydroinformatics: Information Technology and the Aquatic Environment.* Avebury Technical, Aldershot, UK, 145 pp.

Abbott M.B., Bathurst J.C., Cunge J.A. *et al.* (1986) An Introduction to the European Hydrological System – Système Hydrologique Européen, 'SHE', 1: History and philosophy of a physically-based, distributed modelling system. *Journal of Hydrology*, **87**(1), 45–59.

Beven K.J. (1985) Distributed models. in Anderson, M.G. and Burt, T.P. (eds), *Hydrological Forecasting*, eds M.G. Anderson and T.P. Burt. John Wiley & Sons Ltd, Chichester, pp. 405–435.

Beven K.J. and Freer F. (2001) A dynamic TOPMODEL, *Hydrological Processes*, **15**, 1993–2011.

Beven K.J. and Kirkby M.J. (1979) A physically-based variable contribution area model of basin hydrology, *Hydrological Sciences Bulletin*, **24**, 43–69.

Bronstert A. and Plate E.J. (1997) Modelling of runoff generation and soil moisture dynamics for hillslopes and micro-catchments, *Journal of Hydrology*, **198**(1–4), 177–195.

BWK (1999) *Hydraulische Berechnung von naturnahen Fließgewässern, Merkblatt 1/1999.*

Ciliberti S.A., Gomez M., Macchione F. *et al.* (2008) *2D analysis for local flooding assessment in a new square of Barcelona during storm events. 11th International Conference on Urban Drainage*, Edinburgh, Scotland, UK, p. 9.

Cunge J. A. (1975) Two-dimensional modeling of flood plains. in *Unsteady Flow in Open Channels* (ed. K. Mahmood & V. Yevjevich), Water Research Publications, Chapter 17.9, pp. 731–750.

Cunge J.A. (2003) Of model and data. *Journal of Hydroinformatics*, **5**(2), 75–98.

DHI (Danish Hydraulic Institute) (1998a) *Mike 11 – User and Reference guides*, Horsholm, Denmark.

DHI (Danish Hydraulic Institute) (1998b) *Mike 21 – User and Reference guides*, Horsholm, Denmark.

Durairaju Kumaran R., Clews E., Tigli J.Y. *et al.* (2010) Sensors Networks, SOA and Web Based Approach for Fresh Water Environmental Monitoring, 9th International Conference on Hydroinformatics (HIC), Tianjin, China, 7–11 September 2010.

Giammarco P.D., Todini E. and Lamberti P. (1996) A conservative finite element approach to overland flows. The control volume finite element formulation, *Journal of Hydrology*, **175**(1), 267–291.

Gomez M., Macchione F. and Russo B. (2011) Methodologies to study the surface hydraulic behaviour of urban catchments during storm events. *Water Sciences and Technologies*, **63**(11), 9.

Gourbesville P. (2009) Data and hydroinformatics: new possibilities and new challenges. *Journal of Hydroinformatics*, **11**(3–4), 330–343.

Gourbesville P. (2011) ICT for water efficiency in *Environmental Monitoring*, (ed. E.O. Ekundayo) Intech, pp. 411–426, ISBN: 978-953-307-724-6 (http://www.intechopen.com/articles/show/title/ict-for-water-efficiency).

Guinot V. and Gourbesville P. (2003) Calibration of physically based models: back to basics? *Journal of Hydroinformatics*, **5**(4), 233–244.

Henonin J., Russo B., Mark O. and Gourbesville P. (2013) Real-time urban flood forecasting and modelling – a state of the art. *Journal of Hydroinformatics*, **15**(3), 717–736.

Jonch-Clausen T. (2004) *IWRM and Water Efficiency Plans by 2005: Why, What and How?, Global Water Partnership Technical Committee, Background Paper no.10.*

Kirkby M.J., Ed. (1978) *Hillslope Hydrology*, John Wiley & Sons Ltd, Chichester.

Lafarge F., Descombes X., Zerubia J. and Pierrot-Deseilligny M. (2010) Structural approach for building reconstruction from a single DSM. *PAMI*, **32**(1), 14.

Lafarge F. and Mallet C. (2011) *Building large urban environments from unstructured point data in Proceedings of the IEEE International Conference on Computer Vision (ICCV), Barcelona, Spain*, **2011**, p. 8.

Lawton G. (2004) Machine-to-machine technology gears up for growth. *Computer*, **37**(9) 12–15.

Lippert K. (2005) Analyse von Turbulenzmechanismen in naturnahen Fließgewässern und ihre mathematische Formulierung für hydrodynamische Modelle, Hamburger Wasserbau-Schriften, Heft 4, Hamburg.

Mastin A., Kepner J. and Fisher J. (2009) Automatic registration of LIDAR and optical images of urban scenes. inn *IEEE Conference on Computer Vision and Pattern Recognition (Miami, FL, June 2009)*, IEEE, pp. 2639–2646.

Musialski P., Wonka P., Aliaga D. *et al.* (2012) *A survey of urban reconstruction. Eurographics 2012*, State of the Art Reports, pp. 1–28.

Mattern F. and Floerkemeier C. (2010) *From the internet of computers to the internet of things*. in From Active Data Management to Event-based Systems and More, (eds K. Sachs, I, Petrov and P. Guerrero) Springer-Verlag, Berlin, pp. 242–259.

Malcherek A. (2001) *Hydrodynamik der Fließgewässer*, Habilitation, Universität Hannover.

Papazoglou M.P. and Heuvel W.J. (2007) Service oriented architectures: approaches, technologies and research issues, The VLDB Journal, **16**(3), 389–415.

Pasche E. (1984) Turbulenzmechanismen in naturnahen Fließgewässern und die Möglichkeiten ihrer mathematischen Erfassung, Mitteilungen Lehrstuhl und Institut für Wasserbau und Wasserwirtschaft, Nr. 52, RWTH Aachen.

Pasche E., Brüning C., Plöger W. and Teschke U. (2005) *Möglichkeiten der Wirkungs-analyse anthropogener Veränderungen in naturnahen Fließgewässern, Hamburger Wasserbau-Schriften, Heft 4*, Hrsg. Erik Pasche, Technische Universität Hamburg-Harburg.

Pasche E. (2007) *Mannings versus Darcy Weisbach law for flood modelling in rivers*, Proceedings IAHR Conference, Venice.

Price R.K. and Vojinovic Z. (2008) Urban flood disaster management. *Urban Water Journal,* **5**, 259–276.

Remondino F., Barazzetti L., Nex F., Scaioni M., Sarazzi D. (2011) *UVA photogrammetry for mapping and 3D modelling -Current status and future prespectives. Conference on Unmanned Aerial Vehicle in Geomatics. 2011. International Archives of the Photogrammetry*, Remote Sensing and Spatial Information Sciences: Zurich, Switzerland, p. 7.

Rodi W. (1993) *Turbulence Models and their Application in Hydraulics*, International Association for Hydraulic Research (IAHR), Delft, 3rd edition.

Rouvé G. (1987) *Hydraulische Probleme beim naturnahen Gewässerausbau*, VCH, Weinheim.

Sherman L.K. (1932) Streamflow from rainfall by unit-graph method, *Eng. News Record*, **108**, S. 501–505.

Tsubaki R. and Fujita I. (2010) Unstructured grid generation using LiDAR data for urban flood innundation modelling, *Hydrological Processes*, **24**, 17.

Wenka T. (1992) *Numerische Berechnung von Strömungsvorgängen in naturnahen Fließgewässern mit einem tiefengemittelten Modell*, Dissertation Universität Karlsruhe.

World Meteorological Organization (WMO) (2009) *Integrated Flood Management – Concept paper*, Geneva, 32 pp. ISBN 978-92-63-11047-3.

Zanobetti D., Lorgéré H., Preissmann A., Cunge J.A. (1968) Le modèle mathématique du Delta du Mékong (Mathematical model of Mekong Delta). *La Houille Blanche*, **23**(1, 4 and 5).

3.2
Drought: How to be Prepared for the Hazard?

Henny A.J. Van Lanen

Hydrology and Quantitative Water Management Group, Wageningen University, Wageningen, The Netherlands

3.2.1 Introduction

Drought is one of the most extreme weather-related natural hazards. Millions of people across the world are annually exposed to droughts that seriously affect economic development and environment. Poor economies are more impacted than more developed ones. For example, in 2011, a severe drought affected the entire East Africa region and contributed to a severe food crisis. It threatened the livelihood of more than 12 million people in Djibouti, Ethiopia, Kenya and Somalia (UN, 2011). In 2012, about 25% of the victims of natural disasters worldwide were because of drought (Guha-Sapir, Hoyois and Below, 2013). Droughts in developing countries may cause fatalities, which does not happen in the more prosperous countries. However, in Europe almost 80 000 people died due to associated heatwaves and forest fires over the period 1998–2009. Overall losses were estimated to be as high as €4940 billion over the same period (EEA, 2010). Estimated economic impact from the drought in some parts of the United States in 2012 reached US$20 billion and was the highest reported for the decade (Guha-Sapir, Hoyois and Below, 2013). The United States Department of Agriculture's Farm Service Agency distributed an estimated US$1.6 billion for drought relief from 2008 to 2011 through the Livestock Forage Disaster Program (USDM, 2013).

Drought will worsen in parts of the world because of climate change (e.g., Bates *et al.*, 2008; Seneviratne *et al.*, 2012; Prudhomme *et al.*, 2014; Forzieri *et al.*, 2014), although confidence in projections is still rather low, in particular for the early twentyfirst century (IPCC, 2013). However, vulnerability will increase in numerous regions as a response to population growth and, thus, higher demand for water. Water and food security, in particular under drought conditions, are at risk (e.g., Falkenmark, Rockström and Karlberg, 2009; WWDR, 2009; Gerten *et al.*, 2011; Haddeland *et al.*, 2014; Schewe *et al.*, 2014). Large-scale drought can also have a large impact on ecosystems. The 2010 Amazon drought showed that repeated drought may largely influence tropical forests that can shift from buffering CO_2 increase into accelerating it, which can have important

Hydrometeorological Hazards: Interfacing Science and Policy, First Edition. Edited by Philippe Quevauviller.

decadal-scale impacts on the global carbon cycle (Lewis *et al.*, 2011). Drought can also lead to more peat fires in tropical wetlands that affect the ecosystem (e.g. Anshari *et al.*, 2004) and the global climate system through the carbon cycle (e.g. Page *et al.*, 2002; Murdiyarso, Hergoualch and Verchot, 2010).

Drought differs from other hydrometeorological extremes in several ways. It develops gradually and usually over large areas (transnational), mostly resulting from a prolonged period (from months to years) of below-normal precipitation, and drought can occur nearly anywhere on the globe. The influence of precipitation or temperature anomalies propagates through the hydrological cycle (Figure 3.2.1) from its origin as a meteorological drought into soil moisture depletion to the point where crops or terrestrial ecosystems are impacted, and eventually into a hydrological drought. Hydrological drought refers to a prolonged period with below-normal water availability in rivers and streams, and lakes, or groundwater bodies due to natural causes. From an impact and drought management point of view, it is important to distinguish between the different types of drought, and underlying drought-generating processes. In this context, trends in observed drought hazard support understanding of how drought frequency and intensity may develop in future and offer the opportunity to be better prepared for drought.

Various overview papers and textbooks (e.g. Wilhite and Glantz, 1985; Wilhite, 2000; Tallaksen and Van Lanen, 2004) address one or more aspects of the multifaceted drought phenomenon. Recently, overviews have been added by Mishra and Singh (2010) and Sheffield and Wood (2011). This chapter adds knowledge on the drought hazard from recently finished and on-going EU-funded research programmes. It covers a description of recent achievements on: (i) processes that generate the drought hazard and how to identify, (ii) trends in past drought hazard, (iii) projections of drought

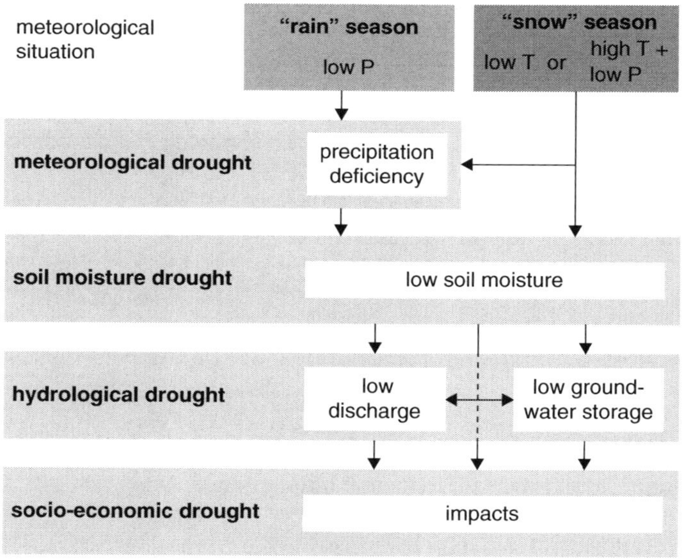

Figure 3.2.1 Different drought types and their development, including precipitation and temperature control. *Source:* Van Loon, 2013, Fig. 8.1, p. 140. Reproduced with permission of the author. See plate section for colour representation of this figure.

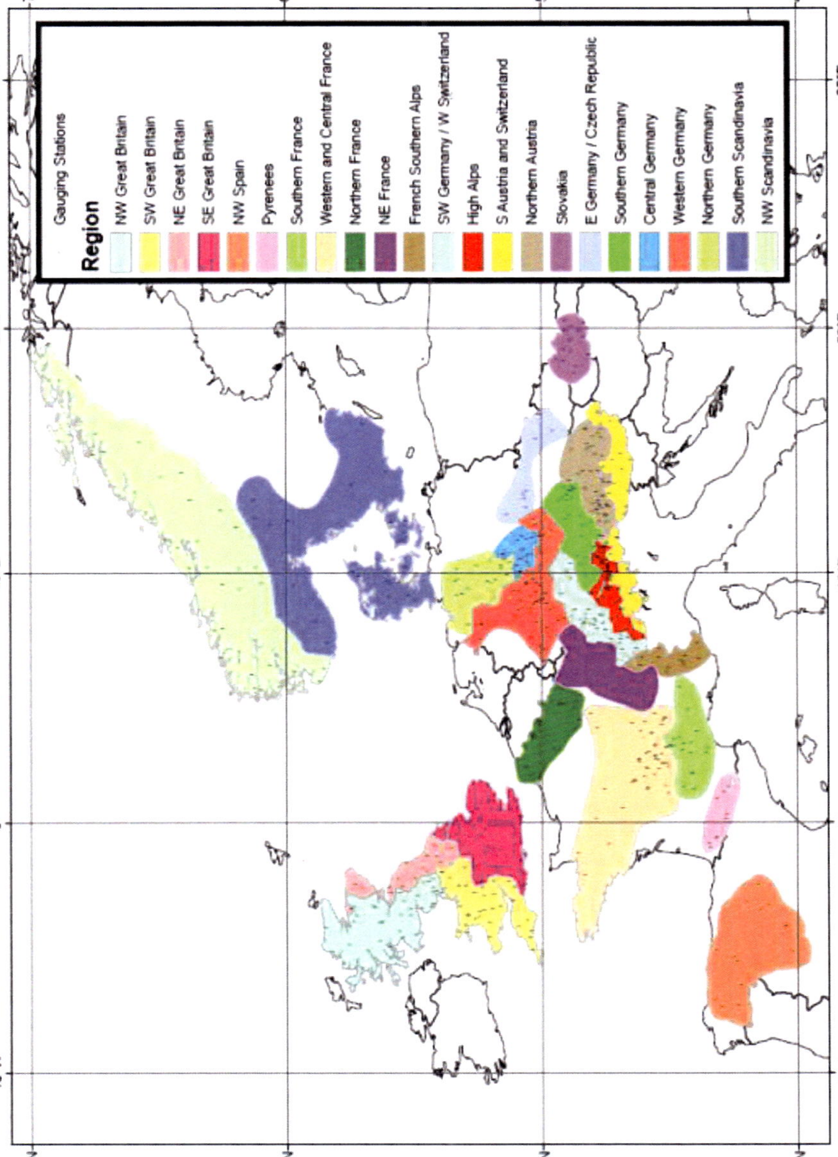

Figure 1.2.3 Regions used in the drought catalogue for Europe. *Source:* Parry *et al.*, 2011.

Hydrometeorological Hazards: Interfacing Science and Policy, First Edition. Edited by Philippe Quevauviller.

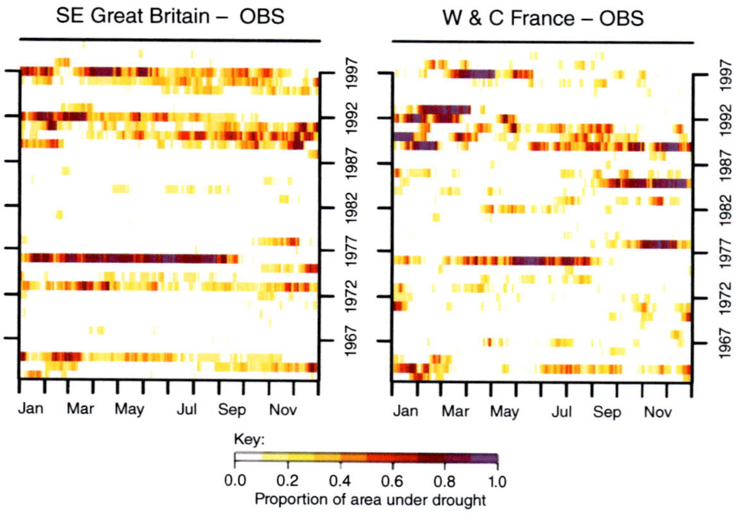

Figure 1.2.4 Drought catalogue derived from observed river flow gauges for two contrasting regions of Europe: Southeast Great Britain and Southwest Germany and West Switzerland. The months of the year are shown on the *x*-axis, years on the *y*-axis. Colour show the Regional Deficiency Index, a measure of the proportion of the region experiencing a flow deficiency. *Source*: Parry *et al.*, in press.

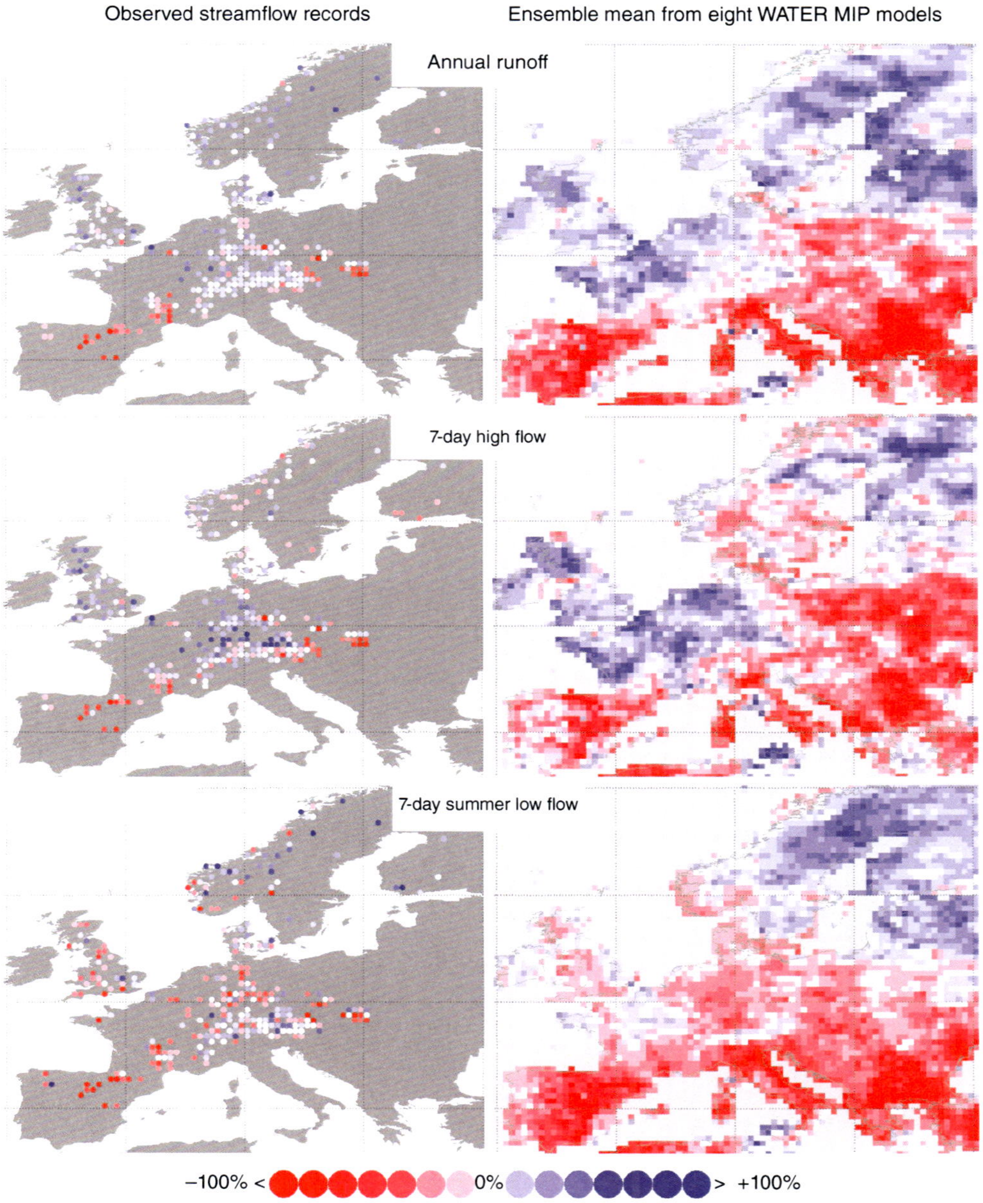

Figure 1.2.5 Trends in annual run-off in Europe: (a) observed and (b) as simulated by a multi-model ensemble of eight large-scale hydrological models. *Source*: Stahl *et al.*, 2011. ©American Meteorological Society.

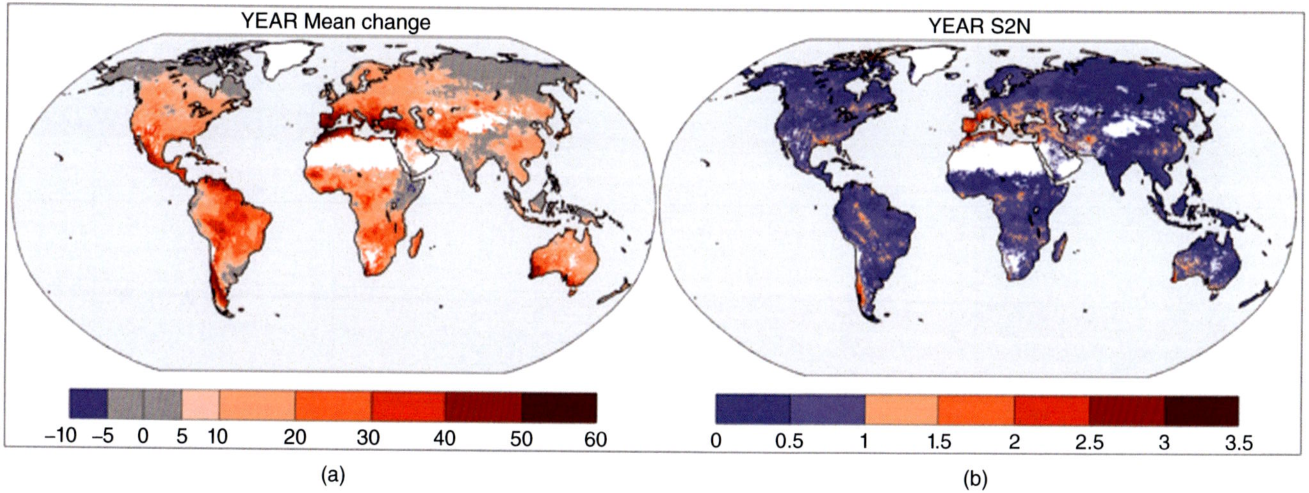

Figure 1.2.6 Percentage change in the occurrence of days under drought conditions for the period 2070–2099 relative to 1976–2005, based on a multimodel ensemble MME experiment under RCP8.5 from five global climate models and seven global impact models: (a) MME mean change and (b) associated signal-to-noise ratio (S2N, MME mean change divided by its inter-quartile range). *Source*: Prudhomme *et al.*, 2013b, Fig. 1.

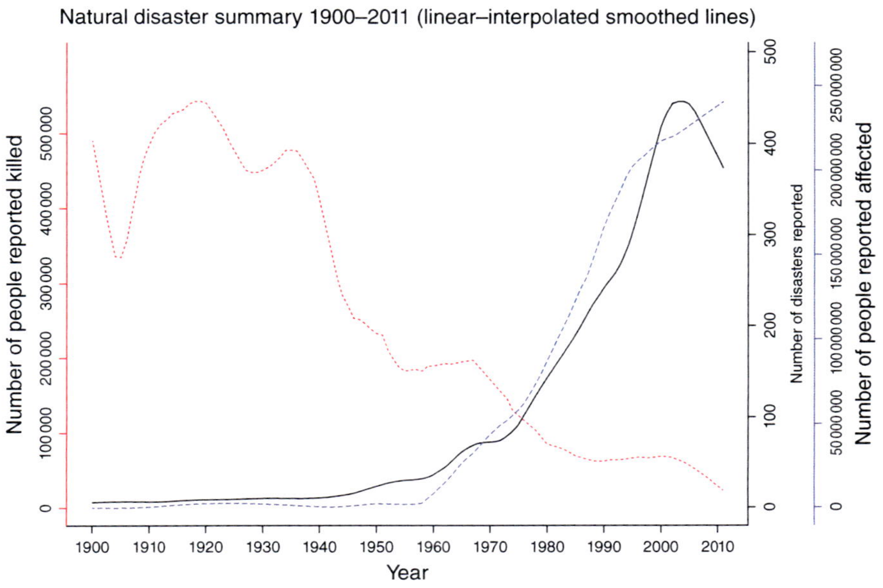

Figure 1.3.1 Trends of disasters and affected people in the period 1990–2011. From EM-Data: The OFDA/CRED International database – www.emdat.be.

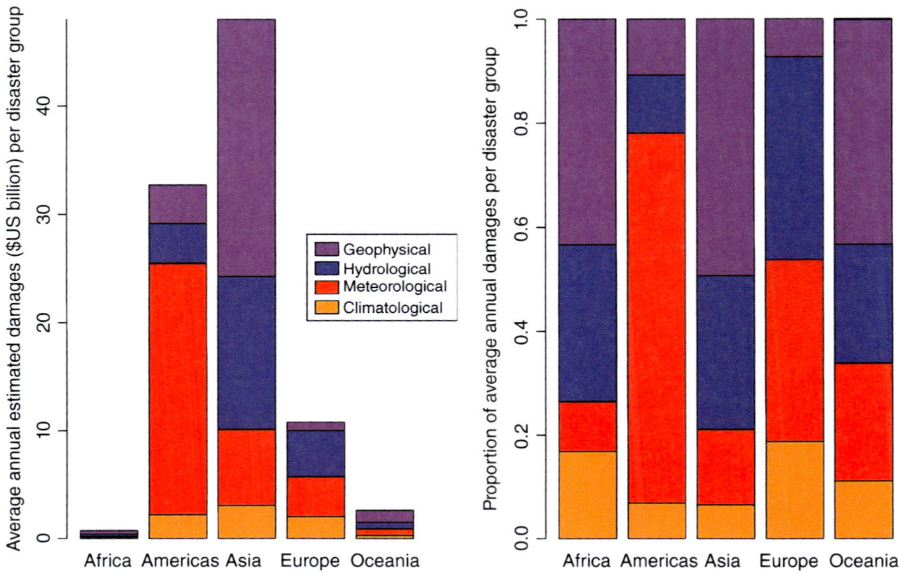

Figure 1.3.2 Average annual damage ($US billion) caused by reported natural disasters in the period 1990–2011. From EM-Data: The OFDA/CRED International database – www.emdat.be.

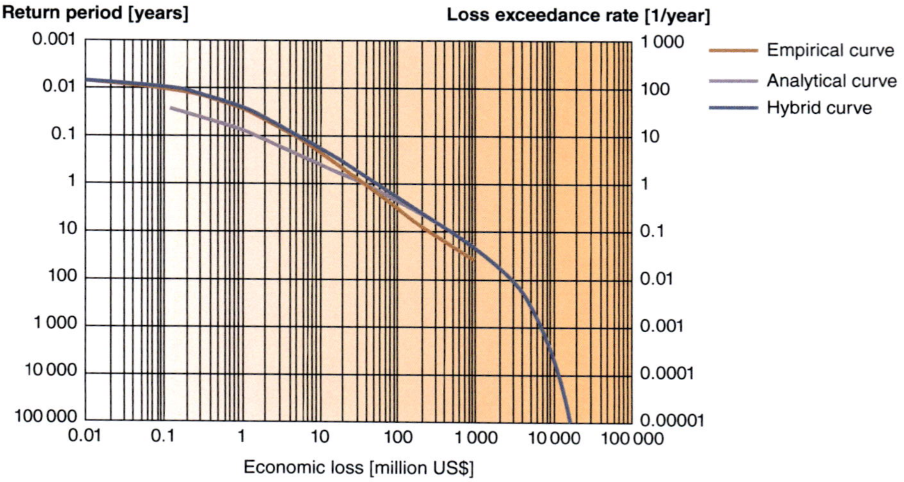

Figure 2.1.1 Hybrid risk model: loss exceedance curve for Colombia. *Source*: Data from ERN-AL in the UNISDR 2011 Global Assessment Report on Disaster Risk Reduction.

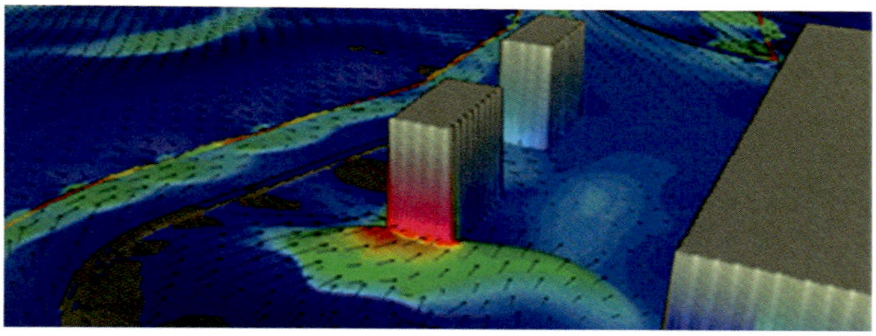

Figure 3.1.3 Two-dimensional hydraulic modelling in an urban environment – velocity vectors and water depth.

Figure 3.1.5 Example of two-dimensional hydraulic modelling based on a LIDAR data set (DSM).

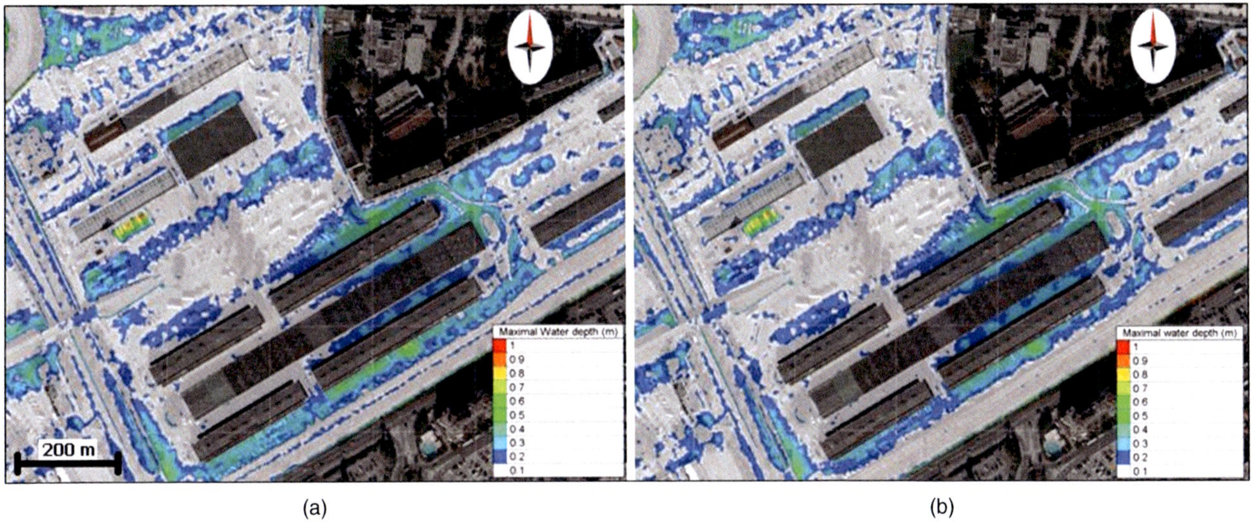

Figure 3.1.6 Maximum water depth using photogrammetry combined with LIDAR DSM with (a) a structured (Mike 21) and (b) a flexible mesh (TELEMAC-2D).

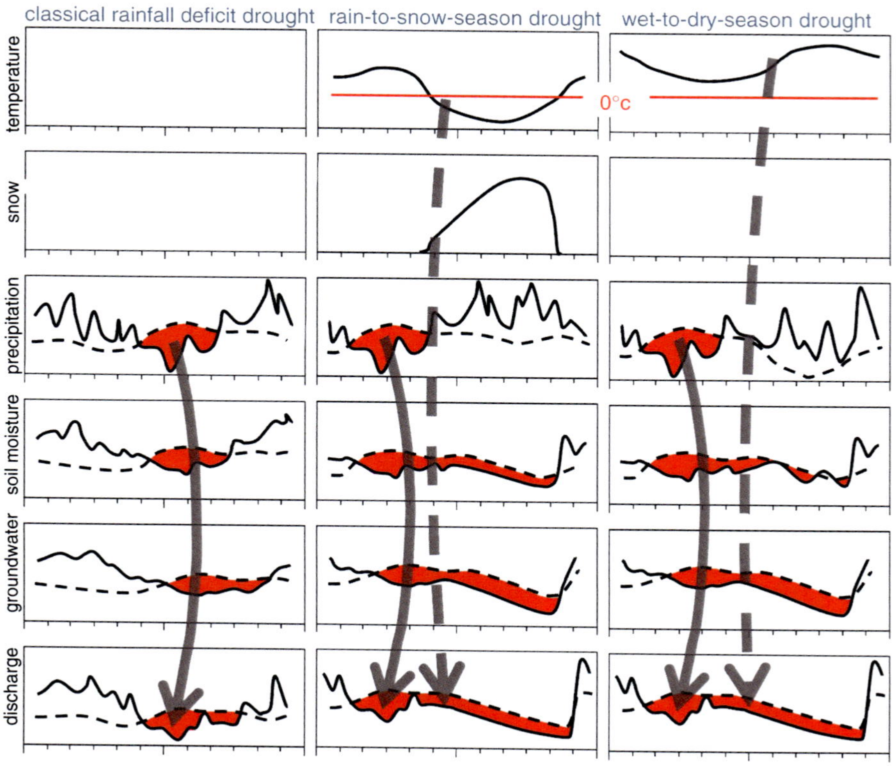

Figure 3.2.2 Drought typologies caused by precipitation and temperature control. *Source*: Van Loon, 2013, adapted from Fig. 8.2, p. 141. Reproduced with permission of the author.

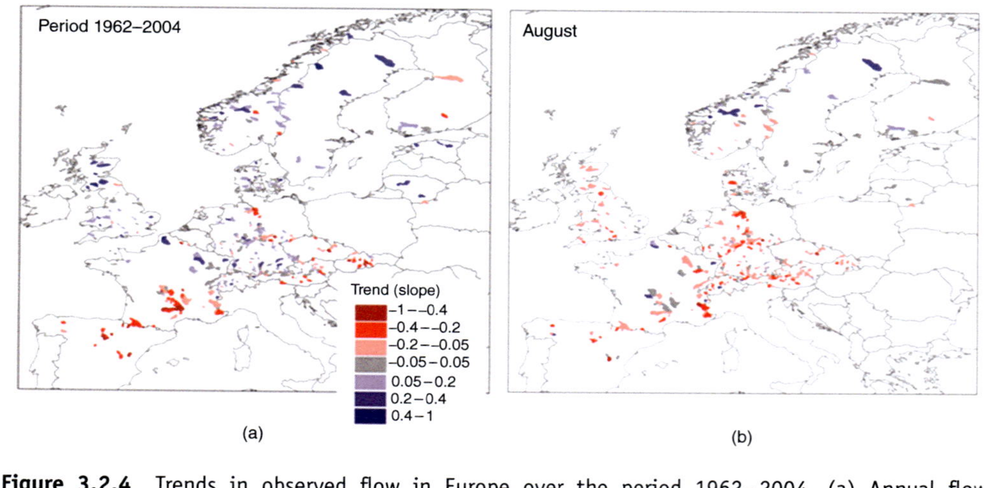

Figure 3.2.4 Trends in observed flow in Europe over the period 1962–2004. (a) Annual flow, (b) August flow. *Source*: Stahl *et al.*, 2010, adapted from Fig. 1, p. 2373 and Fig. 2, p. 2374. Reproduced with permission of the authors. Blue are negative trends, implying increasing flow, towards wetter conditions and red indicates the opposite (positive trend, decreasing flow, towards drier conditions).

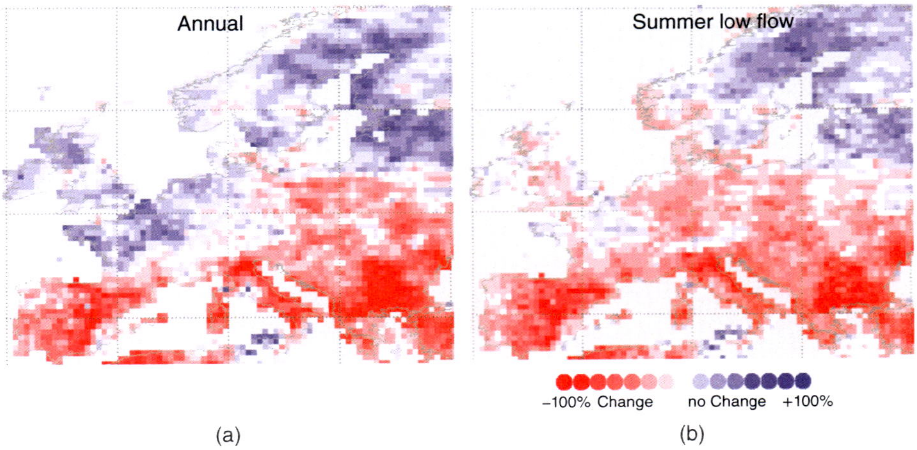

Figure 3.2.5 Trends in simulated flow (multi-model ensemble mean) in Europe over the period 1963-2001. (a) Annual flow, (b) summer low flow. *Source*: Stahl *et al.*, 2012, Fig. 4, p. 2041. Reproduced with permission of authors.

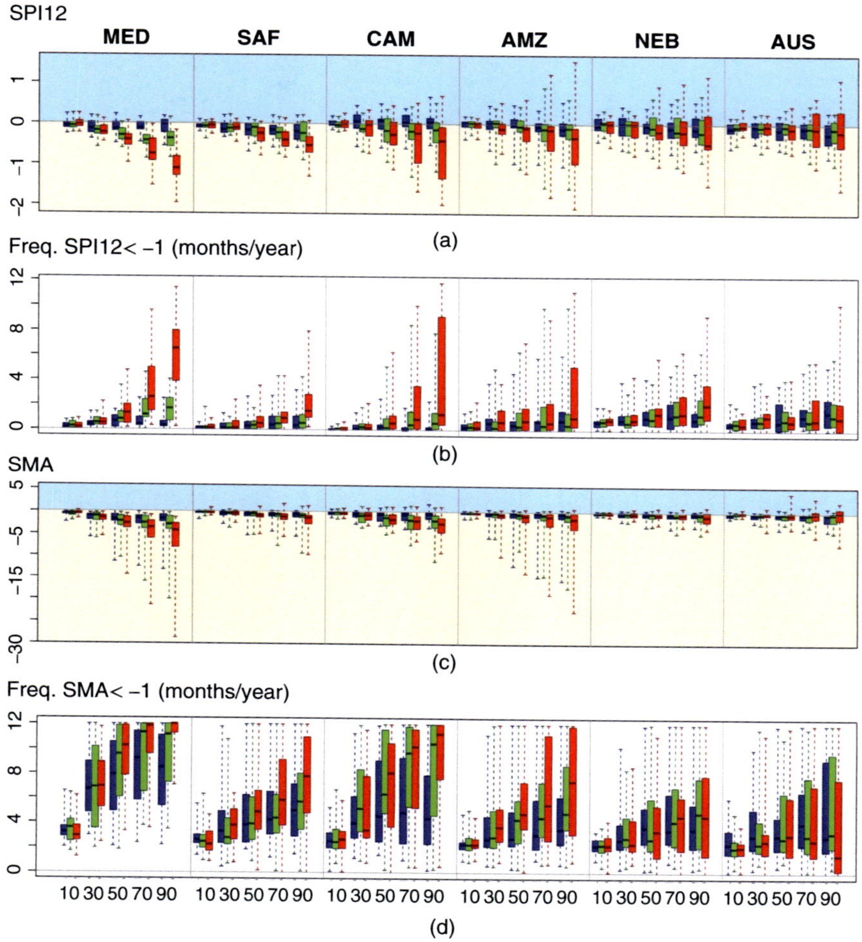

Figure 3.2.7 Drying regions in the twentyfirst century (regions where the median SPI12 decreases): (a) 20 year average SPI12, (b) months per year in which SPI12 < −1, (c) and (d) as (a) and (b), but for SMAs. Changes are given as standard deviations with respect to 1979–2009 in (a), mm in (c) and as months per year in (b) and (d), The numbers on the lowest x-axis denote the central years in the twentyfirst century of the 20 year windows. Colours indicate the respective GHG concentrations scenario, blue: RCP2.6, green: RCP4.5 and red: RCP8.5. *Source*: Orlowski and Seneviratne 2013, Fig. 8, p. 1776. Reproduced with permission of the authors.

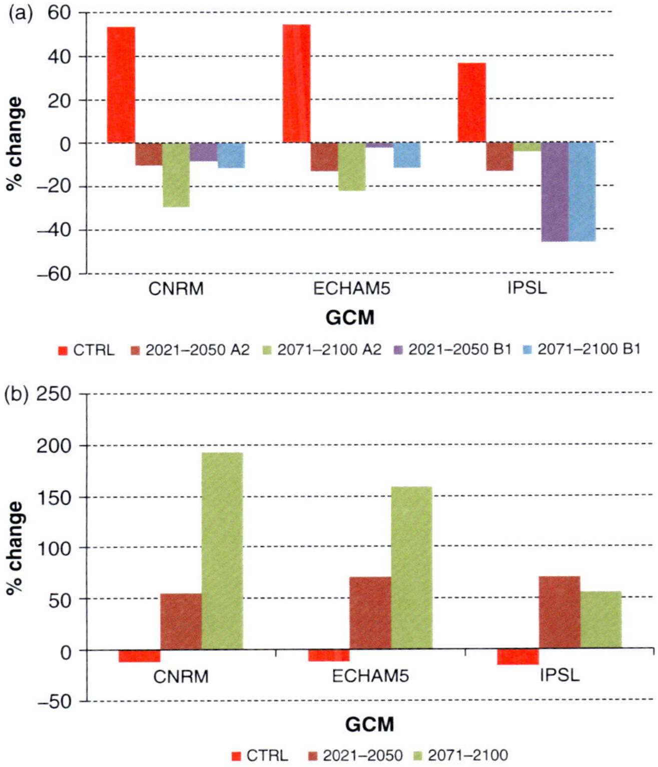

Figure 3.2.8 Projected change in future mean annual run-off (a) and average drought duration (b) derived from a multi-hydrological model ensemble for three GCMs (CNRM, ECHAM5 and IPSL), intermediate future (2021–2050) and far future (2071–2100). CTRL indicates model uncertainty. *Source*: Data from Alderlieste and Van Lanen, 2013.

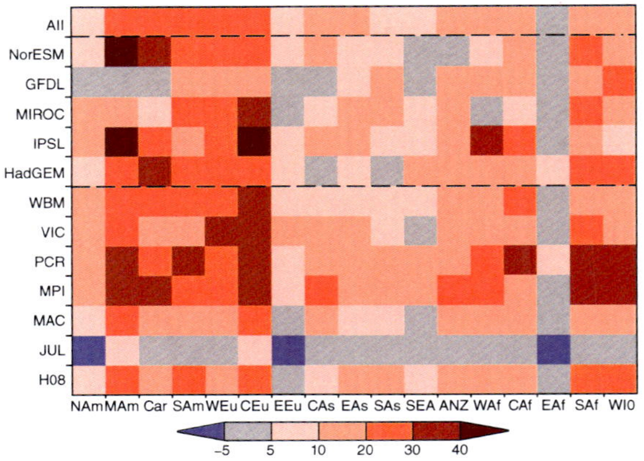

Figure 3.2.9 Mean percentage changes in hydrological drought (i.e. regional deficit index, RDI) in the future (2070–2099) relative to the control period (1976–2005) under RCP8.5 for 17 world regions (*x*-axis). Changes are averaged over all of the seven large-scale hydrological models and five GCMs (All, upper row), per GCM averaged over all seven large-scale hydrological models (rows 2–6), and per large-scale hydrological models averaged over all GCMs (rows 7–13). Prudhomme *et al.*, 2013, Fig. 5. Reproduced with permission of the authors.

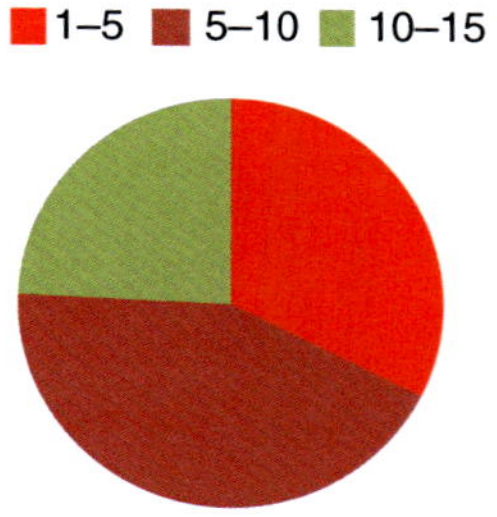

Figure 3.2.10 Number of model ensemble members (large-scale hydrological model and GCM) that passed performance criteria in the control period derived from observed low flow (derived from Van Huijgevoort *et al.*, 2013).

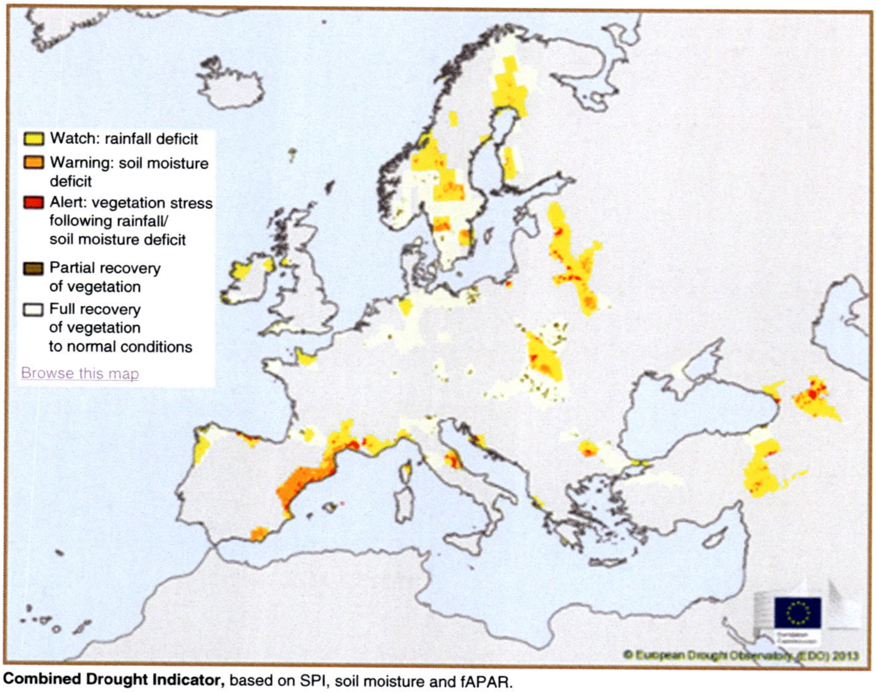

Combined Drought Indicator, based on SPI, soil moisture and fAPAR.

- **Watch:** when a relevant precipitation shortage is observed
- **Warning:** when this precipitation translates into a soil moisture anomaly
- **Alert:** when these two conditions are accompanied by an anomaly in the vegetation condition.

Figure 3.2.11 Pan-European drought situation, 11–20 November 2013. *Source*: European Drought Observatory, JRC. Pan-European drought situation, 11-20 November 2013 (European Drought Observatory, EDO, JRC, 2013). Reproduced with permission of the European Commission, Joint Research Centre, Institute for Environment and Sustainability (EC JRC-IES), Ispra, Italy.

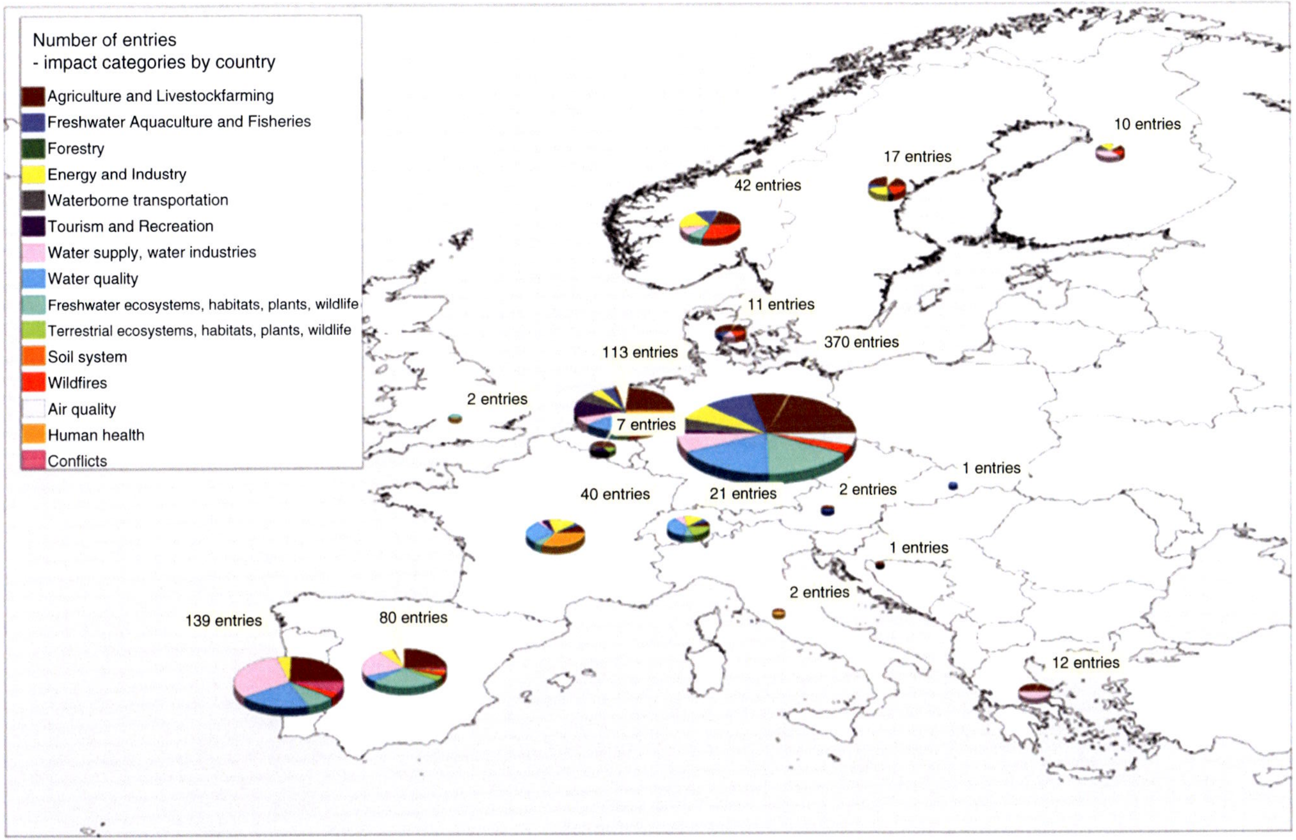

Figure 3.2.12 Different drought impacts across Europe retrieved from the European Drought Impact Inventory (EDII). *Source*: Stahl *et al.*, 2012, Fig. 7, p. 8. Reproduced with permission of the authors.

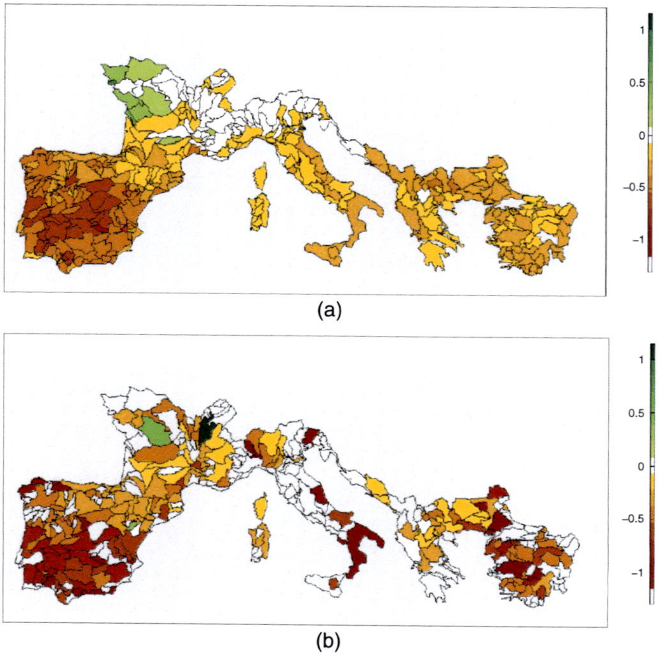

Figure 3.3.4 Per unit changes of run-off (a) and water availability for irrigation (b) in the climate change scenario (2070–2100) with respect to a control run (1960–1990) for the DMI model in Mediterranean European basins.

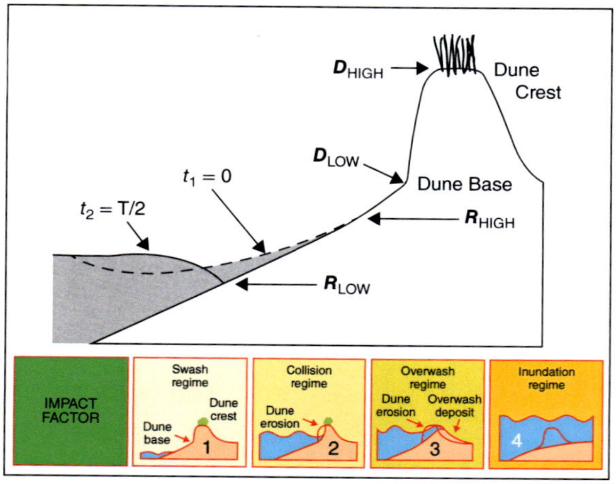

Figure 3.4.1 Sketch describing variables used to scale the impact of storms on barrier islands. *Source*: Adapted from Sallenger, 2000.

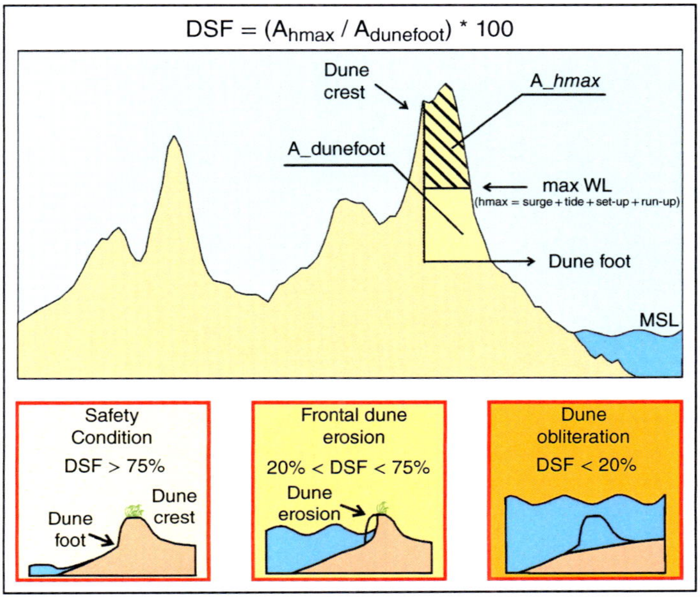

Figure 3.4.3 Sketch describing variables used in scaling the impact of storms on a dune system, the Dune Safety Factor of *Source*: Adapted from Armaroli *et al.* (2012).

		Tangible costs	**Intangible (non-market) costs**
Damage costs	**Direct**	• Physical damage to assets: – buildings – contents – infrastructure	• Loss of life • Health effects • Loss of environmental goods
	Business interruption	• Production interruption because of destroyed machinery	• Ecosystem services interrupted
	Indirect	• Induced production losses of suppliers and customers of companies directly affected by the hazard	• Inconvenience of post-flood recovery • Increased vulnerability of survivors
Risk mitigation costs	**Direct**	• Investment costs • Operation & maintenance costs	• Environmental damage - due to the development of mitigative infrastructure - or due to a change in agricultural practices
	Indirect	• Induced costs in other sectors	

Cost categories applied in this article:

Direct costs | Business interruption costs | Indirect costs | Intangible (non-market) costs | Risk mitigation costs

Figure 4.1.3 Cost categorisation applied in this article with examples. *Source*: Adapted from and extended based on Penning-Rowsell *et al.*, 2003; Smith and Ward, 1998.

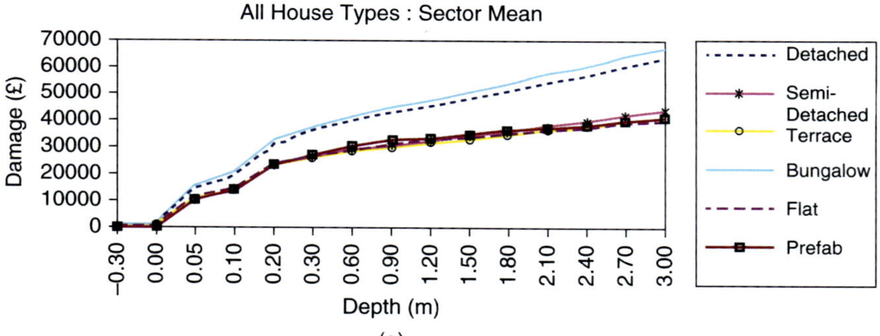

(a)

Figure 4.1.4 (a) Absolute damage function (England and Wales). *Source*: Penning-Rowsell *et al.* 2003. (b) Relative depth damage function. *Source*: Translated from Meyer and Mai 2003.

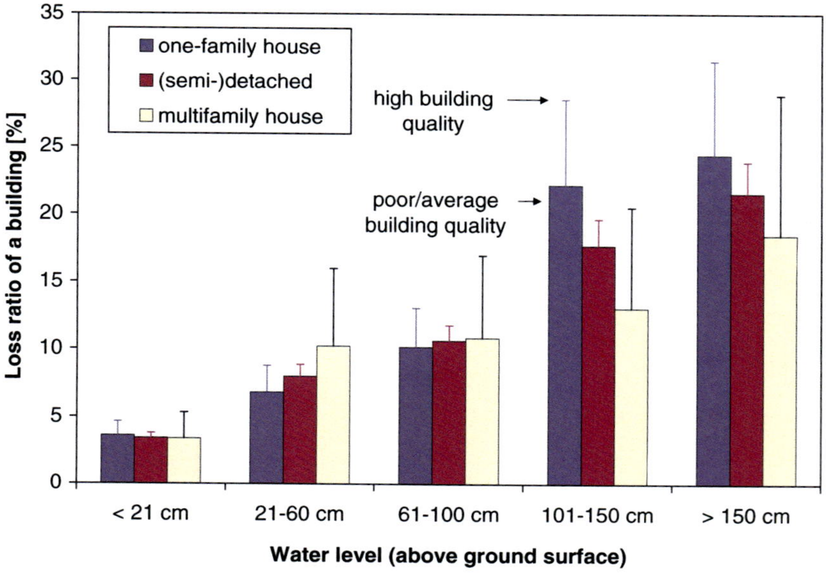

(a)

Figure 4.1.5 (a) Loss ratios for different inundation depth and different building types from FLEMOps. *Source*: Thieken *et al.* 2008. Reproduced with permission of WIT Press). (b) Influence of contamination on losses (from FLEMOps+) (Kreibich *et al.* (2011).

hazard, (iv) monitoring, management and early warning and (v) impacts and policy associated with the drought hazard. Including recent research progress in drought management and drought policy is crucial to become better prepared for the drought hazard, which is associated with building resilience, and hence reducing vulnerability and risk to future drought.

3.2.2 Drought: Generating processes and identification

Understanding of how drought is induced by climate drivers and how it develops through the subsurface part of the hydrological cycle is key information for timely management of water resources and considerable progress has been made over the last 5–10 years. Extreme hydrological droughts and associated heatwaves are likely to be caused by a combination of climate anomalies and already low antecedent catchment storage (e.g. soil moisture, groundwater storage, lake storage). Tallaksen and Fleig (2009) present the main climate drivers of drought in different hydro-climatological regions and for different spatial scales and summarize key challenges. Fleig *et al.* (2010) attempt to use weather and circulation patterns to seek a link between the state of the atmosphere and surface environmental phenomena. They evaluate the potential of tens of weather type classifications (WTCs) and associated weather types (WTs) for analysis of regional hydrological drought development in north-western Europe. Hydrological drought in their study is based on deficits derived from daily river flow series from regions in Great Britain and Denmark. A few WTCs outperformed and the study suggests that explicit information including air flow characteristics of meridionality, zonality and cyclonicity in the definition of WTs is important for a WTC when analysing the development of regional hydrological drought deficits. Fleig *et al.* (2011) conclude on the basis of the five most severe droughts in the period 1964–2001 for the above-mentioned regions that (i) hydrological response time (i.e. the time over which WTs influence drought development) is fundamental in moderating drought response to mesoscale climatic drivers and (ii) severe hydrological droughts may be caused by a complex set of hydroclimatological processes that vary between regions and events. Kingston *et al.* (2006) investigated river flow teleconnections across the northern North Atlantic region (North America and Europe). They identified for the first time, with correlation of monthly inter-regional river flow and composite analysis of large-scale climatic patterns associated with high and low monthly regional river flow, that inverse river flow relationships exist between North America and Europe in some periods of the year.

Climate anomalies propagate in landscape fluxes. Changnon (1987) was the first to describe the propagation of an anomaly in the precipitation (i.e. meteorological drought) into soil moisture, groundwater and streamflow, that is development of hydrological drought. Peters *et al.* (2003) explain with a theoretical example on how a rainfall deficit is amplified into a discharge deficit dependent on groundwater characteristics, and Peters (2003) provides a real-world example for drought propagation in the Pang catchment (UK). Van Lanen *et al.* (2004) systematically elaborated on how land use, soils and groundwater determine how an anomaly in precipitation affects discharge. Recently, Van Loon and Van Lanen (2012) investigated drought-generating processes in a number of catchments with contrasting climate and characteristics using observed and simulated time series of hydrometeorological fluxes and state variables. Based upon anomalies in precipitation and temperature they developed a drought typology that describes different ways of drought propagation. The classical rainfall drought is the most common drought propagation type (Figure 3.2.2a) in the catchments. Drought develops in soil moisture and groundwater as a response to below-normal rainfall. Quick flow processes, if occurring in a catchment, cause earlier drought development in streamflow, but also earlier recovery than groundwater drought. More relevant for water resources management under drought are the drought propagation types where the duration of the precipitation anomaly is rather short compared to the duration of the drought in soil moisture, groundwater and streamflow, that is the rain-to-snow-season drought and the wet-to-dry-season drought (Figure 3.2.2b,c). The rain-to-snow-season drought develops because of below-normal rainfall in the warm season, which does not recover before the cold season commences. In the cold season, drought in soil moisture, groundwater and streamflow cannot recover because precipitation falls as snow. This drought type will not recover before the snowmelt period, which implies that it lasts at least about a year. Drought with a similar duration

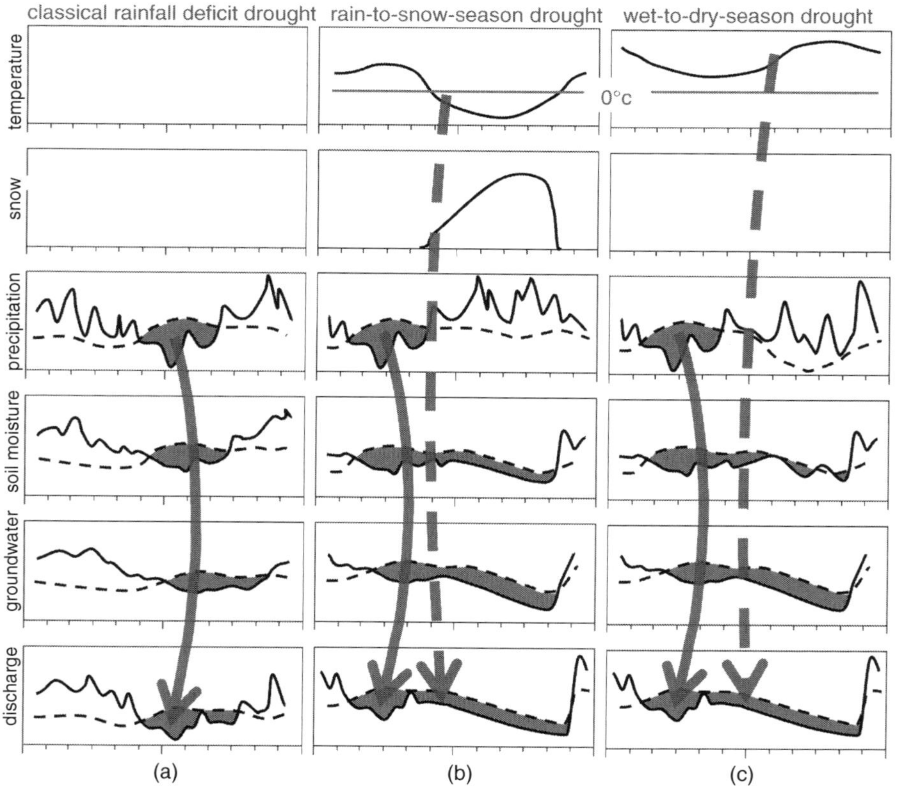

Figure 3.2.2 Drought typologies caused by precipitation and temperature control. *Source:* Van Loon, 2013, adapted from Fig. 8.2, p. 141. Reproduced with permission of the author. See plate section for colour representation of this figure.

belongs to the wet-to-dry-season drought type. Here a drought develops in the wet season because of below-normal rainfall. It cannot recover in the dry season because all rain then is used for evapotranspiration and possibly soil moisture replenishment. It may recover in the next wet season.

Many different approaches are available to identify drought types and typologies, which makes it confusing for water managers, stakeholders and policy-makers. Most indices are only relevant for particular impacted sectors. Extensive overviews are provided by, for example, Hisdal *et al.* (2004); Niemeijer (2008); Mishra and Singh (2010); Wanders *et al.* (2010) and Sheffield and Wood (2011). An important group of indices is based on identifying a probability density function through data of a sufficiently long time series (i.e. 30 years or longer). In the next step the function is transformed to a normal probability function. Moderate drought occurs when the variable of interest is 1 to 1.5 standard deviations lower than normal and severe drought when it is more than 2 standard deviations lower. McKee, Doesken and Kleist (1993) have introduced this approach for precipitation, that is the Standardized Precipitation Index (SPI). Later it was also developed for groundwater heads (groundwater resource index, GRI) (Bloomfield and Marchant, 2013) and streamflow (standardized run-off index, SRI) (e.g. Shukla and Wood, 2008). The challenge for the GRI and SRI is to find candidate probability density functions that adequately describe the groundwater or streamflow data. It is important to make a distinction between precipitation, groundwater and streamflow because these reflect different drought types, which is extremely relevant for water resources management under drought conditions. Some people believe that the SPI for longer periods, for example 6 or 12 month precipitation (SPI-6, SPI-12) is a good proxy for hydrological drought (groundwater or surface water). However, Wanders, Van Lanen and Van Loon (2010) and Bloomfield and Marchant (2013) prove that generally correlations are low and that it is recommended to use specific groundwater or surface water drought indices for water resources management under drought.

A second group of indices or drought characteristics is based on the so-called threshold approach, which was for the first time proposed by Yevjevich (1967). This approach starts with fitting a cumulative frequency distribution through the data from a time series of any hydrometeorological variable (e.g. precipitation, soil moisture, groundwater head, streamflow), which preferably covers 30 years or more. A particular quantile is selected, e.g. X_{80}, which says that this quantity of the variable of interest (i.e. threshold) is equalled or exceeded 80% of the time. Eventually, the time series of the variable is compared with the threshold. If the variable is below the threshold the drought starts (onset) and it will last until the variable equals or exceeds the threshold. Then the duration of the drought is known and the deficit volume can be calculated (cumulative difference between variable and the threshold). Hisdal *et al.* (2004) provide a comprehensive description of the threshold approach to identifying drought characteristics (onset, duration, deficit volume and the drought intensity, being the ratio of the deficit volume and duration). The deficit volume only has a physical meaning for fluxes (precipitation, streamflow). For state variables (soil moisture, groundwater heads) usually the maximum departure of the variable against the threshold is determined. Fleig *et al.* (2006) apply the threshold method to a series of catchments around the world and describe the impact of methods to merge dependent droughts and to exclude minor droughts. If variables have a seasonal behaviour and for some purposes (e.g. recharge of aquifers) then a variable threshold is more relevant than a fixed threshold (i.e. constant throughout the year). Hisdal *et al.* (2004) describe different variable thresholds varying

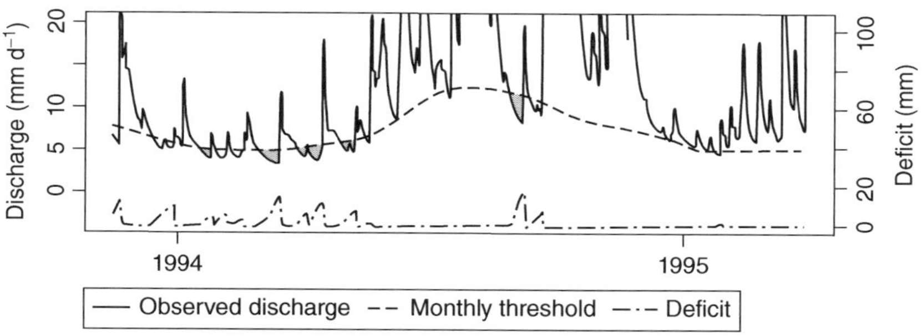

Figure 3.2.3 Variable threshold approach using a smoothed monthly threshold. *Source:* Van Loon *et al.*, 2010, Fig. 2, p. 192. Reproduced with permission of IAHS Press, CEH Wallingford, Oxfordshire OX10 8BB, UK.

from daily to seasonal. Van Loon *et al.* (2010) elaborate a daily smoothed monthly variable threshold (Figure 3.2.3).

The above-mentioned indices are single indices but they can be combined. The most well-known combined index is probably the US Drought Monitor (DM), which is a synthesis of multiple indices and impacts that represents a consensus of federal and academic scientists and is produced weekly. It combines SPI, percent of normal rainfall, Palmer Drought Severity Index (PDSI), soil moisture percentiles and streamflow percentiles (Svoboda *et al.*, 2002). The US National Drought Mitigation Centre (NMDC)[1] also publishes on the web a short-term blend and long-term blend of indices to address possibly impacted sectors. The short-term blend approximates drought-related impacts that mainly respond to precipitation on time scales ranging from a few days to a few months, such as wildfire danger, non-irrigated agriculture, topsoil moisture, range and pasture conditions and unregulated streamflows. The long-term blend approximates drought-related impacts that respond to precipitation on time scales ranging from several months to a few years, such as reservoir stores, irrigated agriculture, groundwater levels and well water depth. The European Drought Observatory (EDO)[2] regularly provides a map with the Combined Drought Index, CDI, (Sepulcre-Canto *et al.*, 2012), which is based upon the SPI-3, soil moisture anomaly (SMA) and the fraction of absorbed photosynthetically active radiation (fAPAR) that is obtained from remote sensing. The CDI distinguishes three phases when a drought progresses. First the SPI-3 will indicate a meteorological drought watch phase, subsequently followed by a soil moisture drought (SMA) (warning phase) and eventually the vegetation will be impacted (fAPAR) (alert phase). Recently, for research purposes combined indices have also been developed. For instance, Van Huijgevoort *et al.* (2012) show that drought characteristics derived from the threshold approach can be combined with the consecutive dry day method to consistently investigate drought across the globe, which includes both areas with permanent and non-permanent run-off. Van Lanen *et al.* (2013) give joint probability distributions of drought duration and deficit volume to demonstrate that subsurface storage processes, in particular aquifer responsiveness, are as important for development of hydrological drought as differences in climate.

[1] http://droughtmonitor.unl.edu/

[2] http://edo.jrc.ec.europe.eu/edov2

The indices described above are all at-site indices, which mean that these give point values (e.g. precipitation, groundwater head) or a number at the outlet of a catchment (streamflow), which might be plotted on a map and interpolated to generate continuous spatial images. Based on such information, areal indices are determined. The most common areal index is the area in drought, which may refer to a river basin, a continent or the whole world. Associated areal indices are average drought duration and average deficit volume for the area in drought. Andreadis *et al.* (2005) introduce severity–area–duration (SAD) curves to relate the area of each drought to its severity. They used SAD curves to analyse the drought history in the Conterminous United States over most of the twentieth century. Sheffield *et al.* (2009) constructed similar SAD curves for continents and the globe for the second part of the twentieth century. Peters *et al.* (2006) and Tallaksen, Hisdal and Van Lanen (2009) elaborated for the Pang basin (UK) on how to derive the above-mentioned areal indices and show how the area in drought is linked to drought propagation. The results illustrate, for instance, that the whole river basin may suffer from a meteorological drought, whereas hydrological drought never covers the entire area.

The wide-range of drought indices, and also the different ways one can calculate these, have led to enormous confusion and to the conclusion by the IPCC that definitional issues substantially contribute to the limited confidence in the outcome of trend studies around the world as to whether drought has become less or more severe (Senevitrane *et al.*, 2012). One recent example, on how the calculation of a particular drought index (PDSI), can lead to significant confusion among the general public is presented in Nature. Dai (2012) mentions increasing drought under global warming in observations, which is claimed to be reproduced in models that try to capture this. Somewhat later, on the contrary, Sheffield, Wood and Roderick (2012) mention that there is little change in drought in the previous 60 years, although both papers use the PDSI as index. The reason for the difference is the way the potential evapotranspiration was calculated. Dai (2012) used a temperature-based method (Thornthwaite), which is more sensitive to global warming than the Penman–Monteith method applied by Sheffield *et al.* (2012).

In summary to be better prepared for the drought hazard, it is prerequisite for adequate water management to implement current knowledge on drought generating processes and carefully investigate which drought type and drought typology apply to the river basin of interest and which drought indices are associated and fit best the possibly impacted sectors.

3.2.3 Trends in drought

3.2.3.1 Trends in observed drought

Detection and attribution of emerging trends in the different drought types is essential for long-term proactive actions. It is vital for: (i) identifying promising water management and policy responses, (ii) verifying current climate and hydrological simulations and (iii) having a "baseline" against which both climate change projections and observed future climate changes can be compared (Stahl *et al.*, 2010).

Many national and regional studies have been carried out over the last decade to explore if streamflow has become lower or higher and whether drought has become less or more severe. Stahl *et al.* (2010) refer to these trend studies in the Czech Republic, France, southern Germany, Slovakia, Spain, Switzerland and the United States. More recently, Wilson, Hisdal and Lawrence (2010) finalized a trend study in the Nordic countries covering three different historic periods. Trends of increased streamflow dominate annual values and the winter and spring seasons. In all three historic periods, a signal towards earlier snowmelt floods was clear, as was the tendency towards more severe summer droughts in southern and eastern Norway.

These national and regional trend studies all describe large spatial variability in the changes detected in streamflow, and results are dependent on the selected flow variable, trend detection method, and the time period chosen. Hence, it is challenging to synthesize the outcome of such regional and national trend studies into a consistent pan-European account of hydrological change. Therefore, Stahl *et al.* (2010) initiated a comprehensive pan-European trend study that uses a consistent methodology on a consolidated "best possible" European reference dataset (European Water Archive, EWA, UNESCO FRIEND-Water[3]). Trends were fitted for annual flow, monthly flow, high flow and low flow characteristics. Daily flow data were from over 400 catchments with no or minor disturbance and covered the period 1962–2004. Trends in annual flow are presented in Figure 3.2.4a. There is a kind of SW–NE line over Europe. Generally, north of the line it became wetter and south of the line drier (Mediterranean, SE Europe). Trends in the August flow are given in Figure 3.2.4b. The map reveals a remarkable picture, because in vast areas of Europe August flow decreased. The Nordic countries only show wetter conditions by the end of the summer. The map for the trend in the 7-day minimum flow (MA(7)) is similar (Stahl *et al.*, 2010).

3.2.3.2 Trends in simulated historic drought

The map with catchments (Figure 3.2.4) reveals that the hundreds of gauging stations are not equally distributed over Europe. Large white spots occur in the southern Mediterranean and southeast Europe. As a follow-up of the 2010 study, Stahl *et al.* (2012) investigated whether the white spots can be infilled using simulated gridded run-off from eight large-scale hydrological models. All these models used the same climate input (WATCH Forcing Dataset, Weedon *et al.*, 2011) that covered the period 1963–2000. Trends in simulated daily run-off in each grid cell were computed for each individual model and for the multi-model ensemble mean run-off. The trend estimates using the model run-off could be verified in 293 cells across Europe with the trends in the observed run-off (Stahl *et al.*, 2010). Overall the trends derived from the multi-model ensemble agreed better than one of the individual models. The trend in the multi-model ensemble mean annual run-off is given in Figure 3.2.5a. Comparison with the map with trends in observed flow (Figure 3.2.4a) shows that the ensemble mean reasonably well reproduces these trends. Maps of trends in simulated annual run-off show a pronounced continental dipole pattern of increasing flow in western and northern Europe and decreasing flow in southern and parts of Eastern Europe, which is similar to what has been found in the observations. There is a widespread decrease in summer low flow in large parts of Europe (Figure 3.2.5b). The map with trends

[3] http://ne-friend.bafg.de/servlet/is/7413/

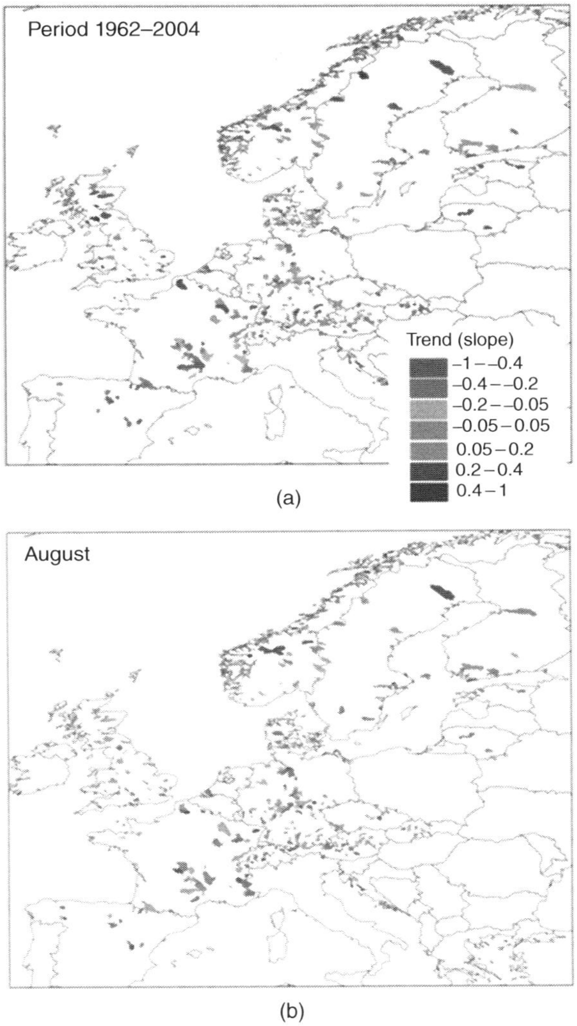

Figure 3.2.4 Trends in observed flow in Europe over the period 1962–2004: (a) Annual flow, (b) August flow. Blue are negative trends, implying increasing flow, towards wetter conditions and red indicates the opposite (positive trend, decreasing flow, towards drier conditions). *Source:* Stahl *et al.*, 2010, adapted from Fig. 1, p. 2373 and Fig. 2, p. 2374. Reproduced with permission of the authors. See plate section for colour representation of this figure.

in simulated run-off (Figure 3.2.5a) and the map with the location of the catchment (Figure 3.2.4a), also reveal that strong negative trends occur, particularly in European regions where no readily available flow data are available. The different models agreed on the predominant continental-scale pattern of trends, but in some areas disagreed on the magnitude and even the direction of trends, particularly in transition zones between regions with increasing and decreasing run-off trends. The use of models to explore changes in hydrological characteristics should therefore be viewed with caution due to higher uncertainty.

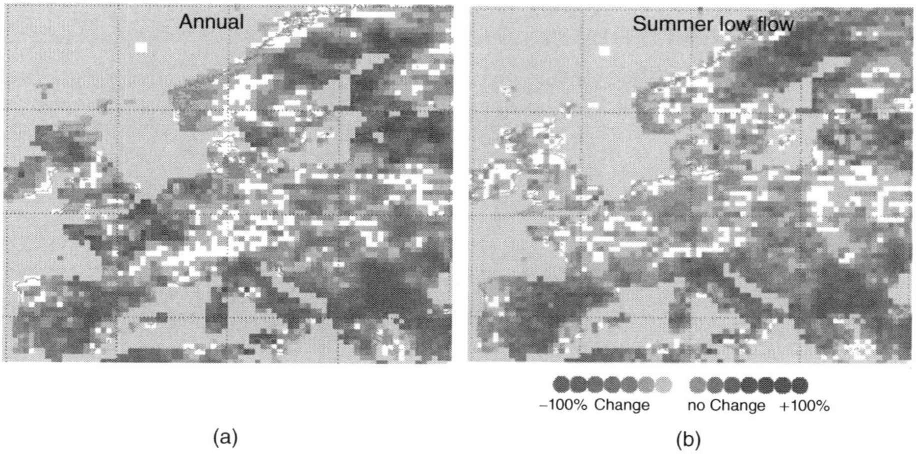

Figure 3.2.5 Trends in simulated runoff (multi-model ensemble mean) in Europe over the period 1963-2001: (a) Annual flow, (b) summer low flow. *Source:* Stahl *et al.*, 2012, Fig. 4, p. 2041. Reproduced with permission of authors. See plate section for colour representation of this figure.

Alderlieste and Van Lanen (2012) determined trends in drought characteristics (e.g. duration, deficit volume) for selected catchments in Europe rather than in simulated flow as done by Stahl *et al.* (2012). Over the period 1963–2001 the trends in the number of droughts over the period 1963–2001 showed a close to zero trend for most selected catchments. The trends in average drought duration show a distinct pattern. Most of the catchments (Spain, Italy, Greece, the Netherlands) reveal clear positive trends (i.e. longer drought duration), whereas others (Swiss Alps and large parts of Portugal) had decreasing trends (i.e. shorter drought duration). The trend magnitudes are much stronger over the period 1963–2001 than over the whole twentieth century. Trends in the average deficit volume follow the same pattern as drought duration.

Garnier *et al.* (2011) carried out a historical drought analysis beyond the instrumental record for the Besancon region and put twentieth century and more recent events in a long-term historical perspective. A similar analysis for the Valencia region in southern Spain unmistakably reveals an increase in droughts since the 1940s with a stronger increase in the years 1980–1990. However, when this was put in the perspective of a longer period, then it becomes clear that the eighteenth century and the first half of the nineteenth century faced outbreaks of droughts much greater than those of the twentieth century.

Hannaford *et al.* (2013) convey an important message to take into account interdecadal variability in trend studies. They collected long records of minimally anthropogenically influenced streamflow data from over 100 gauging stations in northern and central Europe. They adopted an alternative "multi-temporal" approach, whereby trends were computed for every possible combination of start and end years in a record. Through, among others, heat maps they show that the magnitude and even direction of short-term trends are heavily influenced by interdecadal variability. The proposed multi-temporal approach should help future studies to contextualise short-term trends.

Drought trend studies should carefully check if the time series of hydrometeorological data is not affected by water scarcity. Drought is a natural hazard, caused by large-scale climatic variability, and cannot be prevented by water management, whereas water scarcity refers to long-term unsustainable use of water resources, which water managers can influence. Both drought and water scarcity lead to lower than normal soil moisture, groundwater level or streamflow. Making the distinction between drought and water scarcity is not trivial, because both phenomena often occur simultaneously and interlinked. Schmidt and Benítez-Sanz (2013) look at the similarities and differences between the interrelated concepts of drought and water scarcity and provide a framework for defining the two terms. Jaeger *et al.* (2013) propose a simple but robust definition of water scarcity and illustrate it with examples of the many biophysical and socio-economic factors that interact within a broader system to determine water scarcity. Van Loon and Van Lanen (2013) propose an observation-modelling framework to separate natural (drought) and human (water scarcity) effects on the hydrological system. The basis of the framework is simulation of the "naturalised" situation (conditions without human influence), using a hydrological model. The resulting time series of naturalised state variables and fluxes are then compared to observed time series. Secondly, anomalies (i.e. deviations from a threshold, Section 3.2.2) are determined from both time series and compared. Application of this observation-modelling framework in the Upper-Guadiana catchment in Spain suggests that the impact of groundwater abstraction on the hydrological system was, on average, four times as high as the impact of drought during the period 1980–2000.

In summary, the trend studies do not reveal a consistent pattern for the whole of Europe towards drying or wetting. The annual flow in the second part of the twentieth century became higher in western and northern Europe and lower in southern and eastern Europe. However, summer flows decreased and associated hydrological drought became more severe in extended European regions, including western Europe. Additional studies, which are more limited in spatial coverage, reveal that the magnitude of the trend and even the direction is dependent on the period because of interdecadal variability. These studies also address the distinction to be made between the interlinked phenomena of drought and water scarcity. Historical studies on drought, using proxy data for instrumental records, show that more severe droughts have occurred than those derived from hydrometeorological data collected in the last century. The trend studies show that proactive measures to adapt to the impacts of drought should be considered in a long-term context.

3.2.3.3 Future drought

Prediction of future drought is of paramount importance for water management that is prepared for potential environmental conditions. Basically, two approaches can be followed: (i) extrapolation of current trends and (ii) modelling using a suite of climate and hydrological models. Both approaches have their pros and cons and preferably should be used complementarily. Extrapolation of current trends is a strong method, particularly when observation-based and if the patterns are consistent (Section 3.2.3.1), but cannot take us far in the future and interdecadal variability requires consideration.

Alternatively, models can simulate the far future, but results should be interpreted with care because of model uncertainty.

When modelling is used different approaches can be used to explore future drought. The simplest one is to use directly the run-off simulated by a GCM (Figure 3.2.6a approach 1). A well-known example is Milly, Dunne and Vecchia (2005), who show that an ensemble of 12 GCMs have skills in simulating observed regional patterns of twentieth century multidecadal changes in streamflow. These models project 10–40%

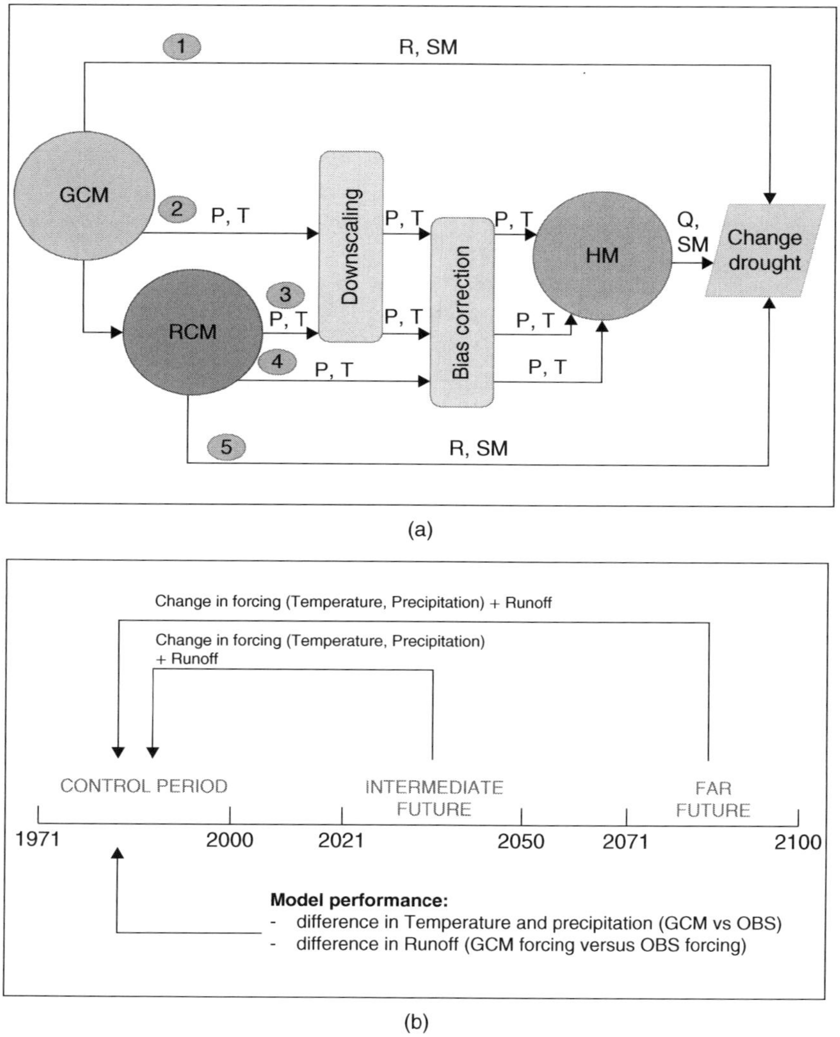

Figure 3.2.6 Schematic presentation on how to assess change in hydrometeorological variables due to climate change: (a) Model chain and (b) procedure for different time windows. GCM: general circulation model, RCM: regional climate model, HM: hydrological model, P: precipitation, T: temperature, R: run-off, SM: soil moisture and Q: streamflow.

increases in run-off in eastern equatorial Africa, and 10–30% decreases in run-off in southern Europe by the year 2050. Dai (2012) analysed model-simulated soil moisture from over 10 GCMs (Figure 3.2.6a, approach 1), which suggest increased drought risk in the twentyfirst century. He concluded that these changes are very likely, because the observed global drying changes up to 2010, which were presented as global area in drought, are consistent with model predictions. Sheffield *et al.* (2012), however, show that the drying was strongly affected by the way the drought indicator (i.e. the potential evapotranspiration) was calculated (Section 3.2.2).

A more sophisticated approach (Figure 3.2.6a, approach 5) is to feed the output from the GCM into a RCM and then to compare the simulated run-off for future periods with the control period (Figure 3.2.6b). Using output directly from a GCM or RCM is, however, too broad for most hydrological drought studies. Hydrological models (HMs) that are forced with the outcome from either GCMs or RCMs are applied to overcome lack of detail. Downscaling and bias correction methodologies are common to generate realistic forcing data (precipitation, temperature) at the land surface (Figure 3.2.6a, approaches 2, 3 and 4). The downscaled and bias-corrected GCM temperature and precipitation can be intercompared with observed weather data for the control period (Figure 3.2.6b) to obtain a measure for GCM performance. Additionally, the downscaled and bias-corrected GCM temperature and precipitation is also used as input for a hydrological model to obtain time series of hydrological variables (GCM-forced HM), which subsequently are the basis for the drought analysis. The same hydrological model is also forced with observed meteorological data or re-analysis data (e.g. Weedon *et al.*, 2011) for the control period (Figure 3.2.6b) to create a reference set with hydrological variables (OBS-forced HM). Theoretically, the droughts derived from the time series of the GCM-forced HM should be identical to those derived from the OBS-forced HM. Differences can be interpreted as underperformance of the GCM-forced HM (model noise). Alderlieste and Van Lanen (2013) carried out such an analysis for streamflow characteristics and drought characteristics of selected catchments in Europe for the control period (Figure 3.2.6b). They conclude that the noise is rather high for flow characteristics and much smaller for drought characteristics, which is caused by drought characteristics being anomalies, whereas flow characteristics are absolute numbers.

Prudhomme *et al.* (2011 present a methodology for exploring the ability of hydrological models (HMs) to reproduce large-scale low flow events, as described by a historical drought catalogue (using the Regional Deficiency Index, RDI), which was derived from river flow measurements across Europe (Hannaford *et al.*, 2011). They used the same methods to compute the RDI from the run-off simulated by three global hydrological models (JULES, WaterGAP and MPI-HM) run with the same meteorological input (Weedon *et al.*, 2011) at the same spatial 0.5° grid. Results show that all three models can broadly reproduce the spatio-temporal evolution of hydrological drought in Europe to varying degrees. Van Huijgevoort *et al.* (2014) use model agreement against observations to capture hydrological drought across the world as a measure for model performance. Drought duration and spatial extent of major historic drought events differed between models, mainly because of the too fast response to rainfall of some models. By using a model ensemble, this fast run-off response was partly overcome and delay in drought propagating from meteorological drought to drought in run-off was included.

Orlowsky and Seneviratne (2013) carried out a comprehensive analysis of future meteorological drought (SPI-12) and soil moisture anomaly (SMA) (Figure 3.2.6a, approach 1). They used simulated monthly data from over 30 GCMs of the 5th phase of the Coupled Model Intercomparison Project (CMIP5) for three future Representative Concentrations Pathways experiments (RCP2.6, RCP4.5 and RCP8.5). Drought hot spot regions were identified based on impact and land use data, as well as projections of future drought changes. They observed hot spot regions that will become wetter (increase of median SPI-12): namely East Asia, South Asia, East Africa, Central North America, Sahel and Central Europe. Other hot spot regions are projected to become drier (median SPI-12 decreases, Figure 3.2.7). These include: Mediterranean (MED), South Africa (SAF), Central America/Mexico (CAM), Amazon (AMZ), North East Brazil (NEB) and Australia (AUS). Figure 3.2.7a exhibits the evolution of the SPI-12 over the twentyfirst century from 2010 to 2090 for these regions. For each period of 20 years three bars are given that present the three RCPs and the box plots give the spread of the ensemble of GCMs. For example, for the Mediterranean the median SPI-12 is expected to decrease by one standard deviation over the twentyfirst century for the RCP8.5. The spread varies from −0.5 to −2.0 standard deviations by the end of the century for the RCP8.5. The frequency (month/year) that the SPI-12 becomes lower than −1 (drought situation) is given in Figure 3.2.7d. The median frequency for the Mediterranean can go up to 5 months per year by the end of the century (RCP8.5). Figure 3.2.7c presents the increase in the soil moisture anomaly, whereas Figure 3.2.7d presents the frequency (month/year) provided that the SMA is lower than one standard deviation from normal.

In some regions (Mediterranean, Central America/Mexico) more than half of the year a soil moisture drought is expected to happen already from 2030. Orlowski and Seneviratne (2013) include a thorough uncertainty analysis. Separation of different sources of uncertainty in drought projections reveals that, for the near future, internal climate variability is the dominant source, while the model structure of Global Circulation Models (GCMs) for soil moisture drought becomes the dominant source of uncertainty by the end of the twentyfirst century. The uncertainty in RCPs (greenhouse gas concentrations scenarios) appears to be small for most hot spot regions.

Different drought types need to be distinguished for adequate management of water resources management under climate change (Section 3.2.2). Projections of hydrological drought, which is key information for water resources management, are not, by definition, equal to the above-described meteorological drought and soil moisture drought, although these clearly trigger the hydrological drought. A hydrological model (HM, approaches 2, 3 and 4) is needed to consider storage processes in the soil and groundwater (e.g. Van Lanen *et al.*, 2013).

Wanders and Van Lanen (2013) developed a conceptual hydrological model to investigate drought in groundwater discharge to the streams in the major climates around the world. The model was forced by downscaled and bias-corrected outcome from three General Circulation Models for the A2 emission scenario (Figure 3.2.6a, approach 2). For the near future (2021–2050) and far future (2071–2100), a decrease in drought frequency for all major climates was found. However, this goes hand in hand with an increase in both average drought duration and deficit volume of the remaining drought events. The largest decrease in hydrological drought frequency is expected in cold (D-)climates where global warming results in a decreased length of the snow season and an increased precipitation. In the dry B-climates the smallest decrease in drought frequency is projected to happen, which probably will lead to even

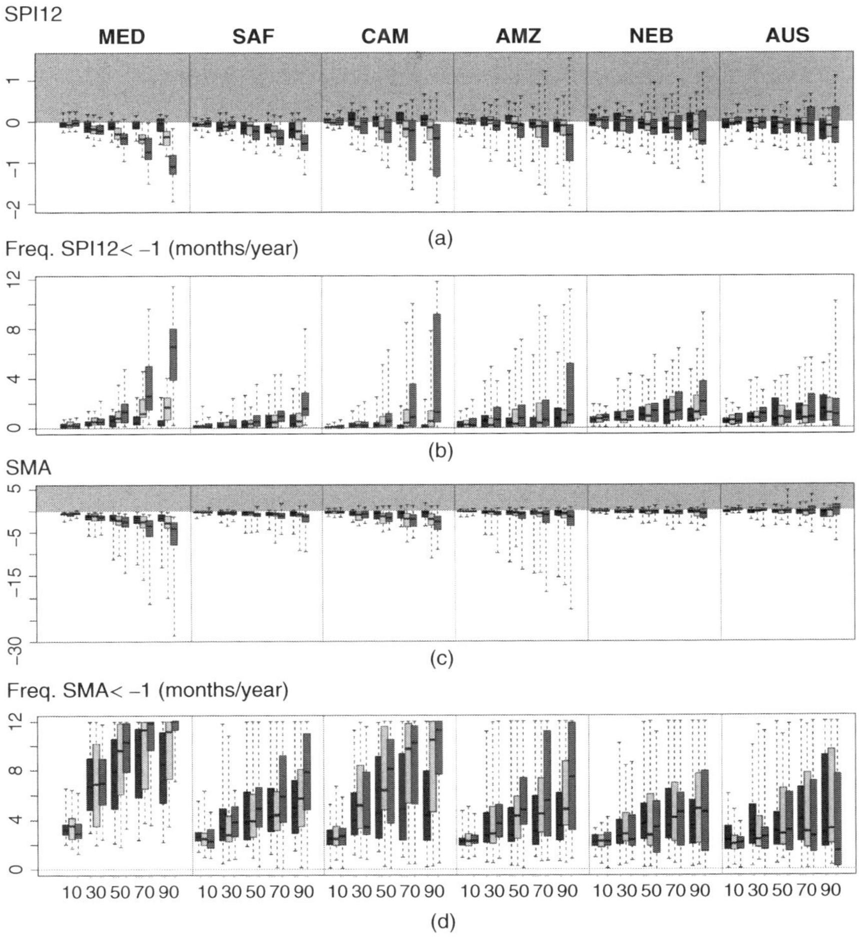

Figure 3.2.7 Drying regions in the twentyfirst century (regions where the median SPI12 decreases): (a) 20 year average SPI12, (b) months per year in which SPI12<−1, (c) and (d) as (a) and (b), but for SMAs. Changes are given as standard deviations with respect to 1979–2009 in (a), mm in (c) and as months per year in (b) and (d), The numbers on the lowest *x*-axis denote the central years in the twentyfirst century of the 20 year windows. Colours indicate the respective GHG concentrations scenario, blue: RCP2.6, green: RCP4.5 and red: RCP8.5. *Source:* Orlowski and Seneviratne 2013, Fig. 8, p. 1776. Reproduced with permission of the authors. See plate section for colour representation of this figure.

more severe water scarcity. However, in the extreme climate regions the analysis for the control period showed that projections are most uncertain in these regions.

Alderlieste and Van Lanen (2013) investigated the change in future run-off for selected case study areas in Europe, which was simulated with six large-scale hydrological models that were forced with three GCMs (Figure 3.2.6a, approach 2). This was done for two emission scenarios (A2 and B1) and two future time windows (intermediate future and far future). The six large-scale models were also run with re-analysis data

(Weedon *et al.*, 2011) for the control period (1971–2000) (Figure 3.2.6b). This enabled an intercomparison for the control period of the multi-model mean annual run-off obtained with the reanalysis data (reference) and the run-off of each of the three GCMs, which gives an indication of how well GCM-forced hydrological models can reproduce past annual run-off (model noise, indicated with CTRL in Figure 3.2.8). The model noise in the simulated annual run-off is rather large for all three GCM forcings (40–50%, red bars in Figure 3.2.8a). The modelling experiment reveals that the annual run-off is projected to decrease (all three GCMs, two emission scenarios, two future periods). Multi-model flow projections from CNRM and ECHAM5 forced hydrological models agree more or less, whereas IPSL leads to an unexpected larger decrease for the B1 emission scenario. The most important information in Figure 3.2.8a, however, is that the impact of climate change on mean annual run-off (climate signal) is smaller than the model noise, which implies that no significant conclusions on change of annual run-off for the selected catchments can be drawn. Alderlieste and Van Lanen (2013) also study the change in drought characteristics (duration, deficit volume) that were

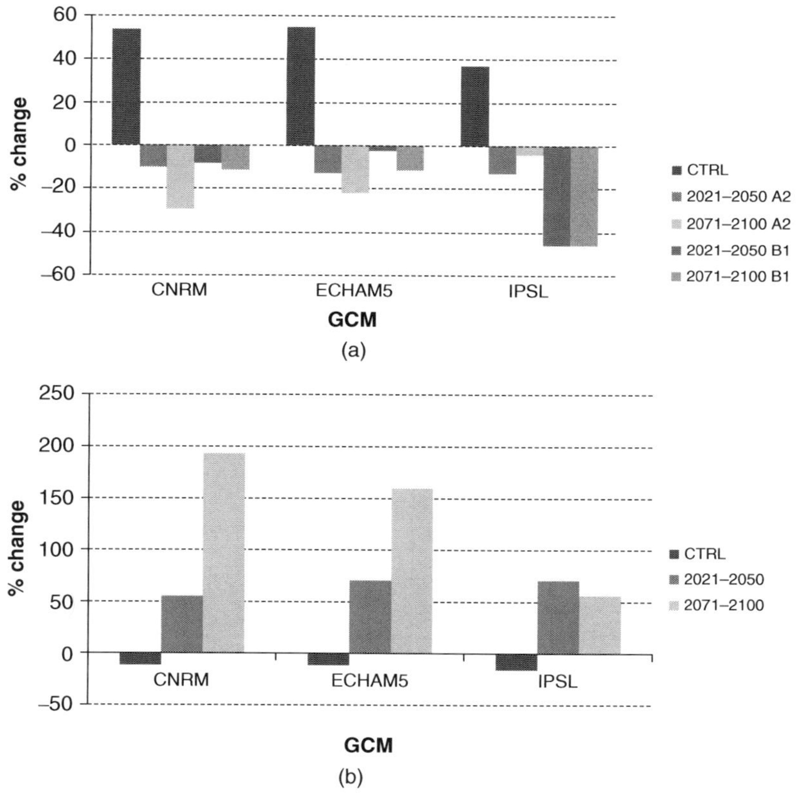

Figure 3.2.8 Projected change in: (a) future mean annual run-off for A2 and B1 scenarios, and (b) average drought duration for the A2 scenario derived from a multi-hydrological model ensemble for three GCMs (CNRM, ECHAM5 and IPSL), intermediate future (2021–2050) and far future (2071–2100). CTRL indicates model uncertainty. *Source:* Data from Alderlieste and Van Lanen, 2013. See plate section for colour representation of this figure.

derived from the simulated multi-hydrological model run-off for A2 scenario using the threshold approach (Section 3.2.2). Remarkably, the model noise is substantially lower (5–10%, red bars in Figure 3.2.8b) when anomalies in flow are used rather than the flow itself (Figure 3.2.8a). The average drought duration is projected to increase by 50–60% in the intermediate future according to all three GCM forcings. For the far future the expected drought duration according to the GCM forcing is more unclear. The average drought duration is expected to increase relative to the control period, but it is unclear if it increases relative to the intermediate future. According to the CNRM and ECHAM5 forcings the average drought is projected to substantially increase (>150% relative to control period), whereas the IPSL forcing projects a slightly smaller duration than for the intermediate future.

Recently, a comprehensive study on future hydrological trends was concluded (Prudhomme *et al.*, 2014). They assessed the impact of climate change on hydrological drought in a multi-model set-up that included seven large-scale hydrological models forced by bias-corrected climate from five GCMs (i.e. 35 members) under four RCPs (Figure 3.2.6, approach 2). Drought was assumed to occur when total daily simulated run-off is less than a given threshold (Section 3.2.2). Future drought was studied by comparing the temporal and spatial patterns of simulated run-off in the years 2070–2099 (RCP forcings) with those from the control period 1976–2005 (historical forcings). A run-off deficit index (DI) was introduced for each land cell; under drought conditions DI = 1, and is zero otherwise. The DIs were spatially aggregated for the globe and 27 regions (GDI and RDI, respectively), and temporally for years and seasons. Drought severity was defined as the fraction of land under drought conditions. The analysis shows a likely increase in the global severity of hydrological drought at the end of the twentyfirst century. Under the most extreme emission scenario (RCP8.5), a GDI>0.4 was found by almost 50% of the ensemble members. Figure 3.2.9 gives the mean percentage change in regional deficit index (RDI) between the far future (2070–2099) relative to the control period for 17 regions under RCP8.5. RDI is presented as the average for all ensemble members (All), for each of the individual GCMs (averaged over seven large-scale hydrological models), and for each of the large-scale hydrological models (averaged over five GCMs). In all regions the fraction of land under drought (severity) is expected to increase by the end of the century, except for East Africa (Figure 3.2.9, upper row). All GCMs predict a higher severity in South America, Western Europe, Central Europe, South Africa and Western Indian Ocean (Figure 3.2.9, rows 2–6). All large-scale hydrological models also project higher severity for these regions, if Jules is excluded. Prudhomme *et al.* (2014) investigate uncertainty as well through an ensemble of different GCMs, RCPs and large-scale hydrological models. They conclude that the uncertainty due to large-scale hydrological models is greater than that from GCMs, particularly if the dynamic response of plants to CO_2 and climate is included in the large-scale hydrological model.

The climate and hydrological communities debate if model performance in the control period (e.g. Figure 3.2.8) is a measure for uncertainty of predicted future drought. For example, Orlowski and Seneviratne (2013) and Prudhomme *et al.* (2014) do not analyse model performance in the past, but choose an ensemble approach to address uncertainty of future drought assessments. Van Huijgevoort *et al.* (2014) studied model performance of an ensemble of five large-scale models and three GCMs in the control period (1971–2000) against observed low flows from 41 river basins across the world

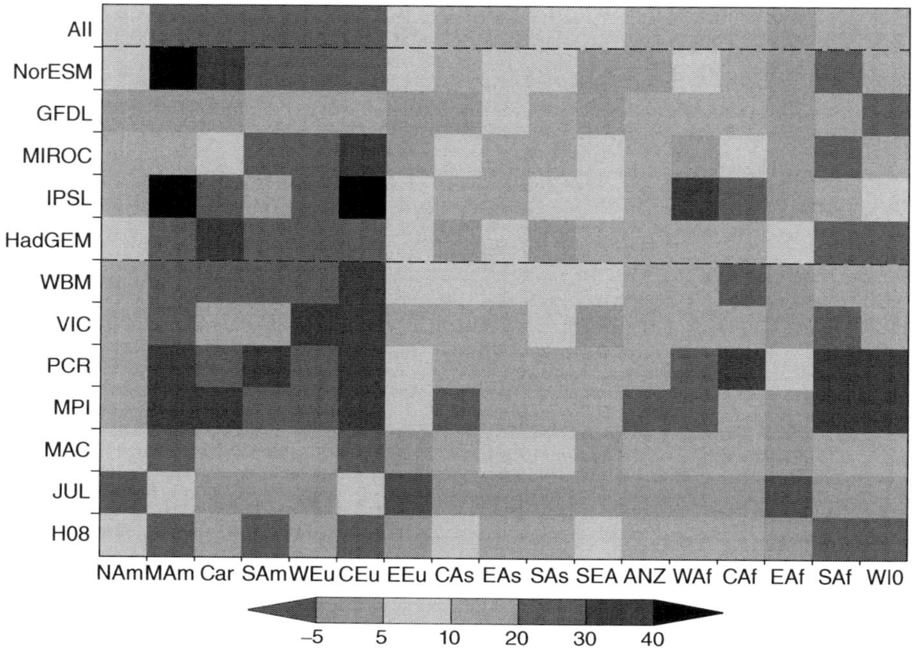

Figure 3.2.9 Mean percentage changes in hydrological drought (i.e. regional deficit index, RDI) in the future (2070–2099) relative to the control period (1976–2005) under RCP8.5 for 17 world regions (*x*-axis). Changes are averaged over all of the seven large-scale hydrological models and five GCMs (All, upper row), per GCM averaged over all seven large-scale hydrological models (rows 2–6), and per large-scale hydrological models averaged over all GCMs (rows 7–13). *Source*: Prudhomme *et al.*, 2014, Fig. 5, p. 3265. Reproduced with permission of the authors. See plate section for colour representation of this figure.

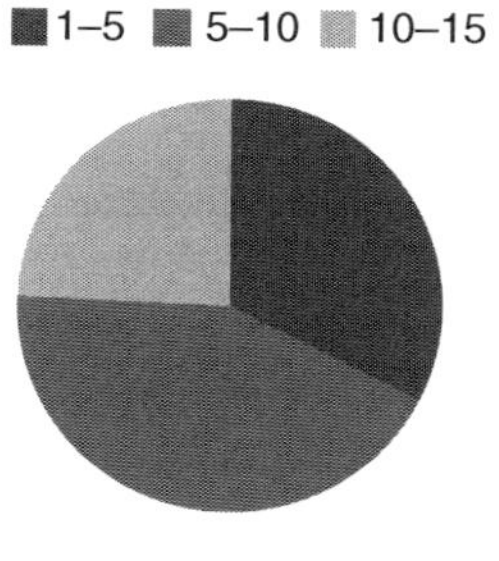

Figure 3.2.10 Number of model ensemble members (large-scale hydrological model and GCM) that passed performance criteria in the control period derived from observed low flow (Data from Van Huijgevoort *et al.*, 2014). See plate section for colour representation of this figure.

(Figure 3.2.6b). Ensemble members were only selected to project future drought if these passed objective performance criteria. Four river basins had no ensemble member that passed the criteria. In about 40% of the basins 5–10 ensemble members were used for drought projections, and in not more than 25% more than 10 ensemble members passed the criteria (Figure 3.2.10).

In summary, drought projections can follow different approaches (combination of climate models, downscaling, bias-correction, hydrological models, Figure 3.2.6*) which affect the outcome. Furthermore, a distinction needs to be made between different drought types, because projections deviate. Future hydrological drought is key information for the management of water resources. Uncertainty needs to be considered when dealing with future drought. It is recommended to use a multi-model ensemble for the future (GCMs, emission scenarios, large-scale hydrological models). Additionally, assessment of model performance against observations in the control period supports building confidence in modelling.*

3.2.4 Monitoring, management and early warning

Proactive measures to reduce drought vulnerability and to build resilience operate at a number of temporal scales, for example for the intermediate and far future (Section 3.2.3), but also at shorter time scales (monthly, seasonal, multi-year). Drought monitoring and drought forecasting as part of early warning are essential at shorter time scales. Drought forecasting is urgently required for drought management as part of integrated water management, but skills are still very limited for the seasonal and multi-year scale, except for regions strongly controlled by El Niño and La Niña. However, on the basis of the current drought conditions it is feasible to provide an outlook, as done by the US National Drought Mitigation Centre. They generally include a description of what the primary physical effects are and label "S", "L" or "SL" on the map where S is short-term, typically less than 6 months (e.g. agriculture, grasslands), and L is long-term, typically more than 6 months (e.g. hydrology, ecology). The outlook should consider storage in the river basins (e.g. occurrence of extensive groundwater bodies, lakes).

When monitoring drought, it is important to monitor the relevant hydrometeorological variables, to address the right drought types, drought typologies and to present the relevant drought indicators (Section 3.2.2) associated with the impacted sectors. In this context it is important to make a distinction between the scale. Awareness indicators, such as the SPI, Standardized Precipitation–Evapotranspiration Index (SPEI) orPDSI are appropriate for broad scales, for example the pan-European scale, but these are too general for the river basin scale where operational drought indicators are more appropriate[4].

The European Drought Observatory (EDO) is responsible for drought awareness monitoring. An example is presented in Figure 3.2.11 for mid November 2013. The monitoring uses the CDI (Sepulcre-Canto *et al.*, 2012). The CDI combines the SPI-3, SMA and fAPAR (Section 3.2.2). The drought monitoring map also gives drought recovery information, which is very important, although it is often forgotten after the first rains give relief. Recovery of hydrological drought, particularly in slowly-responding groundwater bodies and groundwater-fed rivers can last very long (e.g. seasons, years).

Operational drought indicators that are being used at the river basin scale, and sometimes at the national scale, are usually integrated into a comprehensive water information system. These indicators are fine-tuned to the local conditions and

[4] Recommendation by the WFD-CIS Expert Group on Water Scarcity and Drought, 2012.

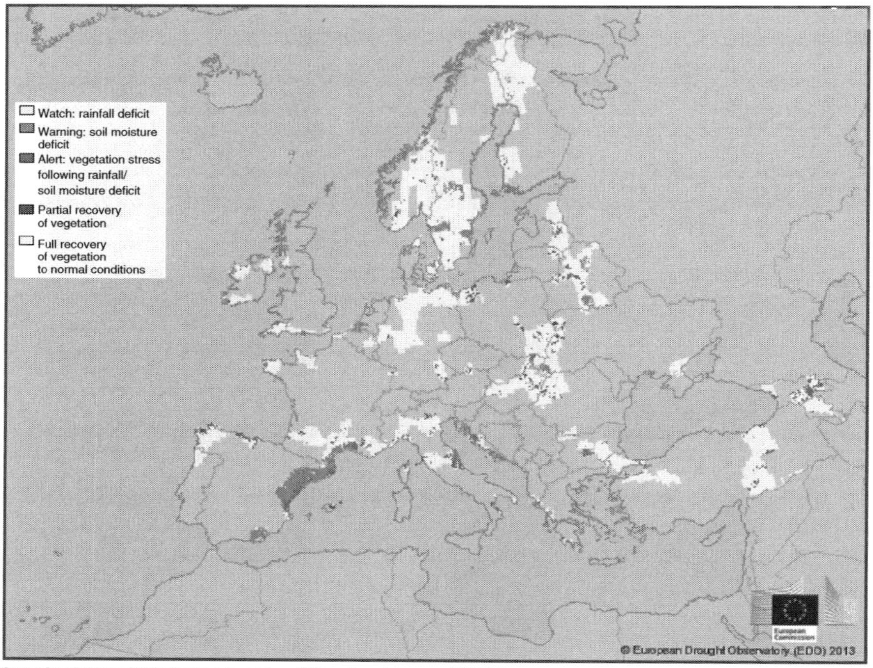

Combined Drought Indicator, based on SPI, soil moisture and fAPAR.
- **Watch:** when a relevant precipitation shortage is observed
- **Warning:** when this precipitation translates into a soil moisture anomaly
- **Alert:** when these two conditions are accompanied by an anomaly in the vegetation condition.

Figure 3.2.11 Pan-European drought situation, 11–20 November 2013. *Source:* European Drought Observatory, EDO, JRC, 2013. Reproduced with permission of the European Commission, Joint Research Centre, Institute for Environment and Sustainability (EC JRC-IES), Ispra, Italy. See plate section for colour representation of this figure.

impacted sectors. Information systems are, for instance, operational at the national scale in the Netherlands and at the basin scale in the Jucar (Spain). The indicators feed a system that analyses river basin management under different drought conditions (emerging, ongoing, recovery). Scenarios for future water flows and storages (temporal and spatial) are simulated, as well as water management scenarios. The probabilistic outcome is then translated into a risk assessment, which is discussed with stakeholders. A decision-support system often facilitates the decision-making. New management alternatives can be proposed by stakeholders and then the water management scenarios are again simulated. Identification of the most appropriate measures is usually the outcome of an iteration process (i.e. iterative risk management strategies). In the Netherlands, a priority ranking has been negotiated prior to the drought and it is used as a basis for proactive measures to be taken to reduce impacts.

In summary, early warning of drought in Europe is hampered because drought forecasting skills are still very limited. However, on the basis of current drought conditions, drought type and river basin characteristics, a short-term and long-term outlook can be provided. Drought monitoring should consider different drought types and typologies and choose the appropriate indicators. Awareness indicators are useful at the broad scale (pan-European scale, national scale). Operational drought indicators are required for the river basin scale. Preferably, these are fine-tuned to local conditions and impacted sectors, and integrated into an information system that stakeholders support to identify in an iterative way the most appropriate proactive measures to reduce drought impacts.

3.2.5 Drought impacts and policy

Drought has a wide range of direct and indirect impacts. Wilhite (2000) and Tallaksen and Van Lanen (2004) provide extensive overviews. The connection between drought impacts and the natural hazard is complex and no straightforward relationships exist. For instance, Dieker, Van Lanen and Svoboda (2010) tried to link the US Drought Monitor (Section 3.2.2) and the Drought Impact Reporter. Lag correlations provided some insight, but no generic information could be drawn that would apply to all drought events across selected US states.

Stahl *et al.* (2012) summarize the establishment of a pan-European inventory of drought impact reports (European Drought Impact Inventory, EDII) for the purpose of cross-disciplinary research on drought vulnerability and risk. Impacts of the 2003 summer drought in Europe and the 2004–2007 multi-year drought on the Iberian Peninsula appeared to be already well documented. The inventory reveals a large variety of diverse drought impacts and response measures (Figure 3.2.12) and, although not yet statistically representative, the initial analysis suggests some geographical differences across Europe that may be important to EU policy development. The inventory will be populated with more data in the near future and aims to become an online source to the public later.

Collated information indicates that in North and West Europe reported impacts on ecosystems, agriculture and forestry dominate, whereas impacts on water supply and agriculture were more often mentioned in South Europe (Stahl *et al.*, 2012).

Recent increase in knowledge about drought as a natural hazard and its wide-faceted impacts need to be communicated to stakeholders and policy-makers, They will put it in the wider context of possible water scarcity and drought and will synthesise this in integrated water and land management measures. It will also support further implementation and review of EU policies (e.g. Water Framework Directive (WFD), Common Agricultural Policy (CAP), Regional Development) and UN actions (e.g. ISDR Hyogo Framework for Action, UNCCD United Nations Convention to Combat Desertification). Science Policy Briefs (SPBs) in a readily-usable format and targetted to specific policies have been used as a start to facilitate communication between policy-makers and scientists.

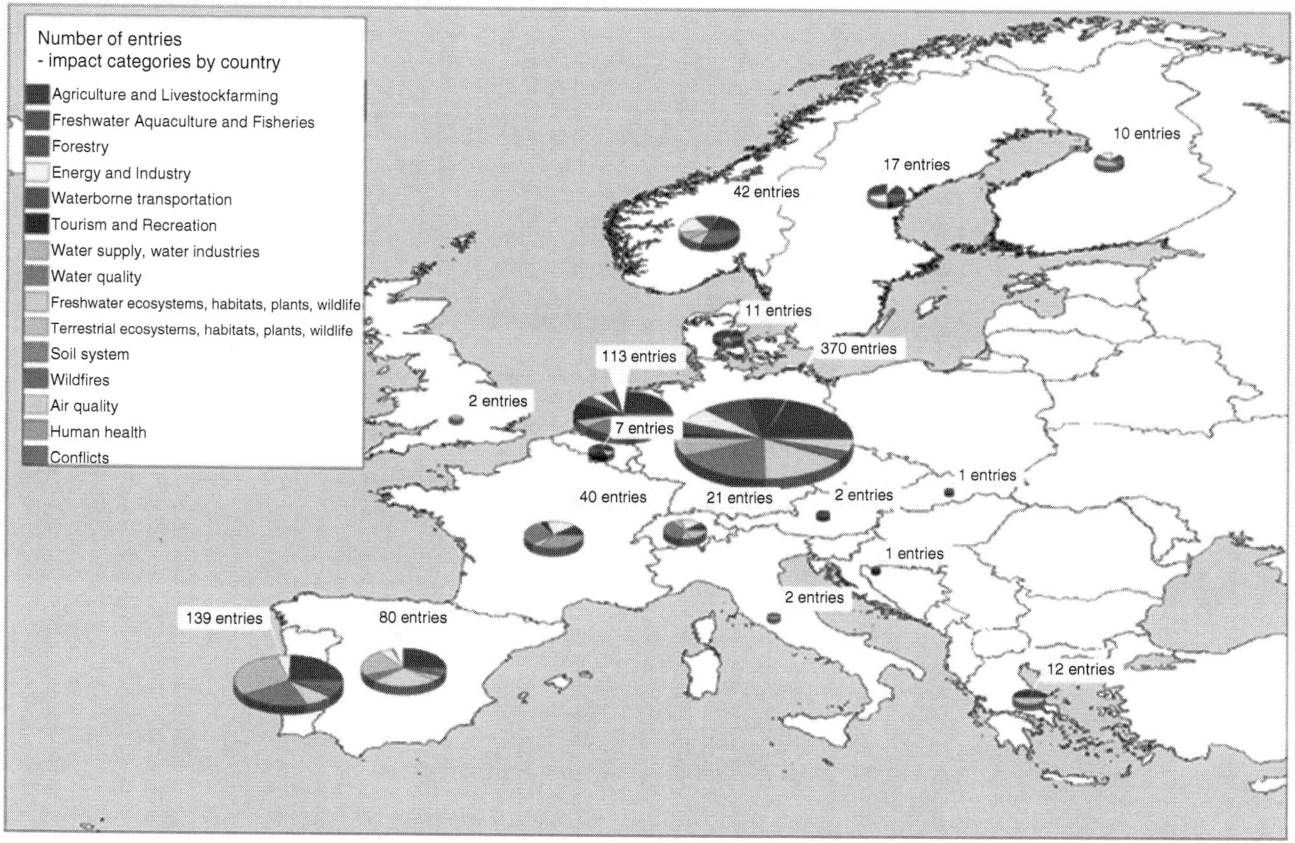

Figure 3.2.12 Different drought impacts across Europe retrieved from the European Drought Impact Inventory (EDII). *Source:* Stahl *et al.*, 2012, Fig. 7, p. 8. Reproduced with permission of the authors. See plate section for colour representation of this figure.

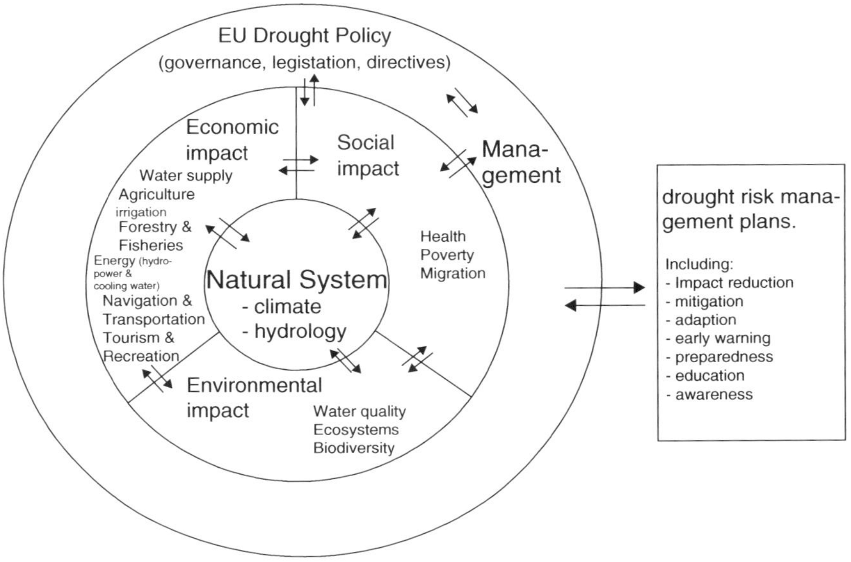

Figure 3.2.13 Integration of relevant aspects towards a future EU Drought Policy.

In the final phase, the EU FP7-project XEROCHORE[5] (Greek: "dry land", 2008–2010) has compiled five SPBs in four languages (English, French, Italian, Spanish). Each SPB is linked to a particular WFD article, and identifies: (i) policy focus, (ii) purpose, (iii) policy milestones and relevant XEROCHORE key outputs, (iv) limitations identified by XEROCHORE and (v) main recommendations (Quevauviller *et al.*, 2012). The schematic project outline on how to arrive from knowledge of the natural physical system (core), via impacts, to management and policy (outer ring) is presented in Figure 3.2.13.

The ongoing EU FP7-project DROUGHT-R&SPI[6] will reduce Europe's future vulnerability and risk to drought by investigating drought in six case study areas in currently water-stressed regions (e.g. Mediterranean, The Netherlands) with drought analyses at the pan-European scale. In the project, science–policy interfacing has been taken on board directly from the beginning of the project. A dialogue has been implemented with actors in the water field at different scales. A Case Study Drought Dialogue Forum (CS-DDF) has been established in each of the six case study areas. In steps, the wide-ranging aspects of drought are being discussed, that is: (i) analysis of past natural hazards, socio-economic and environmental impacts, mapping underlying causes, and identification of responses by decision-makers and local users to mitigate drought impacts and enhance local resilience, (ii) exploration of future scenarios (future drought occurrences, socio-economic, environmental and policy factors that

[5] XEROCHRE: http://www.feem-project.net/xerochore/index.php

[6] DROUGHT-R&SPI: Fostering European Drought Research and Science–Policy Interfacing, http://www.eu-drought.org/

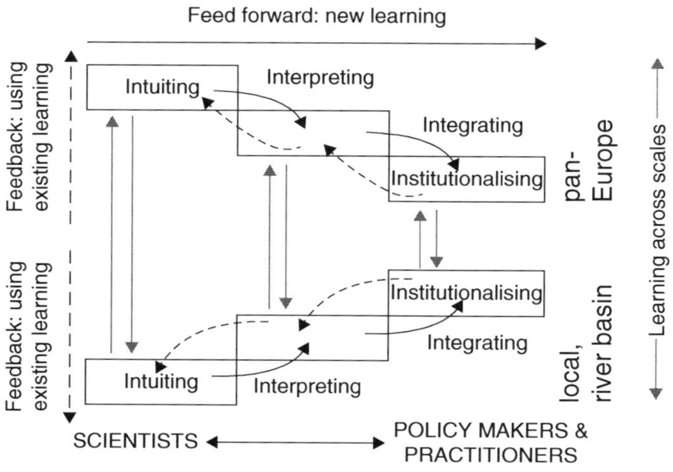

Figure 3.2.14 From research to stakeholders and policy-makers: an organisational learning perspective, according to the Crossan–Lane–White Organisational Learning Framework (solid black arrows: feed forward learning, dashed black arrows: feedback using existing learning and vertical solid arrows: learning across scales).

could affect drought vulnerability), and policies on potential options for long-term risk mitigation, and (iii) assessment of options for long-term drought risk mitigation (effectiveness, cost, benefits and potential trade-offs, incorporating uncertainty). Parallel and interlinked with the CS-DDFs, pan-European Drought Dialogue Fora (pan-EU DDF) have been held (Nicosia, 2012, associated with the Cyprus EU Presidency, and Brussels, 2013), others will be organised with policy-makers, stakeholders, water managers, NGOs and leading scientists (i.e. coordinators of international projects) to ensure that the design and review of EU water-related (including drought) policy-making is fine-tuned to policy implementation at the national and the river basin scale (standard operation procedures). Science–policy interfacing across these scales is paramount because vulnerability is context-specific (e.g. physical, environmental, socio-economic, cultural, legal, institutional) which requires analyses on detailed scales (e.g. river basins) (Kampragou *et al.*, 2011), whereas international policies and drought-generating climate drivers and land surface processes are operating on a (sub)continental scale (e.g. European Union) (Figure 3.2.14).

Additionally to the contribution of DROUGHT-R&SPI to science–policy interfacing, the project output will support the development or further improvement of drought management planning in the framework of the 2nd cycle of WFD River Basin Management Plans (2015). The policy options identified in the Blueprint to safeguard Europe's water resources (EC, 2012) will be important guidance. In this communication from the

European Commission to the European Parliament, the following actions with respect to drought (and water scarcity) are proposed:

- water pricing
- water metering also requires cost-recovery for water services taking into account the polluter pays principle
- water reduction in agriculture
- reduction of illegal abstraction/ impoundments
- awareness of water consumption
- green infrastructure for water storage
- reduction of leakage from water distribution networks
- water re-use for irrigation or industrial purposes
- water accounting
- ecological flows
- target setting
- reduction of drought risk
- improvement of knowledge base.

The European Commission clearly considers that one-size-fits-all actions for the whole of the European Union do not exist and that actions should be site and context specific. The DROUGHT-R&SPI project has published its 1st SPB that presents the evaluation of the responses to drought implemented in the six DROUGHT-R&SPI Case Studies during two past drought episodes (Figure 3.2.15).

The DROUGHT-R&SPI research and science-policy interfacing will be embedded in the expert network of the European Drought Centre (EDC)[7] and linked with the European Drought Observatory (EDO)[8] ensuring that the outcome will be consolidated beyond the project's lifetime.

In summary, drought has a wide range of impacts, which cannot straightforwardly be linked to the natural hazard. A database (European Drought Impact Inventory) has been established that suggests that drought impacts are partly different for northern and western Europe relative to those reported for southern and eastern Europe. Science Policy Briefs (SPBs) facilitate communication between scientists and policy-makers. SPBs need to be timely and targetted to the specific policy topic. A recent pan-European project on drought illustrates that science-policy-making has to be started directly from the project start. The science-policy-making should seek integration across scales, that is from the river basin scale to the pan-European scale and vice versa. Mid-term project key findings are presented and reflect actions proposed in the EU Blueprint to safeguard Europe's water resources.

---

[7] European Drought Centre: http://www.geo.uio.no/edc/

[8] European Drought Observatory: http://edo.jrc.ec.europa.eu/edov2

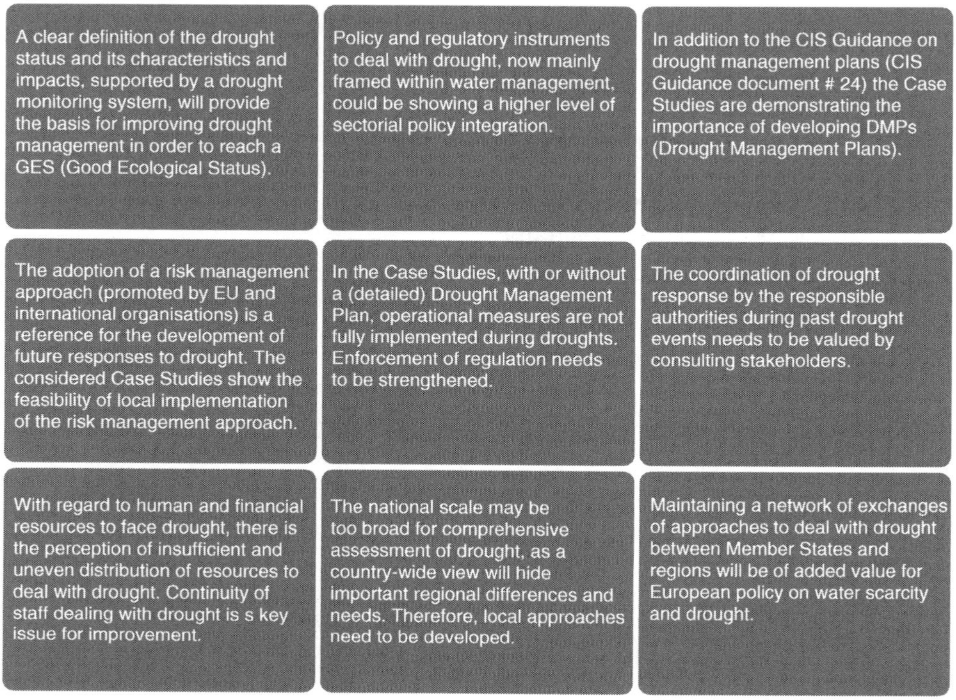

Figure 3.2.15 Mid-term key findings of the DROUGHT-R&SPI project obtained from the Case Study Drought Dialogue Fora. *Source*: DROUGHT-R&SPI project, www.eu-drought.org/sciencepolicybriefs. Reproduced with permission of the author.

Acknowledgements

Most of the research reported in this chapter has been financed by EU drought projects, that is ARIDE (ENV4-CT97-0553), ASTHyDA (EVK1-2001-00166), WATCH (036946), XEROCHORE (211837) and DROUGHT-R&SPI (282769). The study also contributes to research programmes of the UNESCO FRIEND-Water project group on Low Flows and Drought and the Wageningen Institute for Environment and Climate Research (WIMEK-SENSE).

References

Alderlieste M.A.A. and Van Lanen H.A.J. (2012) Trends in low flow and drought in selected European areas derived from WATCH forcing data and simulated multi-model mean runoff. DROUGHT-R&SPI Technical Report No. 1, available online at: http://www.eu-drought.org/technicalreports/3.

Alderlieste M.A.A. and Van Lanen H.A.J. (2013) *Change in future low flow and drought in selected European areas derived from WATCH GCM Forcing Data and simulated multi-model runoff*. DROUGHT-R&SPI Technical Report No. 5, available at: http://www.eu-drought.org/technicalreports/2.

Andreadis K.M., Clark E.A., Wood A., Hamlet A.F., Lettenmaier D.P. (2005) Twentieth-Century Drought in the Conterminous United States. *Journal of Hydrometeorology,* **6**, 985–1001.

Anshari G., Kershaw A.P., Van der Kaars S. and Jacobsen G. (2004) Environmental change and peatland forest dynamics in the Lake Sentarum area, West Kalimantan, Indonesia. *Journal of Quaternary Science,* **19**(7), 637–655.

Bates B., Kundzewicz Z., Wu S. and Palutikof J. (2008) *Climate Change and Water. Technical Paper of the Intergovernmental Panel on Climate Change*, Intergovernmental Panel on Climate Change, Geneva.

Bloomfield, J.P. and Marchant, B.P. (2013): Analysis of groundwater drought building on the standardised precipitation index approach. Hydrology and Earth System Sciences, **17**(12), 4769–4787, doi:10.5194/hess-17-4769-2013.

Changnon Jr., S.A. (1987): *Detecting drought conditions in Illinois*. Illinois State Water Survey Circular **164-87**.

Dai A. (2012) Increasing drought under global warming in observations and models, *Nature Climate Change*, doi:10.1038/nclimate1633.

Dieker E., Van Lanen H.A.J. and Svoboda M. (2010) *Comparison of Three Drought Monitoring Tools in the USA*. WATCH Technical Report No. 25, available online at: http://www.eu-watch.org/publications/technical-reports/3.

EC (2012) *A Blueprint to Safeguard Europe's Water Resources*, COM(2012), 673 final.

EEA (2010) *Mapping the impacts of natural hazards and technological accidents in Europe*. An overview of the last decade, EEA Technical report No 13/2010, Copenhagen. Brussels.

Falkenmark M., Rockström J. and Karlberg L. (2009) Present and future water requirements for feeding humanity, *Food Security*, **1**, 59–69.

Fleig, A.K., Tallaksen, L.M., Hisdal, H. and Demuth, S. (2006) A global evaluation of streamflow drought characteristics. *Hydrology and Earth System Sciences*, **10**(4), 535–552, doi:10.5194/hess-10-535-2006.

Fleig A.K, Tallaksen L.M., Hisdal H., Stahl K. and Hannah D.M. (2010) Inter-comparison of weather and circulation type classifications for hydrological drought development. *Physics and Chemistry of the Earth,* **35**, 507–515.

Fleig A.K, Tallaksen L.M., Hisdal H. and Hannah D.M. (2011) Regional hydrological drought in north-western Europe: linking a new Regional Drought Area Index with weather types. *Hydrological Processes*, **25**, 1163, doi: 10.1002/hyp.7644.

Forzieri, G., Feyen, L., Rojas, R., Flörke, M., Wimmer, F. and Bianchi, A. (2014) Ensemble projections of future streamflow droughts in Europe. *Hydrology and Earth System Sciences*, **18**, 85–108, doi:10.5194/hess-18-85-2014.

Garnier E., Daux V., Yiou P. and García de Cortázar-Atauri, I. (2011) Grapevine harvest dates in Besançon (France) between 1525 and 1847: Social outcomes or climatic evidence? *Climatic Change*, **104**(3-4), 703–727.

Gerten D., Heinke J., Hoff H., Biemans H., Fader M. and Waha K. (2011) Global water availability and requirements for future food production, *Journal of Hydrometeorology*, **12**, 885–899, doi:10.1175/2011JHM1328.1.

Guha-Sapir D., Hoyois Ph. and Below R. (2013) Annual *Disaster Statistical Review 2012: The Numbers and Trends*, CRED, Brussels.

Haddeland, I., Heinke, J., Biemans, H., Eisnere, S., Flörke, M., Hanasaki, N., Konzmann, M., Ludwig, F., Masaki, Y., Scheweb, J., Stacke, T., Tessler, Z.D., Wada, Y. and Wisser, D. (2014) Global water resources affected by human interventions and climate change. *PNAS*, 111(9), 3251-3256, doi/10.1073/pnas.1222475110.

Hannaford J., Lloyd-Hughes B., Keef C. *et al.* (2011) Examining the large-scale spatial coherence of European drought using regional indicators of precipitation and streamflow deficit, *Hydrological Processes* **25**, 1146–1162, DOI:10.1002/hyp.7725.

Hannaford J., Buys G., Stahl K. and Tallaksen L.M. (2013) The influence of decadal-scale variability on trends in long European streamflow records. *Hydrology and Earth System Sciences*, **17**, 2717–2733, doi:10.5194/hess-17-2717-2013.

Hisdal, H., Tallaksen, L.M., Clausen B., Peters, E. and Gustard, A. (2004) Hydrological drought characteristics. in Tallaksen, L.M., Van Lanen, H.A.J. (Eds.), *Hydrological Drought Processes and Estimation Methods for Streamflow and Groundwater* (eds L.M. Tallaksen and H.A.J. Van Lanen), Developments in Water Science, Vol. 48, Elsevier Science B.V., Amsterdam, The Netherlands, pp. 139–198.

IPCC (2013) *Summary for Policymakers Working Group I*, Contribution to the IPCC Fifth Assessment Report Climate Change 2013: The Physical Science Basis, Geneva.

Jaeger W.K., Plantinga A.J., Chang H., Dello K., Grant G., Hulse D., McDonnell J.J., Lancaster S., Moradkhani H., Morzillo A.T., Mote P., Nolin A., Santelmann M. and Wu J. (2013) Toward a formal definition of water scarcity in natural-human systems, *Water Resources Research*, doi: 10.1002/wrcr.20249.

Kampragou, E., Apostolaki, S., Manoli, E., Froebrich, J. and Assimacopoulos, D. (2011) Towards the harmonization of water-related policies for managing drought risks across the EU. *Environmental Science and Policy* **14**, 815–824, doi:10.1016/j.envsci.2011.04.001.

Kingston D.G., McGregor G.R., Hannah D.M. and Lawler D.M. (2006) River flow teleconnections across the northern North Atlantic region. *Geophysical Research Letters*, **33**, L14705, doi:10.1029/2006GL026574.

Lewis S.L., Brando P.M., Phillips O.L. *et al.* (2011) The 2010 Amazon Drought, *Science*, **331**, 554.

McKee T.B., Doesken N.J. and Kleist J. (1993) *The relationship of drought frequency and duration to time scales*. in: Eighth Conference on Applied Climatology. 17-22 January, Anaheim, California.

Milly P.C.D., Dunne K.A. and Vecchia A.V. (2005) Global pattern of trends in streamflow and water availability in a changing climate. *Nature,* **438**, doi:10.1038/nature04312.

Mishra K.K. and Singh V.P. (2010) A review of drought concepts, *Journal of Hydrology*, **391**, 202–216.

Murdiyarso D., Hergoualch K and Verchot L.V. (2010) Opportunities for reducing greenhouse gas emissions in tropical peatlands. *PNAS*, **107**(46): 19655–19660.

Niemeijer S. (2008) *New drought indices. in Proceedings of the 1st International Conference on Drought Management: Scientific and Technological Innovations*, Zaragoza, Spain, 12–14 June 2008, pp. 267–274.

Orlowsky B. and Seneviratne S.I. (2013) Elusive drought: uncertainty in observed trends and short- and long-term CMIP5 projections. *Hydrology and Earth System Sciences*, **17**, 1765–1781, doi:10.5194/hess-17-1765-2013.

Page S.E., Siegert F., Rieley J.O., Boehm H-D.V., Jayak A. and Limink S. (2002) The amount of carbon released from peat and forest fires in Indonesia during 1997. *Nature,* **420**, 61–65.

Peters E. (2003) *Propagation of drought through groundwater systems – illustrated in the Pang (UK) and Upper-Guadiana (ES) catchments*. PhD thesis Wageningen University, available online at: http://edepot.wur.nl/121490.

Peters E., Torfs P.J.J.F., Van Lanen H.A.J. and Bier G. (2003) Propagation of drought through groundwater – a new approach using linear reservoir theory. *Hydrological Processes,* **17**(15) 3023–3040.

Peters E., Bier G., Van Lanen H.A.J. and Torfs P.J.J.F. (2006) Propagation and spatial distribution of drought in a groundwater catchment. *Journal of Hydrology,* **321**, 257–275.

Prudhomme C., Parry S., Hannaford J., Clark D.B., Hagemann S. and Voss F. (2011) How well do large-scale models reproduce regional hydrological extremes in Europe?, *Journal of Hydrometeorology*, **12**(6), 1181–1204, doi:10.1175/2011JHM1387.1.

Prudhomme C., Giuntoli I., Robinso E.L., Clark D.B., Arnell N.W., Dankers R., Feketee B.M., Franssen W., Gerten D., Gosling S.N., Hagemann S., Hannah D.M., Kim H., Masaki Y., Satoh Y., Stacke T., Wada Y. and Wisser D. (2014): Hydrological droughts in the 21st century: Hotspots and uncertainties from a global multimodel ensemble experiment. *PNAS*, 111(9), 3262–3267, doi/10.1073/pnas.1222473110.

Quevauviller Ph., Barceló D., Beniston M., Djordjevic S., Harding R.J., Iglesias A., Ludwig R.A., Navarra A., Navarro Ortega A., Mark O., Roson R., Sempere D., Stoffel M., Van Lanen H.A.J. and Werner M. (2012) Integration of research advances in modelling and monitoring in support of WFD river basin management planning in the context of climate change. *Science of the Total Environment,* **440**, 167–177, 10.1016/j.scitotenv.2012.07.055.

Schewe, J., Heinke, J., Gerten, D., Haddeland, I., Arnell, N.W., Clark, D.B., Dankers, R., Eisner, S., Fekete, B.M. Colón-González, F.J., Gosling, S.N., Kim, H., Liu, X., Masaki, Y., Portmann, F.T., Satoh, Y., Stacke, T., Tang, Q., Wada, Y., Wisser, D., Albrecht, T., Frieler, K., Piontek, F., Warszawski, L. and Kabat, P. (2014) Multimodel assessment of water scarcity under climate change. *PNAS*, 111(9), 3245–3250, doi/10.1073/pnas.1222460110.

Schmidt, G. and Benítez-Sanz, C. (2013) *How to distinguish water scarcity and drought in EU water policy? GWF Discussion Paper 1333*, Global Water Forum, Canberra, Australia, available online at: http://www.globalwaterforum.org/2013/08/26/.

Seneviratne S.I., Nicholls N., Easterling D., Goodess C.M., Kanae S., Kossin J., Luo Y., Marengo J., McInnes K., Rahimi M., Reichstein M., Sorteberg A., Vera C. and Zhang X. (2012) Changes in climate extremes and their impacts on the natural physical environment, in *Managing the Risks of Extreme Events and Disasters to Advance Climate Change Adaptation*, eds C.B. Field, V. Barros, T.F. Stocker *et al.*, A Special Report of Working Groups I and II of the Intergovernmental Panel on Climate Change (IPCC), Cambridge University Press, Cambridge, pp. 109–230.

Sepulcre-Canto G., Horion S., Singleton A., Carrao H. and Vogt J. (2012) Development of a Combined Drought Indicator to detect agricultural drought in Europe. Natural Hazards and Earth System Sciences, **12**, 3519–3531, doi:10.5194/nhess-12-3519-2012.

Sheffield, J. and Wood, E.F. (2011) *Drought: Past Problems and Future Scenarios*. Earthscan.

Sheffield, J., Wood, E.F. and Roderick, M.L. (2012): Little change in global drought over the past 60 years, *Nature*, **491**, 435–438, doi:10.1038/nature11575.

Sheffield, J., Andreadis, K.M., Wood, E.F. and Lettenmaier, D.P (2009) Global and continental drought in the second half of the twentieth century: Severity–area–duration analysis and temporal variability of large-scale events. *Journal of Climate*, **22**, 1962–1981.

Shukla, S. and Wood, A.W. (2008) Use of a standardized runoff index for characterizing hydrologic drought, *Geophysical Research Letters*, **35**, L02405, doi:10.1029/2007GL032487.

Stahl, K., Hisdal, H., Hannaford, J., Tallaksen, L.M., van Lanen, H.A.J., Sauquet, E., Demuth, S., Fendeková, M. and Jódar, J. (2010) Streamflow trends in Europe: evidence from a dataset of near-natural catchments. *Hydrology and Earth System Sciences* **14**, 2376-2382, doi:10.5194/hess-14-2367–2010.

Stahl, K., Tallaksen, L.M., Hannaford, J. and Van Lanen, H.A.J. (2012a) Filling the white space on maps of European runoff trends: estimates from a multi-model ensemble, Hydrology and Earth System Sciences, **16**, 2035–2047, doi:10.5194/hess-16-2035-2012.

Stahl, K., Blauhut, V., Kohn, I., Acácio, V., Assimacopoulos, D., Bifulco, C., De Stefano, L., Dias, S., Eilertz, D., Frielingsdorf, B., Hegdahl, T.J., Kampragou, E., Kourentzis, V., Melsen, L., Van Lanen, H.A.J., Van Loon, A.F., Massarutto, A., Musolino, D., De Paoli, L., Senn, L., Stagge, J.H., Tallaksen, L.M., and Urquijo, J. (2012b) *A European Drought Impact Report Inventory (EDII): Design and Test for Selected Recent Droughts in Europe*. DROUGHT-R&SPI Technical Report No. 3, available online at: http://www.eu-drought.org/technicalreports/2.

Svoboda, M., LeComte, D., Hayes, M., Heim, R., Gleason, K., Angel, J., Rippey, B., Tinker, R., Palecki, M.A., Stooksbury, D., Miskus, D. and Stephens, S. (2002) The Drought Monitor. *Bulletin of the American Meteorological Society*, **83**(8), 1181–1190.

Tallaksen L.M. and van Lanen H.A.J., ed. (2004) *Hydrological Drought. Processes and Estimation Methods for Streamflow and Groundwater*, Developments in Water Science, 48, Elsevier Science B.V., Amsterdam.

Tallaksen L.M. and Fleig A. (2009) *Atmosphere: climatic drivers of drought*. Chapter 3 in: Wipfler L. and Van Lanen H.A.J (Eds.) Extended Guidance Document on the Natural System & Drought, XEROCHORE Technical Report, Wageningen University, pp. 7–16.

Tallaksen L.M., Hisdal H. and Van Lanen H.A.J. (2009) Space-time modeling of catchment scale drought characteristics. *Journal of Hydrology* **375**, 363–372, doi:10.1016/j.jhydrol.2009.06.032.

UN (2011): *Humanitarian Requirements for the Horn of Africa Drought 2011*. United Nations Report. Office for the Coordination of Humanitarian Affairs (OCHA), New York and Geneva, available online at: http://reliefweb.int/sites/reliefweb.int/files/resources/Full report 216.pdf.

USDM (2013) United States Drought Monitor [http://droughtmonitor.unl.edu/AboutUSDM .aspx, accessed: 11 November 2013].

Van Huijgevoort, M.H.J., Hazenberg, P., Van Lanen, H.A.J. and Uijlenhoet, R. (2012): A generic method for hydrological drought identification across different climate regions. *Hydrology and Earth System Sciences,* **16**(8), 2437–2451, doi:10.5194/hess-16-2437-2012.

Van Huijgevoort M.H.J., Hazenberg P., Van Lanen H.A.J., Teuling R., Clark D., Folwell S., Gosling S., Hanasaki N., Heinke J., Koirala S., Stacke T., Voβ F., Sheffield J. and Uijlenhoet R. (2013) Global multi-model analysis of hydrological drought in the second part of the 20th century (1963-2000). *Journal of Hydrometeorology*, **14**, 1535–1552, doi:10.1175/JHM-D-12-0186.1.

Van Huijgevoort M.H.J., Van Lanen H.A.J., Teuling A.J. and Uijlenhoet R. (2014) Identification of changes in hydrological drought characteristics from a multi-GCM driven ensemble constrained by observed discharge. *Journal of Hydrology*, **512**, 421–434, 10.1016/j.jhydrol.2014.02.060.

Van Lanen H.A.J., Fendeková M., Kupczyk E., Kasprzyk A. and Pokojski W. (2004) Flow generating processes. in *Hydrological Drought. Processes and Estimation Methods for Streamflow and Groundwater* (eds L.M. Tallaksen and H.A.J. Van Lanen) Developments in Water Science, 48, Elsevier Science B.V., Chapter 3, pp. 53–96.

Van Lanen H.A.J., Wanders N., Tallaksen L.M. and van Loon A.F. (2013) Hydrological drought across the world: impact of climate and physical catchment structure. *Hydrology and Earth System Sciences*, **17**, 1715–1732, doi:10.5194/hess-17-1715-2013.

Van Loon A.F. (2013) *On the propagation of drought. How climate and catchment characteristics influence hydrological drought development and recovery*. PhD thesis, Wageningen University [available at: http://library.wur.nl/WebQuery/wda/2027352].

Van Loon A.F., Van Lanen H.A.J., Hisdal H., Tallaksen L.M., Fendeková M., Oosterwijk J., Horvát O. and Machlica A. (2010) Understanding hydrological winter drought in Europe. in: Servat E., Demuth S., Dezetter A., Daniell T., Ferrari E., Ijjaali M., Jabrane R., van Lanen H.,

Huang Y. (Eds.) *Global Change: Facing Risks and Threats to Water Resources*, (eds E. Servat, S. Demuth, A. Dezetter *et al.*) IAHS Publication No. 340, IAHS press, pp. 189–197.

Van Loon A.F. and Van Lanen H.A.J. (2012) A process-based typology of hydrological drought, *Hydrology and Earth System Sciences* **16**, 1915–1946, doi:10.5194/hess-16-1915-2012.

Van Loon A.F. and Van Lanen H.A.J. (2013) Making the distinction between water scarcity and drought using an observation-modeling framework, *Water Resource Research*, **49**, 1483–1502, doi:10.1002/wrcr.20147.

Wanders N. and Van Lanen H.A.J. (2013) Future discharge drought across climate regions around the world modelled with a synthetic hydrological modelling approach forced by three General Circulation Models. *Natural Hazard Earth System Sciences and Discussions* **1**, 7701–7738, doi:10.5194/nhessd-1-7701-2013.

Wanders, N., Van Lanen, H.A.J. and Van Loon, A.F. (2010) *Indicators for drought characterization on a global scale*. WATCH Technical Report 24 (http://www.eu-watch.org/publications/technical-reports, accessed: 6 June 2014).

Weedon G.P., Gomes S., Viterbo P., Shuttleworth W. J., Blyth E., Österle H., Adam J.C., Bellouin N., Boucher O. and Best M. (2011) Creation of the WATCH Forcing Data and its use to assess global and regional reference crop evaporation over land during the twentieth century, *Journal of Hydrometeorology*, **12**, 823–848, doi:10.1175/2011JHM1369.1, 2011.

Wilhite D. A. (2000) *Drought: A Global Assessment*, Routledge, London, New York.

Wilhite D.A. and Glantz M.H. (1985) Understanding the drought phenomenon: the role of definitions. *Water International,* **10**, 111–120.

Wilson D., Hisdal H. and Lawrence D. (2010) Has streamflow changed in the Nordic countries? – Recent trends and comparisons to hydrological projections. *Journal of Hydrology,* **394**, 334–346.

WWDR (2009) World Water Assessment Programme 2009, *The United Nations World Water Development Report 3: Water in a Changing World*, UNESCO, Paris and Earthscan, London.

Yevjevich, V. (1967) *An objective approach to definition and investigations of continental hydrologic droughts*. Hydrology papers 23, Colorado State University, Fort Collins, USA.

3.3 Drought in the Light of Climate Change in the Mediterranean Area

Ana Iglesias and Luis Garrote

Department of Agricultural Economics and Social Sciences, Technical University of Madrid, Madrid, Spain

3.3.1 Introduction

Drought can have serious effects on the economy and the environment in Mediterranean countries and on the population's well being. Major drought events in the last 20 years affected almost 10 times more people than were affected by floods in the Mediterranean during the last 50 year period. The economic damage caused by drought in the Mediterranean during the last 20 years is about 5 times more than in the entire United States (EM-DAT, 2013). Drought events affect water supplies for irrigation, urban and industrial use, the ecosystem's health, and give rise to conflicts among users that limit coherent integrated water resource management. The reduction of coastal wetlands has been in part related to recurrent drought episodes and surface water scarcity, and amplified by the excessive groundwater pumping to compensate for these problems. In addition to water scarcity, droughts also cause water quality problems, since water quality parameters deteriorate during drought due to lack of dilution and the water may not be acceptable for human consumption.

As climate change drought impacts become more noticeable and costlier in the Mediterranean region, some current water management strategies will not be useful. According to the research findings, the areas with limited water resources will increase in the coming decades with major consequences for the way we produce food and protect ecosystems. Based on these projections this chapter discusses water policy priorities for climate change adaptation in the Mediterranean.

The geographical focus of this work is the Mediterranean region, including countries in Europe and North Africa; the aim is to provide some insight into the policy priorities for reducing the current impacts of drought on water resources today and in the light of climate change. We think two questions are relevant: Does the understanding of

Hydrometeorological Hazards: Interfacing Science and Policy, First Edition. Edited by Philippe Quevauviller.

current drought vulnerability provide us with policy insights to reduce drought damage in the region? Can improved drought management be an adaptation policy in the light of climate change? We address these questions by evaluating the limits of rainfall in the region, drought hazard and water scarcity (Section 3.3.2), estimating drought vulnerability (Section 3.3.3), looking into the future (Section 3.3.4) and establishing policy priorities (Section 3.3.5).

3.3.2 The limits of rainfall

3.3.2.1 Drought and water scarcity: Overlapping challenges in the region

Drought, water scarcity and aridity are overlapping issues in Mediterranean countries. Water resources are limited, scarce and highly irregular in space and time. The potential use of surface water under the natural regime is less than 10% of total natural resources. The availability has increased to 40% due to the intensive development of hydraulic infrastructure during the twentieth century. Groundwater use is also intensive in many areas of the region and contributes to an additional 10% of the total available resources. Water use is mainly for agriculture (over 50% of water demand), but other economic and social water demands are rapidly increasing, such as tourism (current urban demand is 13%) and ecosystem services. With limited and scarce water resources and demand rising due to demographic shifts, economic development and lifestyle changes, water management problems are significant, even without drought events. Climate change is a great challenge.

Water demand already exceeds water availability in many areas of the Mediterranean region, imposing a strain on social development and ecosystems (Iglesias *et al.*, 2009). The average annual potential water availability per capita considering the total freshwater resources is less than 1000 m^3 per capita per year, the smallest of the European and African averages. In some basins demand is above the available resources, and water scarcity crises are common (Tables 3.3.1 and 3.3.2).

Under current conditions all Mediterranean countries also face significant problems due to the unbalanced distribution of water resources and conflicts among users and between countries. It seems most likely that climate change will lead to an intensification of these problems

The difficulty in forecasting highly variable rainfall multiplies the challenges faced by water resource managers and increases the likelihood of water conflicts.

The Mediterranean climate is characterized by hot and dry summers and cold and wet winters. In general, precipitation is concentrated between October and April (southern regions) or May (northern regions). The average seasonal temperature decreases with increasing altitude and latitude. The weather is characterized by large inter-annual variability. Summer precipitation (June, July and August) is well below 100 mm in most regions, except in the northwest, thus requiring supplementary irrigation during part of the crop cycle in order to obtain adequate production levels. In general, the dry period also shows the largest inter-annual variability. The coefficient of variation (CV) of precipitation varies from 20 to 60%, implying a high risk of rain-fed crop failures and the

Table 3.3.1 Water resource indicators: 2008 total freshwater resources, available resources, use, and water availability in selected Mediterranean countries. (Source of data: FAO 2011).

Country	Total area (x 10^3 km^2)	Population (million)	Rainfall (mm yr^{-1})	Internal usable water resources[a] (km^3 yr^{-1})	Usable water resources[b] (km^3 yr^{-1})	Internal groundwater[c] (km^3 yr^{-1})	Total water use (km^3 yr^{-1})	Total water use (% renewable)	Potential total usable water resources per capita (m^3/capita/year)
Algeria	2, 382	32	89	13.90	14.32	1.70	5.74	40	473
Egypt	1, 001	72	51	1.80	58.30	1.30	61.70	106	859
France	552	60	867	178.50	203.70	100.00	35.63	17	3439
Greece	132	11	652	58.00	74.25	10.30	7.99	11	6998
Israel	22	6.5	435	0.75	1.67	0.5	1.63	103	254
Italy	301	57	832	182.50	191.30	43.00	43.04	22	3325
Libyan Arab J.	1, 770	6	56	0.60	0.60	0.50	5.73	954	113
Morocco	447	31	346	29.00	29.00	10.00	12.23	42	971
the Mediterranean	506	41	636	111.20	111.50	29.90	35.90	32	2794
Syrian Arab R.	180	18	252	7	26.26	4.2	20.6	100	1403
Tunisia	164	10	313	4.15	4.56	1.45	2.58	57	482
Turkey	770	71	593	227	213	69	37	18	2800

[a]The values refer to both regulated and unregulated water. Real available water resources in all cases are a fraction of these values.

[b]These values include transboundary water.

[c]A proportion of these values is included in the total renewable water resources.

Table 3.3.2 Summary of the drought management actions in the Mediterranean. *Source:* Iglesias *et al.* (2009).

Concept	Cyprus	Greece	Italy	Morocco	Tunisia	the Mediterranean
Water law	Includes drought	Includes drought	Includes drought	Includes drought	Includes drought	Includes drought
River basin authorities	Managed at central level Developed	Developed	Developed	Development	Partially Developed	Developed
Relation among institutions	High	Low	Low	Medium	High	Medium
Public participation in water management	Low	Medium	High	Low	Low	High
Drought contingency plan	Developed	In development	Sub-national	In development	National	River basin
Drought monitoring system	Partially developed	Partially developed	River basin	National	National	River basin

National: developed at country level. Sub-national: developed at a level smaller than the country, such as a province or district. River basin level: refers to the portion of the river basin within the country.

need for supplementary irrigation to meet full crop water requirements and achieve potential yield.

The region's overall socio-economic model places available water resources under considerable stress. In many cases, agriculture is responsible for water imbalances because it accounts for more than 50% of water use in most countries (FAO 2011). Other economic uses of water – urban, energy and tourism – impose further challenges for meeting ecosystem services and increase conflicts among the affected parties. Some of the potential solutions to these problems – such as changes in infrastructure or limitations of irrigation – are not accepted by all social sectors. Water resource managers face the challenge of ensuring the future sustainability of water resources while maintaining strategic agricultural, social and environmental targets. Climate change imposes an additional challenge and understanding its implications and policy requirements is a complex process, as we shall see.

As mentioned above, the average annual potential water availability per capita considering the total freshwater resources in southern Mediterranean countries is less than 1000 m^3 per capita per year (Table 3.3.1). Usable water resources are always less than potential water resources in all countries. For example in the Mediterranean area available water resources are less than half of the total freshwater resources and the potential use of surface water under the natural regime is only 7% (Iglesias *et al.*, 2007). In most areas of all countries water demand is above the available resources (Table 3.3.1), and water scarcity crisis is common. Water allocation decisions could be made in advance and contribute to decreasing water conflicts. However, this would require forecasting the highly variable seasonal and inter-annual rainfall and this still remains a challenge.

Drought events in the Mediterranean have been more frequent since 1970 (Iglesias *et al.*, 2009).

Figure 3.3.1 shows the time series of aggregated precipitation in Morocco and the corresponding widely used drought index SPI calculated at 24 month intervals. Drought

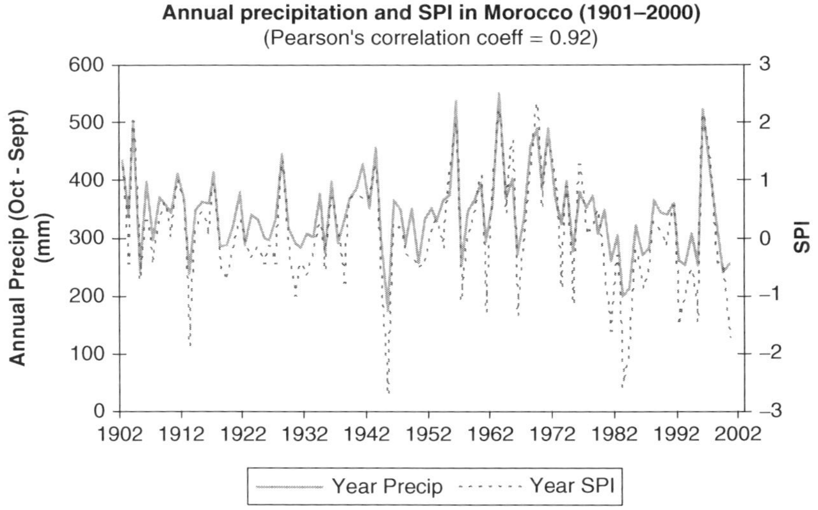

Figure 3.3.1 Time series of aggregated annual precipitation and SPI values (12 month time scale) in Morocco. *Source:* The Tyndall Center database TYN CY 1.1; Mitchell, Hulme and New, 2002.

indices do not correlate well with hydrological drought periods or historical drought impacts, due to the effect of storage (Martin-Carrasco *et al.*, 2014). Many of the more complex indices that take storage and management into account are not easily interpreted across the regions and cannot be validated with the data available over wide geographical areas (Martin-Carrasco *et al.*, 2014). The figure shows at least two periods with different precipitation trends and variability patterns. Precipitation after the 1970s has clearly followed a decreasing trend and provoked further water deficit in many areas in the country (Iglesias *et al.,* 2009).

Pressures and trends Population growth in many southern Mediterranean countries is the major factor affecting water resources, reducing the water availability per capita (Figure 3.3.2a and b, and Table 3.3.1). Urbanization increases urban demands which are of high priority and intensifies conflicts among users. The tourist population in the Mediterranean countries is very significant, and tourist water consumption is about three times higher than local demands (EEA, 2011). The number of international tourist arrivals increases every year; in the Mediterranean countries, Greece, and France; the number exceeds the total population of the countries by about one third (UNWTO, 2006–2013). Tourist consumption is highly seasonal but the industry increases the permanent water demand for facilities and leisure structures. In particular there is a tourist and second-home boom all across the Mediterranean coast and together with the 275 golf courses (with 75 more planned) this represents an increase in demand of about 30 million m^3. This demand is a small fraction of the agricultural demand at the country or regional level (for example in Cyprus tourist demand represents 5% of the total demand compared with about 70% for agriculture), but at the local level in key tourist destinations it is the main demand (for example in Majorca, the Mediterranean; Essex, Kent and Newnham, 2004).

Almost one half of the Earth's land surface is covered by trans-boundary river basins and freshwater scarcity problems are becoming more frequent. Water can lead to political hostilities and many regions with political conflicts are sharing water resources. International Organizations need to address cooperation among nations in order to solve conflicts. Most Mediterranean freshwater and groundwater resources are shared among countries (Iglesias *et al.*, 2009), the Nile River being a key global example. Within the Mediterranean countries, water shared between administrative regions is also common. Disputes exist, especially during drought conditions, which will probably increase as a result of imbalance in the distribution of water resources among the regions. The policies of central government or single basin management cannot resolve issues over shared water bodies and local interests are likely to diverge. International Institutions can play a key role as official and independent mechanisms to deal with water-related conflicts between the regions.

3.3.3 Estimating drought vulnerability

3.3.3.1 Underlying causes of drought risk

The objective of the vulnerability assessment is to identify underlying causes of risk derived from inadequate structures, management and technology, or by economic,

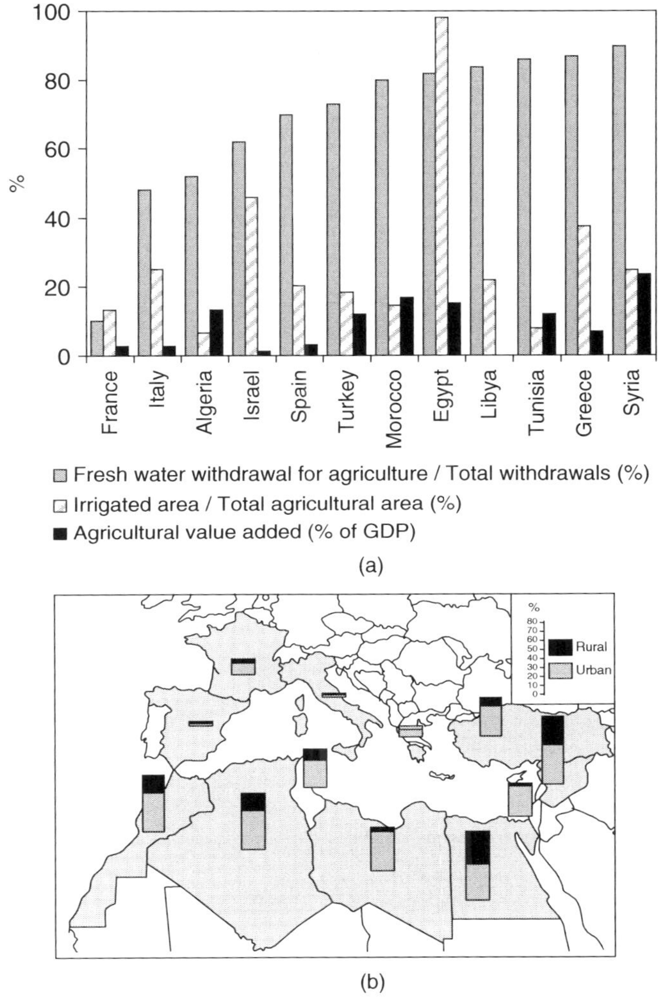

Figure 3.3.2 (a) Water use for irrigation and irrigated agricultural areas in selected Mediterranean countries. (b) Mediterranean countries and population increase projections for 2025. *Source:* FAO, 2012.

environmental and social factors. Vulnerability refers to the characteristics of a group in terms of its capacity to anticipate, cope with, resist and recover from the impact of drought. Vulnerability assessment is to identify characteristics of the systems that modify the level of risk derived from inadequate structures, management and technology, or by economic, environmental and social factors.

The consequences of droughts can usually be predicted, but drought vulnerability – that is the reasons that make a system suffer – is poorly understood. Therefore preventive action is frequently absent or insufficient to mitigate serious impacts in many regions. Here we aim to understand vulnerability in order to provide information for the design of early warning systems.

This section examines Mediterranean patterns of drought vulnerability through the mapping of several drought-related indicators – either at a country level or at regular grid scales. The proposed methods for mapping drought vulnerability at different spatial scales include characteristics that cover various aspects of droughts – from meteorological and hydrological drought hazard to social vulnerability.

Drought has multiple and severe social and economic impacts in any region, therefore drought vulnerability is hard to quantify because the vulnerability aspects are more complex than in other natural phenomena. The magnitude of drought vulnerability is determined, in addition to the natural hazard and exposure, by the level of development, population density and structure, demands on water and other natural resources, government policies and institutional capacity, technology and the political system, among others.

Vulnerability indices help to identify and target vulnerable regions or populations, to raise awareness, and form part of a monitoring and adaptation strategy.

While the research and mapping of disaster risks, water scarcity, climate change and related subjects have been significant, there has been little, if any, attempt to date to comprehensively describe and map various aspects and impacts of a drought as an individual natural disaster and as a global multifaceted phenomenon (Iglesias *et al.*, 2009). Some studies (Iglesias *et al.*, 2011) suggest that lack of a common understanding of what drought is stands in the way of cohesive anti-drought action. Beyond semantics, it is clear that Mediterranean countries and societies are among the most vulnerable in the world, and there are many ongoing efforts to define the causes and consequences of Mediterranean vulnerability to drought.

Dai, Trenberthke and Qian (2004) developed a global monthly dataset of the Palmer Drought Severity Index (PDSI) for 1870–2002 and established that very dry areas had more than doubled since the 1970s, with acute increases in Mediterranean countries. Dettinger and Diaz (2000) used monthly streamflow series to characterize and map global geographic differences in the seasonality and annual variability of streamflow; here again, the Mediterranean is highlighted as a region with very high vulnerability.

The social aspects of vulnerability have also been studied. Iglesias *et al.* (2009, 2011) proposed an index of social vulnerability to drought by combining the indicators that characterize direct exposure to drought with those that characterize social and economic aspects. For example, given a specific farm, the vulnerability is directly related to the intensity of the drought event. In contrast, given a defined drought event, the most vulnerable farming system is the one that has less social and economic resilience; in general, marginal and poor farming systems suffer the largest consequences of drought.

Global drought vulnerability is a key element in several major global projects. The Climate Impact on Agriculture (CLIMPAG) project of the FAO has carried out an analysis of rainfall variability and drought for the period 1961–2002 and presented results through time series maps. The Natural Disaster Hotspots project of the World Bank has assessed the global risks of two disaster-related outcomes – mortality and economic losses – by considering physical exposure and historical loss rates (Iglesias *et al.*, 2009). The Global Water System Project (GWSP) examines global water assessment indicators with links to poverty and food security. The Global Drought Monitor of the University College London, UK, evaluates drought vulnerability based in publicly available datasets. The International Research Institute for Climate and Society of Columbia University, USA (IRI) monitors global drought based on the SPI index and issues climate forecasts and the European Centre for Medium-Range Weather Forecasts (ECMRW) also includes a global meteorological drought analysis. The USGS drought monitor is included in the FEWS early warning systems.

The most comprehensive effort in water resources analysis – including many aspects of drought vulnerability – is carried out by the UNH Water Systems Analysis Group of the University of New Hampshire, USA which has developed a compendium of Earth System and socio-economic databases describing the current state of global water resources, including associated human interactions and pressures. The group cooperates with several UN agencies and NASA to integrate a wide array of satellite-derived and land-based monitoring products from around the world with regional and country-level socio-economic data. This group has produced a digital archive presenting more than 130 thematic global datasets facilitating indicator calculation and mapping (WWI mapping) that are very relevant to drought vulnerability.

Current mapping exercises have barriers and opportunities. A limitation of the mapping efforts and the vulnerability evaluation is the limited in-country information resources; this information shows an alarming severe decline in the last two decades of the twentieth century, particularly in North African countries (Iglesias *et al.*, 2009). At the same time, state-of-the-art data sets from the Earth Systems Sciences, offers an opportunity to vulnerability evaluation, linking social and environmental aspects (Vörösmarty *et al.*, 2010).

Vulnerability in Mediterranean countries depends for instance on access to local water resources and the ability to use them, and several socioeconomic circumstances such as employment opportunities, available household capital and market access. Agriculture is a very specific water user for which also other external factors determine drought vulnerability such as soil properties, tillage practices and the ability to supplement rain fed agriculture by irrigation.

The underlying causes of risk may be related to structural problems, such as lack of adequate hydraulic infrastructures or technology, and also to management, economic and social features that increase the vulnerability of the region, watershed or water supply system under analysis. For example, the direct impact of precipitation deficiencies may be a reduction of crop yields. The underlying cause of this vulnerability, however, may be that the farmers did not use drought-resistant seeds, either because they the cost of the improved drought tolerant seeds was too high or they did not have enough information on improved management choices.

Another example could be farm foreclosure related to drought. The underlying cause of this vulnerability could be many things, such as small farm size because of historical land appropriation policies, lack of credit for diversification options, farming on marginal lands, limited knowledge of possible farming options, a lack of local industry for off-farm supplemental income or government policies.

3.3.3.2 A drought vulnerability index

Drought vulnerability is complex to evaluate since it includes intangible elements, such as the societal capacity to anticipate, respond and recover from drought. Here we estimate these aspects of social vulnerability evaluating the natural resource structure, the economic capacity, the human and civic resources, and aspects of agricultural innovation. The weight of each component of the index is a key determinant of the final value (Iglesias *et al.*, 2009).The sequential steps taken for the quantification of the vulnerability index are: (i) select proxy variables for factors that contribute to the vulnerability; (ii) normalize the proxy variables with respect to some common baseline; (iii) combine the sub-component proxy variables within each vulnerability category by weighted averages and (iv) quantify vulnerability as the weighted average of the components.

The total vulnerability index has been quantified as the weighted average of each of the components. The scores of the vulnerability index range on a scale of 0 to 1, 0 being the least vulnerable and 1 the most vulnerable. The total index is generated as the average of all components. The final value of the index depends on the valuation of each component. Here we present the results of the index under the assumptions that all components are weighted equally. This reflects the assumption that a society with institutional coordination and strengths for public participation may play a very important role in drought mitigation.

The indicators selected include characterisation of the drought natural event, water resources indicators, water infrastructure and management, and socio-economic indicators. The socio-economic indicators include adult literacy, life expectancy at birth, population without access to improved water, and GDP, that are among the most favoured indicators of sustainable development, and are used as components of the United Nations Development Programme's (UNDP) Human Development Index. The policy-relevance of the groups of indicators and the individual indicators selected is presented below.

Social characteristics depend to a large extent on the type of policies implemented in the country or region, such as self-sufficiency strategies based on market or protective policies for industrialised nations where agriculture might only play a marginal role. As in the previous case, the level of economic development is a variable of the capacity of a system to make investments in development technologies, food security and income stabilization. Eco-efficiency increases significantly the adaptation potential of a system. The development of agriculture significantly decreases the dependence of this sector on climatic variables and stabilises production, and this evolution is driven both by policies that aim for more productive crops or by private initiatives to increase the revenue from agriculture.

Figure 3.3.3 shows the individual values for the components that integrate the global drought vulnerability index (DVI). From this figure it is clear that for northern countries economic and social capacity are playing a major role in the maintenance of low DVI values, while the natural capital and the climate capital are quite similar or even lower than in southern countries. Technological eco-efficiency is also higher in France and Italy, but Spain shows levels quite similar to those of countries in the South. This kind of information is useful for identifying priority aspects for the definition of adaptation strategies. According to the results from this study it would seem appropriate to develop strategies to improve the economic and social capacity of the countries in the south Mediterranean.

All components, including the DVI itself, show distributions below that equity line, indicating that countries have different levels of contribution. This difference is especially clear in the case of economic and social capacities. In the first case, nearly 60% of the countries contribute around 10% of the total economic capacity of the region, and in the case of social capacity, 40% of the countries contribute more than 65% to the total

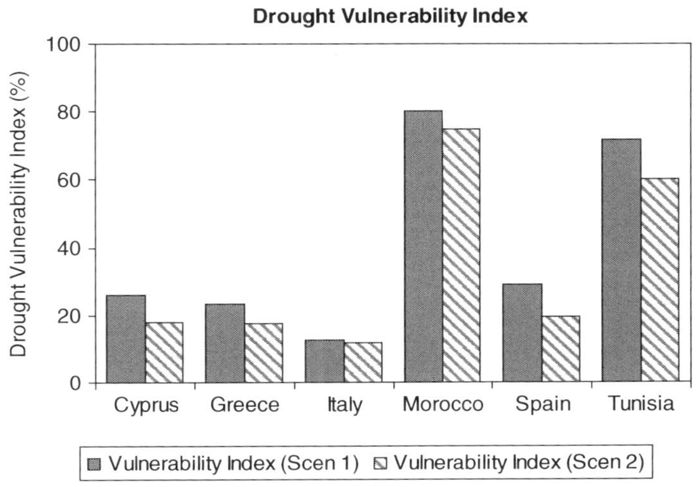

Figure 3.3.3 Evaluation of the DVI and components for selected Mediterranean countries. Scen1: scenario where all components are weighted equally, Scen 2: scenario where the natural resources component is weighted 20% more than the other components.

of the region. The distribution is quite even for the rest of the components, including the DVI itself, which underlines the importance of developing a partial analysis of the components.

Although, Mediterranean countries have quite similar DVI distributions overall, countries differ greatly in their economic and social capacity. In contrast, countries have similar technological and eco eco-efficiency capacity, natural capital and climate capital. As we shall see in the next section, this insight is extremely valuable for the development of policies that seek to redress low levels of vulnerability in order to cope with climate change impacts.

When moving from large scales to national and local scales, the local conditions will strongly influence whether a dry period is experienced as a drought, that is to what degree a specific local activity or group of water users is vulnerable to droughts (Wilhite, 2005).

3.3.4 From drought vulnerability to drought management

3.3.4.1 Policies, actions and examples

Water scarcity and drought have multidimensional implications for society and therefore no single management action, legislation or policy can respond to all aspects and demand objectives (Table 3.3.2). Mediterranean countries have developed legislation with different perspectives and levels of integration into the overall water management policy. When water is managed at the river basin level and the institutional responsibilities are clearly defined, drought legislation is more effectively applied. Nevertheless, a characteristic of all countries in the region is the weak cooperation among different institutions, and the fragmented roles of the State, the administrative regions and the river basin authorities, that often results in conflicts and impediments for implementation of existing legislation.

The European Water Framework Directive (2000/60/EC) is an example of recent efforts to develop coherent water management legislation and promote creation of institutions responsible for planning, management and control of water resources at the basin level. A main advantage of the explicit linkage of legislation and management to the basin level is the opportunity to address directly the needs and problems of the natural hydrological system and integrate the stakeholders in the decision process. However, the implementation of the Directive leads to difficulties, especially in transboundary basins – between countries or administrative regions of a country – where institutional coordination is required. Neither the current legislation, nor the aforeseen implementation of the Directive, provide explicit regulations for the ecological quality of water bodies or the quality of the discharges in drought situations. This important issue is being left to the discretion and responsibility of the various River Authorities.

3.3.4.2 Linking indicators to drought management actions

In most Mediterranean basins, measures are grouped according to different severity levels. A commonly used ranking describes three levels of severity (i.e. pre-alarm, alarm, and emergency). It is also extremely important to define the "normal" situation, since the plan is optimally developed at this stage. The management plan is considered a proactive measure that defines a protocol for implementing reactive measures when the water scarcity situation occurs. The severity levels are determined by established thresholds of indicators that trigger groups of measures in response to the objective of each level (Table 3.3.3). There are many examples that validate this framework over the past decades in Mediterranean countries, especially in the pre-alarm and alarm levels. In the emergency level, the main priority is to satisfy drinking water demands and all structural and non-structural measures of high economic, social, or environmental cost are designed and taken in order to minimise water restrictions for urban demand.

Monitoring and early warning of potential water quality and quantity is a key component of the plan (Wilhite, 2005). Continuous technological improvements in instrumental monitoring devices (i.e. gauges, piezometers, etc.) play a key role for accounting of resources. Drought indices, adequately calibrated, represent local features of the water resources system of the basin and they can be used as auxiliary tools for drought monitoring and forecasting (Wilhite, 2005). Realistic models appropriate for water management need to be incorporated in monitoring and early warning systems (Rossi *et al.*, 2003). Finally, scientific advances in understanding variations of the climate system offer an opportunity to develop prediction methods.

In the Mediterranean, national governments and the local authorities have responded to extreme drought vigorously, taking emergency measures, but so far the responses have focused on the effects of drought *ex post*, rather than on anticipatory measures *ex ante* (i.e., developing a drought management plan and coherent resource management). In general, these efforts have neglected to build the capacity needed to deal with similar situations in the future. Information on possible longer-term climate forecasts and/or development of plausible scenarios has not yet been incorporated into any specific action plans.

Table 3.3.3 Summary of a planning framework for risk management of water scarcity.

	Preparedness	Pre-alert	Alert	Emergency
Monitoring indicators	Indicators show a normal situation	Indicators show initial stage of danger; no observed impacts (meteorological drought)	Drought is occurring and impacts will occur if measures are not taken (meteorological and hydrological drought)	Drought is persistent and impacts have occurred; water supply is not guaranteed (socio-economic drought)
Objective of the plan in each stage	To ensure that a preparedness and early warning plan is in place	To ensure acceptance of measures to be taken in case of alarm or emergency by raising awareness of the danger of drought	To overcome the drought situation and to guarantee water supply while emergency measures can be put in place	To minimise damage, the priority is drinking water
Measures	Development of a management plan and strategy for revision and review Implementation of a monitoring and early warning system Integration with development and land use policies	Low cost, indirect, voluntary Non-structural directed to influence water demand and avoid worse situations Focus on communication and awareness Intensification of monitoring and evaluation of worse case scenarios	Low cost, direct, coercive, direct impact on consumption costs Non-structural directed to specific water-use groups Water restrictions for uses that do not affect drinking water Changes in management Revision of tariffs Rights Exchanging Centres	High cost, direct, restrictive, approved as general interest actions Structural, new infrastructure, intra-basin, inter-basin and transboundary transfers Non-structural, such as permission for new groundwater abstraction points Water restrictions for all users, including urban demand

The drought management is implemented by theauthorities responsible for water management, in general at the basin level. These planning bodies develop and use management plans that incorporate: demand and supply analysis and projections, contingency and preparedness plans, and scenario analysis for drought. In an optimal situation, permanent monitoring provides indicators that can trigger specific drought management actions. These indicators ideally include: hydrological, socio-economic and environmental aspects. In general, drought management plans include measures that respond to different drought severity levels.

3.3.5 Looking into the future

3.3.5.1 Climate change scenarios

Climate change projections for the region derived from a global climate model driven by socio-economic scenarios (IPCC, 2014) result in an increase in temperature (1.5 to 3.6 °C by the 2050s) and precipitation decrease in most of the territory (about 10 to 20% decrease, depending on the season, by the 2050s). Climate change projections also indicate an increased likelihood of droughts (IPCC, 2014) and variability of precipitation – in time, space, and intensity – that would directly influence the availability of water resources. The combination of long-term change (e.g. warmer average temperatures) and greater extremes (e.g. droughts) can have decisive impacts on water scarcity throughout the region. These new IPCC findings confirm and reinforce previous estimates: under all climate change scenarios, water availability in the Mediterranean region will decrease while irrigation demand increases (Iglesias *et al.*, 2007).

Climate change also affects sea level rise. The effects of sea level rise in North Africa, especially on the coast of the Delta region of Egypt, would impose additional constraints to the use of resources (IPCC, 2014).

The management of the decreasing water resources, as a result of the climatic changes within the Mediterranean region, is a particular challenge, as climate change coincides with high development pressures, increasing populations and high agricultural demands. Evidence for limited capacity to cope with socio-economic and agricultural demands in the Mediterranean region can be documented in recent history. For example, water reserves were not able to cope with extensive droughts in the late 1990s in the Mediterranean, Morocco and Tunisia, causing many irrigation-dependent agricultural systems to cease production. Effective measures to cope with long-term drought and water scarcity are limited and difficult to implement due to the variety of stakeholders involved and the lack of adequate means to negotiate new policies.

The Mediterranean is considered to be a region that will experience large changes in climate mean and variability; that is a climate change "hot-spot" (Giorgi, 2006). Our current understanding of Mediterranean climate leads to projected overall temperature increase from 2 to 4 °C and precipitation changes of 10 to −50% by the 2080s (Alpert *et al.*, 2011). The changes are not equally distributed across the regions or the seasons. The changes are likely to be more pronounced in North Africa, with a temperature increase that reaches +5 °C by the 2080s in some scenarios and an alarming increase in extreme temperature (hot and very hot days); drought periods may increase throughout the Mediterranean (Giorgi and Lionello, 2008; Christensen and Christensen, 2007).

Protecting the world's freshwater resources requires the diagnosis of threats over a broad range of scales (Vorosmarty *et al.*, 2010) and sectors (Table 3.3.2). In the Mediterranean, climate change impacts on water will have a large impact on human water security and biodiversity (Vorosmarty *et al.*, 2010). There are several hundred studies on the potential impacts of climate change on water resources in the Mediterranean which apply many different approaches (European Environment Agency, 2009). According to Gleick and Palaniappan (2010), more and more watersheds appear to have passed the point of "peak water", a concept related to the sustainability of water management. These studies have different focuses – from ecosystems to water pricing to recreational water–, a wide range of time-frames, different scenarios and spatial scales that vary from the local to the global analysis. Although the results are diverse and sometimes contradictory, a common element is that one of the primary impacts of climate change will be a reduction of water availability in the Mediterranean (European Environment Agency, 2007, 2009).

3.3.5.2 Higher drought risk

Climate change will increase drought vulnerability. Burke, Brown and Christidis, (2006) found that at present climate conditions, on average, 20% of the land surface is in drought at any given time while the proportion of land surface in extreme drought is predicted to increase from 1% at present to 30% by the end of the twenty-first century.

Climate change results in intensification of drought and water scarcity in the Mediterranean region which may become increasingly unstable and vulnerable. Drought management in both regulated and unregulated systems will have to adapt to the slow evolution of climate. The human dimension of climate change in the Mediterranean may not stop at the boundaries of a country, since there is the potential for more pronounced water conflicts with neighbouring regions (i.e. transboundary issues in the Nile and in many shared aquifers).

As a result, evapotranspiration rates will increase and soil structure changes will result in increased rates of soil erosion. Climate change may also produce some positive changes in water resources in some areas, giving an adequate adaptive management. The changes may results in risks and opportunities for the water system and the environmental and social systems that depend on water.

These projections may result in reductions of average annual run-off by up to 50%, challenging the whole socioeconomic model which is based largely on water demanding activities: recreation, tourism, and food production. The solution to those problems will imply social changes, a progressive increase in water demand management and a consensus reallocation of water availability to essential users. The agreement on essential uses remains a controversial issue across the region. In this process, policies regulating water usage, water accessibility and hydraulic infrastructure, will play a critical role in making water available to users by overcoming the spatial and temporal irregularities of natural regimes.

A number of studies have shown that under climate change annual river flow is expected to decrease in Southern Europe and increase in Northern Europe; changes are also expected in the seasonality of river flows with considerable differences over the European region (Arnell, 2004; Milly, Dunne and Vecchia, 2005; Alcamo, Floerke and Maerker, 2007). Nevertheless, many of these projections do not take into account the effects of policy. One alternative measure that has been used to include some policy aspects is the water exploitation index (WEI), which is calculated annually as the ratio of total freshwater abstraction to the total renewable resource (Raskin *et al.*, 1997). But even though the WEI can provide additional information regarding run-off, such an analysis still struggles to fully reflect the level of available water resources.

3.3.5.3 Changes in water availability

Water resources management in the Mediterranean is in a time of transition. The ability to project future water availability depends on understanding the formal relationships between the determinants of change and the responses to change. This is complex. Here we evaluate the main categories of factors that influence water availability (i.e., irrigation, economic change and changes resulting from concerns about environmental protection) and the potential responses.

The Water Availability and Policy Assessment model (Garrote *et al.*, 2014) links water supply, demand and management and is used to analyse policy options. The model computes water availability and reliability as result of implementing climate or policy scenarios. WAPA is used to compute water availability and demand-reliability curves, which provide a simple way to evaluate water availability under different policy and climate change scenarios. The model has been applied to evaluate economic decisions of drought policy and water policy in the Mediterranean (Garrote *et al.*, 2014), Figure 3.3.4.

3.3.5.4 Climate change as an opportunity to revise drought management

Water scarcity and drought have multidimensional implications for society and, therefore, no single management action, legislation or policy can respond to all aspects and demand objectives. The Mediterranean stands among the leading countries in the development of legislation with different perspectives and levels of integration into the overall water management policy. The fact that water is managed at the basin level with clearly defined institutional responsibilities ensures that the implementation of drought legislation is potentially effective and that public participation in water management increases rapidly. The basin planning bodies develop and use drought management plans that define a protocol for water allocation under scarcity conditions. Drought management plans incorporate: demand and supply analysis and projections, contingency and preparedness plans, and scenario analysis for drought. In an optimal situation, permanent monitoring provides indicators that can trigger specific drought management actions. These indicators ideally include: hydrological, socio-economic and environmental aspects. Recent drought management plans include a set of measures grouped according to different severity levels. A commonly used ranking describes three levels of severity (i.e. pre-alert, alert and emergency). It is extremely important to also define the "normal" situation, since the plan is optimally developed at this stage. There are many examples that validate this framework over the past decades in the Mediterranean (Garrote *et al.*, 2014), especially in the pre-alert and alert levels. In the

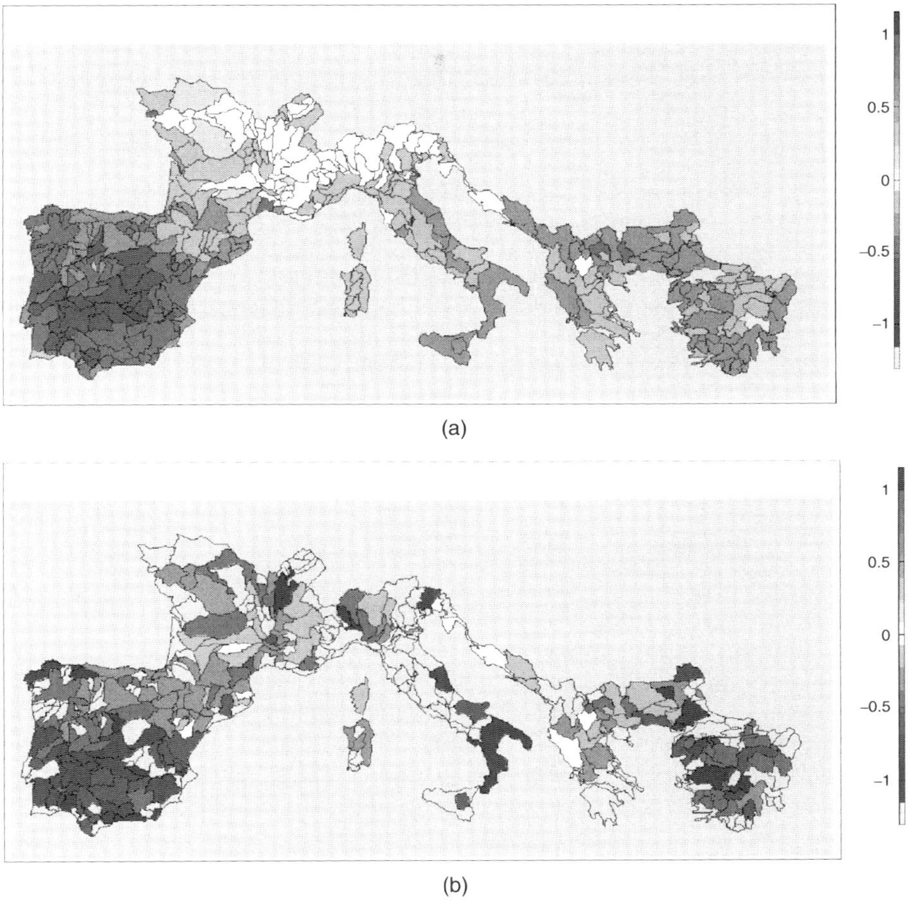

Figure 3.3.4 Per unit changes of run-off (a) and water availability for irrigation (b) in the climate change scenario (2070–2100) with respect to a control run (1960–1990) for the DMI model in Mediterranean European basins. See plate section for colour representation of this figure.

emergency level, the main priority is to maintain drinking water supply, and all structural and non-structural measures of high economic, social, or environmental cost are designed and taken in order to avoid water restrictions for urban demand.

In spite of the many positive aspects of recent drought management plans, these need to continue evolving and expanding to include some key climate change considerations, ecological issues and overall long-term issues related to sustainable development. Information on possible longer-term climate forecasts and/or development of plausible scenarios has not yet been incorporated into any specific action plans. Furthermore, neither the current legislation nor the envisioned implementation of the European Union's Water Framework Directive, provides explicit regulations about how to calculate the ecological discharge during drought situations. This important question is being left to the discretion and responsibility of the various basin authorities. It is surprising that international initiatives such as the United Nations Convention to Combat Desertification (UNCCD, 2000) that provides the global framework for implementing drought mitigation strategies, or the United Nations International Strategy for Disaster Reduction (UNISDR, 2002) that establishes a protocol for drought risk analysis, are not taken into account by local drought management plans.

3.3.5.5 Conservation-oriented policies

The evolution of water use for agriculture reflects changes in the societal values and is guided by the economic and political drivers. The political pressures arising from water scarcity and social change resulted in the transition of the Spanish water management regime towards conservation-oriented policies. The transition was initially resisted by irrigation farmers, who were beneficiaries of the status quo, but these farmers have been gradually adapting to efficient management. In spite of the profound transformation of the sector, most people outside the agricultural population perceive the dedication of a substantial amount of water supplies to irrigation as inequitable and inefficient. This is especially true in areas that compete with urban or environmental water. A recent survey in Castilla-Leon (García de Jalon *et al.*, 2013) shows that urban dwellers consider maintaining the population of the rural areas a top priority. The views of Andalusians are not shared in regions where the farm sector is less important, or their territories less vulnerable to depopulation and aging.

The ability of societies to anticipate and face an external shock is often reflected in their vulnerability. When the external shock is climate change, this vulnerability is estimated by environmental, social and economic factors. At the same time these factors are essential components of a country's development status and of the sustainability of its socio-economic model. In other words, actions that reduce vulnerability and development are closely linked processes that feed and rely on each other. In the case of water the synergies between the two are particularly noticeable.

The reuse of urban water may be included in a group of policies that will need to become increasingly important, since future scenarios project higher population and per-capita water requirement. Other demand-side policies could make use of appropriate water-pricing mechanisms, investments in technology to improve efficiency, upgraded distribution networks and ensuring that agricultural subsidies are linked to efficient use (European Environment Agency, 2009). Efficiency policies may play a major role for improving management. For example, reduction of per-capita or per-hectare water use that always results in an increase of water availability and reliability.

A number of policies may be implemented to overcome temporary water deficits. Water rights exchange programmes may be implemented to overcome temporary deficits and to increase system performance. Proactive drought management measures to increase drought resilience may include improved performance for irrigation demands. Policies that foster communication and education are also needed, since it has been shown that joint participative knowledge is an important factor in facilitating efficient water management (Huntjens Pahl-Wostl and Grin, 2010).

Finally, policies may seek to increase water supply by effectively increasing the regulation volume available for water conservation or a densification of the water

Table 3.3.4 Proposal of potential policy interventions based on water scarcity and vulnerability.

Water scarcity levels	Weakest component of the vulnerability	Policy Recommendations
Low to medium	Social and economic factors	Promote pro-poor management Promote health and education Improve access to water for production and sanitation
	Technological eco-efficiency & natural capital	Focus on environmental mitigation Promote more efficient technologies Reduce water use
	Climate	Promote flexible water storage options Invest in physical infrastructure Develop safety net programmes
Serious to very serious	Social and economic factors	Develop micro-irrigation technologies Integrate ground and surface water management
	Technological eco-efficiency & natural capital	Reform conflict prevention institutions. Promote information sharing and cooperative management
	Climate	Promote policies that help create a paradigm shift Design adaptation policies Develop new alternative sectors

distribution networks. Among other measures this may include water recycling and desalination (European Environment Agency, 2009).

Table 3.3.4 summarizes a proposal of potential policy interventions based on water scarcity and vulnerability. In conditions of low to medium water scarcity, where social and economic drivers of inequality prevail, policies should foment social and economic development policies that will improve policy formulation and implementation in the future, thus minimiaing the risks of climate change impacts. In this case, basic needs such as education and health need to be taken care of. At the same time, in the context of climate change, a longer term strategy that seeks to diversify a country's economy could also help mitigate the effects of climate change impacts.

Where water scarcity remains at low to medium levels and inequality is driven by technological eco-efficiency and natural capital, then environmental mitigation policies are recommended. Some policy options might include the development of more efficient technologies that would reduce water use in key economic activities.

When water scarcity is low or medium and climate drives inequality then current policies should be adjusted for adaptation. Policy options might include fostering greater flexibility in water storage options in order to choose those that will maximise sustainability and effectiveness. Other options could be building or improving physical infrastructure to prevent the risk of extreme events or developing safety net programmes to strengthen livelihood asset availability for vulnerable populations.

If water scarcity levels are serious or very serious and social and economic drivers of inequality prevail, then pro-poor water management policies are recommended. Effective policies might include the development of micro-irrigation technologies, or the promotion of mechanisms that allow an integrated management of groundwater and surface water.

If water scarcity is serious or very serious and technological eco-efficiency or natural capital drive inequality then the most appropriate policies would be those that reform institutions for the resolution of water conflicts. These reforms would imply developing a more holistic approach for the management of shared water resources that ensures that different user needs are taken care of. Increased cooperation and information-sharing at sub-national, national and even regional levels will be required in order to prevent water conflicts from aggravating.

Finally, in conditions of serious to very serious water scarcity where climate drives inequalities then climate-driven policies that allow adaptation and a greater paradigm shift should be put in place. In this case, profound reforms aimed at transforming a country's economic activity should be emphasized. Particularly in agriculture-dependent countries, the strengthening of other economic sectors such as industry or tourism would be in order.

The policies outlined above show that, for the water sector, planned interventions must consider both supply-side and demand-side solutions. On the supply side, adaptation options involve increases in storage capacity or abstraction from water courses; demand-side options, such as increasing the allocated efficiency of water to ensure that economic and social benefit is maximised through use in higher-value sectors, aim to increase value per volume used and to ensure that quality is maintained. All in all it becomes clear that the water sector's importance for numerous other productive and social arenas requires policies and management strategies to be well aware of water's widespread impacts.

3.3.6 Conclusions

The reality of climate change renders it impossible to use the past as an indicator for the future. In this assessment we find reasons to be optimistic, given the important role that science and technology will play in reducing vulnerability and improving water accessibility. There are, however, also reasons to be pessimistic. It remains to be seen whether the current inequalities that exist in the Mediterranean will be successfully redressed, given the high costs associated with technology transfer for less advantaged regions and countries.

Uncertainties in climate change science and long planning horizons add to the complexity of adaptation decision-making. A further important complication is presented by the difficulties in identifying and linking adaptation and development policies in many areas in the Mediterranean where a large proportion of the population does not have access to clean water and sanitation. The uncertainty of the cost and benefits of the various policies suggested is not addressed here; this is a shortcoming of this assessment.

While there is a continuing need to strengthen the climate change knowledge base (through research), improved understanding of climate change science will be insufficient on its own for adaptation policy development and to drive adaptation action. There is a complementary need to engage stakeholders, by developing suitable methodologies for assessment of impacts, vulnerabilities and planning as a pre-requisite for cost-effective adaptation.

Adaptation is unlikely to be facilitated through the introduction of new and separate policies, but rather by the revision of existing policies that currently undermine adaptation and the strengthening of policies that currently promote it. If adaptation is to become "mainstreamed", it will be necessary for relevant policies, such as the CAP and the Water Framework Directive to address the issue more directly.

The Mediterranean region is undergoing rapid socio-economic and technological changes that increase the pressure on its already structural water deficit and question the ability to maintain the current management philosophy. In addition, climate change projections indicate an increased likelihood of droughts. Institutions in the region are evolving to respond to these pressures and to ensure more sustainable management of water resources. There is ongoing progress in many of these countries, which is favoured by the increasing regional cooperation, better monitoring and management systems and, above all, by the awareness of governments. The adoption of emerging technologies for using fresh or unconventional water resources more effectively is crucial for water management. Drought management measures need to be integrated into the long-term strategies for water and land uses and overall development strategies. When water resources are managed at the river basin level, there is an opportunity to respond directly to policy decisions and to the needs and problems of the natural hydrological system. Monitoring and early warning systems continue to improve and are being incorporated into the planning processes. Lastly, strengthened regional cooperation and better understanding of the resource's dynamics and social dimension, and more efficient monitoring systems give hope for alleviating the present pressures on the water resources in the next decades.

Acknowledgements

Funding was provided by the CIRCE project of the 6th FP of the EU and to the DEWFORA and BASE projects from the 7th FP of the EU.

References

Alcamo J., Floerke M. and Maerker M. (2007) Future long-term changes in global water resources driven by socio-economic and climatic changes. *Hydrological Sciences,* **52**(2), 247–275.

Alpert P., Hemmings D., Jin F. *et al.* (2011) The hydrological cycle of the Mediterranean. in *Regional Assessment of Climate Change Impacts in the Mediterranean*. (eds A. Navarra and L. Tubiana) Springer, Dordrecht, The Netherlands.

Arnell N.W. (2004) Climate change and global water resources: SRES emissions and socio-economic scenarios. *Global Environmental Change*, **14**(1), 31–52.

Burke E.J., Brown S.J. and Christidis N. (2006) Modeling the recent evolution of global drought and projections for the twenty-first century with the Hadley Centre Climate Model. *Journal of Hydrometeorology*, **7**, 1125–1113.

Christensen J.H. and Christensen O.B. (2007) A summary of the PRUDENCE model projections of changes in European climate by the end of this century. *Climatic Change*, **81**, 7–30.

Dai A., Trenberthke K.E., Qian T. (2004) A global dataset of Palmer Drought Severity Index for 1870–2002: Relationship with soil moisture and effects of surface warming. *American Meteorological Society*, **5**, 1117–1130.

Dettinger M.D. and Diaz H.F. (2000) Global characteristics of stream flow seasonality and variability. *Journal of Hydrometeorology*, **1**, 289–310.

EM-DAT (2013) E*M-DAT: The OFDA/CRED International Disaster Database* – www.emdat.be, Université Catholique de Louvain, Brussels (Belgium).

Essex S., Kent M. and Newnham R. (2004) Tourism development in Mallorca: is water supply a constraint? *Journal of Sustainable Tourism*, **12**(1), 4–28.

European Environment Agency (2007) *Climate change and water adaptation issues*. EEA Technical Report No. 2/2007, 110 pp.

European Environment Agency (2009) *Water resources across Europe – confronting water scarcity and drought*. EEA Report No 2/2009, available at http://www.eea.europa.eu/publications/water-resources-across-europe.

EEA (2011) *Europe's environment –An Assessment of Assessments*. European Environment Agency, Copenhagen.

FAO (2011) *Aquastat database, Global water information system of the Food and Agriculture Organisation of the United Nations*. http://www.fao.org/nr/water/aquastat/main/index.stm.

FAO (2012) *World Agriculture Towards 2030/2050*. The 2012 Revision. ESA Working Paper No. 12-03. June 2012. Agricultural Development Economics Division. Food and Agriculture Organization of the United Nations. www.fao.org/economic/esa.

García de Jalón S., Iglesias A., Quiroga S. and Bardají I. (2013) Exploring public support for climate change adaptation policies in the Mediterranean region: A case study in Southern Spain. *Environmental Science and Policy*, **29**, 1–11.

Garrote L., Iglesias A., Granados A., Mediero L. and Martin-Carrasco F. (2014) *Quantitative assessment of climate change vulnerability of irrigation demands in Mediterranean Europe. Water Resources Management* (in press).

Giorgi F. (2006) Climate change hot-spots. *Geophysics Research Letters*, **33**, L08707.

Giorgi F. and Lionello P. (2008) Climate change projections for the Mediterranean region. *Global and Planetary Change*, **63**, 90–104.

Gleick P. and Palaniappan M. (2010) Peak water limits to freshwater withdrawal and use. *Proceedings of National Academy of Sciences (PNAS)*: 1004812107v1-8.

Hisdal H., Stahl K., Tallaksen L.M. and Demuth S. (2001) Have streamflow droughts in Europe become more severe or frequent? *International Journal of Climatology*, **21**(3), 317–333.

Huntjens P., Pahl-Wostl C. and Grin J. (2010) Climate change adaptation in European river basins. *Regional Environmental Change*, **10**, 263–284.

Iglesias A., Cancelliere A., Cubillo F., Garrote L. and Wilhite D.A. (eds) (2009) *Coping with Drought Risk in Agriculture and Water Supply Systems: Drought Management and Policy Development in the Mediterranean*. Springer, The Netherlands.

Iglesias A., Garrote L., Diz A. *et al.* (2011) Rethinking water policy priorities in the Mediterranean Region in view of climate change. *Environmental Science and Policy*, **14**, 744–757.

Iglesias A., Garrote L., Flores F. and Moneo M. (2007) Challenges to manage the risk of water scarcity and climate change in the Mediterranean. *Water Resources Management*, **21**, 775–788.

IPCC (2014). *Climate Change 2014: Impacts, Adaptation, and Vulnerability*. Intergovernmental Panel on Climate Change, Geneva.

Milly P.C.D., Dunne K.A. and Vecchia A.V. (2005) Global pattern of trends in streamflow and water availability in a changing climate, *Nature*, **438**(7066), 347–350.

Mitchell, T.D., Hulme, M. and New, M., 2002: *Climate data for political areas*. Area **34**, 109–112.

Raskin P., Gleick P.H., Kirshen P. *et al.* (1997) *Comprehensive assessment of the freshwater resources of the world*. Stockholm Environmental Institute, Sweden. Document prepared for UN Commission for Sustainable Development 5th Session 1997.

Rossi G., Cacelliere A., Pereira L.S. *et al.* (eds) (2003) *Tools for Drought Mitigation in Mediterranean Regions*. Kluver Academic Publishers, The Netherlands. 357 pp.

UNCCD (2000) *United National Convention to Combat Desertification*.

UNISDR (2002) *United Nations International Strategy for Disaster Reduction*.

UNWTO (2006–2013), *United Nations World Tourism Organisation*. World Tourism Barometers (+Statistical Annexes), Madrid, May 2014.

Vorosmarty C.J., McIntyre P.B., Gessner M.O. *et al.* (2010) Global threats to human water security and river biodiversity, *Nature*, **467**(7315) 555–561.

Wilhite D.A. (ed.) (2005) *Drought and Water Crises*. CRC Press, London.

3.4
Prediction of Storm Impacts on Beach and Dune Systems

Paolo Ciavola[1], Oscar Ferreira[2], Ap Van Dongeren[3], Jaap Van Thiel de Vries[4], Clara Armaroli[1] and Mitchell Harley[1]

[1]*Dipartimento di Fisica e Scienze della Terra, Università di Ferrara, Ferrara, Italy*
[2]*FCT, CIMA, Universidade do Algarve, Faro, Portugal*
[3]*Deltares, Delft, The Netherlands*
[4]*Faculty of Civil Engineering and Geosciences, University of Delft, Delft, The Netherlands*

3.4.1 Introduction

A large part of the European coastline consists of sandy beaches and dunes which can undergo dramatic changes during storms. Usually storms are the most important factor controlling short-term shoreline movement with longshore transport gradients dominating the longer-term shoreline position. However, in some cases storm events completely dominate the erosion history of the coastline (Birkemeier, 1979; Thom and Hall, 1991; Russell, 1993; Fenster, Dolan and Morton, 2001; Ciavola *et al.*, 2007; Houser and Hamilton, 2009). Since coastal areas are often densely populated, there is a need to obtain an in-depth knowledge of storm impacts on sandy coastal systems (i.e. dune erosion, overwash and inundation), as these systems ultimately defend the coastline. The study and prediction of storm effects on sandy shores is of great relevance to the comprehensive understanding of coastal dynamics and shoreline evolution, and thus fundamental to carrying out effective coastal planning and management (Ferreira, 2005).

Dune erosion and overwash are fairly complex processes and hitherto the scientific and dune risk assessment problem has been necessarily simplified by making a number of assumptions, including alongshore uniformity of the beach profile and a fully sandy profile without coastal structures or vegetation. These simplifications have led to a number of models of dune erosion and overwash that consider only the cross-shore dimension (e.g. Kriebel and Dean 1985; Vellinga, 1986; Larson and Kraus, 1989; Steetzel, 1993; Larson Wise and Kraus, 2005; Donnelly, Kraus and Larson 2006). Calibration of these models has primarily been undertaken using large-scale laboratory experiments in flumes (see Dette *et al.*, 2002 and, more recently, Gencarelli *et al.*, 2008 or Van Gent *et al.*, 2008). The use of field measurements for model calibration on the other hand still remains limited, thus casting doubts on the reliability of the models'

Hydrometeorological Hazards: Interfacing Science and Policy, First Edition. Edited by Philippe Quevauviller.

predictive capability. A temptation lies for example in tuning the model to match a particular measured profile response. This is quite a dangerous procedure as it is undertaken using empirical coefficients (e.g. the K-transport rate coefficient of S-Beach, see Larson and Kraus, 1989; Larson, Kraus and Byrnes, 1990) which have limited validity having often being developed for laboratory experiments under controlled wave conditions. Another limitation is the fact that the traditional profile-evolution models only consider the dominance of cross-shore processes. This may be a valid assumption for intermediate to dissipative beaches with low-angle nearshore slopes and nearshore bars controlling breaking processes. However, for steeper beaches with wave breaking closer to the shore, the assumption of normally-incident waves is not correct. Additionally, the dune crest may be alongshore irregular in both height and shape, which can result in preferential overtopping and/or overwash hot spots.

This chapter provides an overview of the current knowledge of storm impacts on low-lying coastlines, including beaches and dunes. It first outlines the complexity of identifying a storm event in an objective manner. The impact of storms on dune ridges is then discussed, making a comparison between research recently carried out in both the USA and Europe. Here, considerations are presented on the gaps in current knowledge and which aspects deserve further investigation. The main impacts of storm events are then reviewed, taking into account the development of overwash processes and breaching of the dune system as well as flooding of the hinterland. Thereafter, a detailed overview of qualitative and quantitative models of dune erosion is presented considering the most advanced numerical techniques. The last part of the chapter is dedicated to methods for hazard and risk evaluation, taking into account the emerging role that early warning systems are playing in this field.

3.4.2 Coastal storm definitions

An adequate definition of coastal storms is required in order to isolate individual storm impacts from daily and long-term coastal change, as well as to investigate the spatial and temporal variability in coastal storm frequency, severity and other relevant storm parameters. In a broad sense, a coastal storm can be considered as an anomalous set of meteorological conditions that has the potential to cause damage to the coastal zone and surrounding hinterland. One of the most devastating types of coastal storms is that of a tropical cyclone striking the coast (Stone and Muller, 2005). As demonstrated by Hurricane Katrina and Superstorm Sandy in the USA, the combination of storm surge, large waves and extreme winds generated from tropical cyclones can result in the loss of many lives and significant damage to coastal property and infrastructure (e.g. Fritz *et al.*, 2007; Irish *et al.*, 2013).

Establishing a quantitative and all-encompassing definition of coastal storms is, however, complex for a number of reasons. First, the coastal zone is highly variable and different coastlines respond to the same offshore forcing conditions in vastly different ways. Low-lying coastlines for instance are much more vulnerable to extreme water levels than coastlines backed by large cliffs. Likewise, coastlines that are usually exposed to low wave energy conditions are likely to be more susceptible to wave attack than high energy, dissipative coastlines (Wright and Short, 1984). This introduces the concept of the equilibrium forcing conditions of a beach system, which are the combination of waves and water levels that result in negligible beach change (Wright, Short and Green, 1985; Yates, Guza and O'Reilly, 2009). The further the forcing conditions during a storm event are from the equilibrium conditions, the more beach change (and damage) is likely to occur.

The second reason why coastal storms are difficult to define in a quantitative manner is related to the timing of storms and the ability of sandy beach systems to recover following storm damage. Rather than occurring as one-off events, coastal storms often arrive in clusters (e.g. Lee, Nicholls and Birkemeier, 1998; Ferreira, 2005; Callaghan *et al.*, 2008; Vousdoukas *et al.*, 2012; Coco *et al.*, 2014). This can have the effect of accentuating storm damage, not only by allowing insufficient time in between storms for beach recovery processes to occur, but by successive storms 'exploiting' weak points, such as rip channels formed during the initial storm event. Storm clustering can mean that a series of relatively minor storm events occurring in succession may be just as, or even more significant than, a single more-extreme event (Ferreira, 2005). Recently Coco *et al.* (2014) concluded, on the basis of frequent high-resolution beach surveys, that it is not possible to scale-up single-storm erosion studies into predictions of cluster-storm erosion. A similar conclusion was reached by Karunarathna *et al.* (2014), who point out how the strengths of single storms in a storm cluster alone are not the reason for dramatic beach change, but the time interval between successive storms and the post-storm recovery speed of the beach play important roles.

Finally, a coastal storm does not necessarily have to be locally generated, but may be the product of meteorological conditions that have occurred some thousands of kilometres away. This is particularly the case for open-coast beaches situated on oceanic basins, where long-period swell waves can travel across the ocean and cause major coastal damage, even when local weather conditions are calm (e.g. Terwindt, Hulsbergen and Kohsiek, 1984). These coastlines are therefore susceptible to subtle changes in storm tracks occurring at long distances from the coast, which can in turn result in large shifts in the direction of swell waves arriving at the beach and subsequent beach erosion (Barnard *et al.*, 2011).

A fundamental consideration for meteorological and coastal engineering/management applications is the severity of a coastal storm event. This storm severity is usually framed in the context of its return period, for example a one in one hundred year coastal storm event. Since coastal storms are intrinsically linked to their potential to cause damage for that particular stretch of coastline, a storm of a certain return period should ideally also be indicative of an equivalent damage potential (Corbella and Stretch, 2012). However, for the reasons outlined above, this is not necessarily the case. The following presents a summary of the meteorological and coastal engineering approaches to assess coastal storm severity in terms of the offshore forcing conditions.

3.4.2.1 Meteorological approaches to assessing coastal storm severity

Coastal storms are predominantly driven by atmospheric pressure gradients, which cause winds to blow over the ocean or sea to generate waves or raise water levels locally by means of storm surge. One of the oldest and most basic measures of storm severity is the Beaufort scale, which ranks sea states from a number (known as a

Beaufort number) from 0 to 12 based on visual observations and/or wind speeds. Winds on this scale are deemed of storm strength when they have a Beaufort number of at least 10, equivalent to wind speeds in excess of 55 mph (89 km h^{-1}). The Beaufort scale, however, does not consider whether the winds are associated with a certain type of synoptic system, such as a cold front or tropical cyclone. Nor does this scale take into account the duration and fetch length of the wind, making its applicability to coastal studies relatively limited.

Two of the most important synoptic systems related to coastal storms are those of tropical and extra-tropical cyclones (Stone and Muller, 2005). Tropical cyclones are intense low-pressure systems that have a warm core and are formed predominantly in waters above 26 °C and between latitudes 5° and 15 °N and S. In the north Atlantic and northeastern Pacific they are referred to as hurricanes and in the northwestern Pacific as typhoons. Tropical cyclones generally follow a parabolic track towards the coastline, deflecting to the right in the Northern Hemisphere and to the left in the Southern Hemisphere (May *et al.*, 2013). They usually die out upon reaching land or when entering colder water at higher latitudes.

Depending on the region, a number of different systems exist to rank the severity of tropical cyclones. The most well-known of these systems is the Saffir–Simpson Hurricane Wind Scale, which is a five-category scale applied to tropical cyclones in the north Atlantic and northeastern Pacific. Tropical cyclones in this system are ranked according to their one-minute maximum sustained wind speed. For example, Hurricane Katrina that devastated the area of New Orleans in 2005 reached the maximum of a Category 5 hurricane according to this scale (one-minute maximum sustained wind speed >157 mph/252 km h^{-1}) over the Gulf of Mexico before weakening to a Category 3 cyclone (one-minute maximum sustained wind speed = 111–129 mph/178–208 km h^{-1}) when it made landfall close to New Orleans (Fritz *et al.*, 2007). Other tropical cyclone system scales include the Australian and the Japanese tropical cyclone intensity scales, which are instead based on ten-minute average sustained winds and use slightly different terminology.

In contrast to tropical cyclones, extra-tropical or mid-latitude cyclones reflect a broad class of cold-core cyclones that draw their energy from horizontal temperature gradients in the atmosphere, such as those that occur at cold and warm fronts. They form predominantly in the middle latitudes between 30° and 60 °N and S and, unless interrupted by other synoptic systems, tend to follow a west–east path around the globe. The strongest extra-tropical cyclones usually occur during the winter months when the atmospheric temperature differences are at their greatest (May *et al.*, 2013). Some examples of extra-tropical cyclones that have been observed to cause major coastal damage around the world are nor'easters on the eastern coast of North America (e.g. Dolan and Davis, 1992), east coast lows on the east coast of Australia (e.g. Splinter and Palmsten, 2012) and windstorms in Europe (e.g. Cyclone Xynthia in France).

The more expansive and generally slower-moving nature of extra-tropical cyclones in comparison to tropical cyclones, however, means that a sustained wind speed-based severity scale is often not appropriate for these types of storms (Dolan and Davis, 1992). Storm severity for extra-tropical cyclones has instead been typically assessed based on the historical record of wave and water-levels resulting from these storms. These engineering-type approaches are described in more detail below.

3.4.2.2 Engineering approaches to assessing coastal storm severity

Engineering approaches to assess coastal storm severity at a particular site typically rely on data analysis techniques from local measuring stations, for example wave buoys and/or tidal gauges. One of the most common techniques of defining individual storm events from continuous wave data is the peak-over-threshold (POT) method (Mendez *et al.*, 2006). This method involves defining a certain wave height threshold and minimum storm duration in the dataset from which storm events can be considered. Once these thresholds have been set, individual storm characteristics, such as the peak and duration of an event, can then be determined.

Although there is no clear methodology to choosing a storm threshold for the POT method, it is clearly related to the modal wave conditions for a particular site. For example for the reasonably energetic coastlines of Australia and South Africa, Harley *et al.* (2010) and Corbella and Stretch (2012) adopted significant wave height thresholds of 3.0 and 3.5 m, respectively. On the other hand, for the less-energetic Adriatic and Mediterranean Sea coastlines, Armaroli *et al.* (2012) and Mendoza, Jimenez and Mateo (2011) adopted thresholds of 1.5 and 2.0 m, respectively. To avoid short spikes in wave height being considered a storm event, Mendoza, Jimenez and Mateo (2011) imposed a minimum duration of six hours for storm classification. Likewise, to prevent lull periods of wave heights below the storm threshold during the same event being considered as two separate events, a minimum time period between storm events in the range of six hours (Armaroli *et al.*, 2012) and up to two weeks (Corbella and Stretch, 2012) have been set.

Following storm classification using the POT method, Dolan and Davis (1992) proposed a storm severity criteria based on the total energy content E of a storm event. This is given by the equation:

$$E = \int_{t1}^{t2} H_{\text{sig}}^2 \, \mathrm{d}t \qquad H_{\text{sig}} > H_{\text{thresh}} \tag{3.4.1}$$

where H_{sig} is the significant wave height, H_{thresh} the storm threshold value and t_1 and t_2 are the start and end times of the storm event. This calculation therefore takes into account both the intensity of the event in terms of its wave height and the event's duration. Using a clustering analysis technique based on this equation, Dolan and Davis (1992) and Mendoza, Jimenez and Mateo (2011) established five storm severity classes for Atlantic nor'easters and Mediterranean storms, respectively. The selection of five storm classes was chosen in each case to reflect the well-known Saffir–Simpson Hurricane Wind Scale discussed above. Harley *et al.* (2010) on the other hand used Equation (3.4.1) to rank the top 10 coastal storm events in Sydney Australia over the last 45 years, using ERA-40 wave re-analysis data.

3.4.3 The storm impact scale

The nature of coastal storm response is often based on the storm impact scale proposed by Sallenger (2000), which distinguishes between the swash, collision, overwash

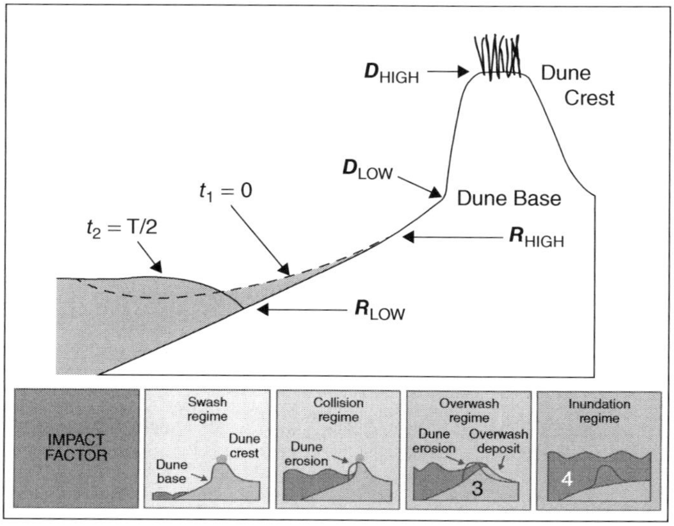

Figure 3.4.1 Sketch describing variables used to scale the impact of storms on barrier islands *Source:* Adapted from Sallenger, 2000. See plate section for colour representation of this figure.

and inundation regimes. Delineation between each regime is based on the elevation of R_{HIGH} (run-up level) and R_{LOW} (run-down level) relative to the elevation of critical geomorphic features on barrier islands. Critical features along barrier islands are the dune base and dune crest, defined as D_{LOW} and D_{HIGH}, respectively. By considering how R_{HIGH} and R_{LOW} vary relatively to D_{HIGH} and D_{LOW} (Figure 3.4.1), the storm impact regimes can be defined. Figure 3.4.2 demonstrates the different forms of dune impacts according to this storm impact scale, described in more detail below.

3.4.3.1 Swash regime

The swash regime occurs when the maximum run-up is confined to the beach face, and does not reach the dune base ($R_{\text{HIGH}}<D_{\text{LOW}}$). On dissipative beaches swash motions are to a large extent a result of wave-group forcing of infragravity waves (Tucker, 1954) or simply of wave-by-wave interactions with the beach face after wave breaking. Depending on the beach configuration and directional properties of the incident wave spectrum, both leaky and trapped infragravity waves contribute to the swash spectrum (Huntley, Guza and Thornton, 1981). Raubenheimer and Guza (1996) show that even when the incident band swash is saturated, infragravity swash is not, meaning that infragravity swash is dominant during storm conditions. Models to predict swash motion and run-up levels range from empirical formulations (e.g. Holman, 1986; Ruggiero *et al.*, 1996; Ruggiero *et al.*, 2001; Stockdon *et al.*, 2006) through to analytical approaches (Schaeffer, 1994, Erikson, Larson and Hanson, 2005) numerical models in 1D (e.g. List, 1992, Roelvink, 1993) and 2DH (e.g. van Dongeren *et al.*, 2003, Reniers *et al.*, 2004, 2006). 2DH wave group-resolving models are well capable of describing low-frequency motions. However, for such a model to be applied for swash, a robust drying/flooding formulation is required.

Figure 3.4.2 Examples of storm impacts on dune systems according to the Sallenger Scale. (a) Swash regime, (b) collision regime, (c) overwash regime, (d) inundation regime. All examples are from the coast near Ravenna, Italy.

3.4.3.2 Collision regime

As R_{HIGH} increases, run-up will eventually collide with the base of the dune. This collision regime occurs when R_{HIGH}/D_{HIGH} exceeds $D_{LOW}/D_{HIGH.}$ In this regime contact with the dune is made and an erosional scarp can develop. The scarp gradually retreats due to episodic slumping of the dune face after it has been undercut by impacting waves (van Gent *et al.*, 2008). The dune scarp erosion process supplies sand to the swash and surf zone, which is subsequently transported seawards by the (infragravity) backwash motions and undertow. The net offshore-directed flows combined with high sediment suspensions result in a strong offshore transport capacity in front of the dune that keeps the erosion process going. Seaward eroded sediment from the dunes settles and creates a new foreshore, which is more efficient in dissipating the incoming waves. As a result the wave loads on the dune face decrease and the dune erosion process slows down as a storm progresses.

3.4.3.3 Overwash regime

As R_{HIGH} continues to increase, overwash of the dune crest occurs (i.e. when $R_{HIGH}>D_{HIGH}$). Overwash is, therefore, a natural storm-related process that occurs when wave run-up exceeds the dune crest. Overwash occurs throughout a singular combination of oceanographic and foreshore geomorphologic factors. Factors controlling the frequency and intensity of overwash, and the resulting morphologies have been summarised by Matias *et al.* (2010) and include marine conditions (e.g. Fisher, Leatherman and Perry, 1974), the orientation of a coast relative to the storm (e.g. Fletcher *et al.*, 1995), nearshore bathymetry (e.g. Ritchie and Penland, 1988), beach topography (e.g. Leatherman, 1976), backbeach elevations (e.g. Morton and Sallenger, 2003), dune morphology (e.g. Donnelly and Sallenger, 2007), engineering structures (e.g. Hayden and Dolan, 1977), location and orientation of footpaths and roads (e.g. Nordstrom and Jackson, 1995), and buildings on the shorefront (e.g. Hall *et al.*, 1990). Intensive overwash acting in association with high sea levels (e.g. storm surge) can lead to dune/barrier breaching and therefore create the potential for large-scale inundation of low-lying coastal areas.

3.4.3.4 Inundation regime

If R_{LOW} exceeds D_{HIGH} then the storm-induced sea level is sufficient to completely submerge a barrier island. In this case the flow over the barrier is no longer a simple overwash, but a complete inundation of the coast (Sallenger, 2000).

3.4.4 Analytical methods of hazard definition

Overwash and inundation are some of the most dramatic consequences of storms on barrier islands, at pristine or developed sandy coasts with dunes and are thus responsible for a substantial part of storm-induced hazards. The occurrence of overwash and inundation in areas of human occupation may cause: loss of or damage to property; loss of access to property; damage to roads and other infrastructure (including their burial); intrusion of sediment into navigation channels; requirement for the removal of washover deposits from public and private property to regain functionality of the property; loss of protection to the mainland afforded by protective barriers or dunes if they are lowered; changes to the backshore environment; shoreline recession and barrier island migration and increased susceptibility to breaching (USACE, 2004). It may also cause intrusion of salt and sand into agriculture soils, interference with back-barrier channel navigation, and loss of human life (Matias *et al.*, 2009). Because overwash and inundation events are important natural sedimentation–erosion processes that promote significant changes to coastal systems, and since coastlines are globally occupied, there is a demand to predict their occurrence and to define the vulnerability of a coastal area to such events.

In order to predict the overwash hazard during a storm, different return periods are often defined. For each return period, the highest run-up elevation (R_{HIGH}) is calculated, according to the impact classification of Sallenger (2000) described in the previous paragraphs:

$$R_{\mathrm{HIGH}} = R_2 + \eta_{\mathrm{mean}} \tag{3.4.2}$$

where R_2 is the run-up that has exceeded 2% of the recorded events and η_{mean} is the mean sea level (the sum of the astronomical tide and storm surge). R_2 can be easily obtained through the empirical parameterizations given by Stockdon *et al.* (2006) or others (e.g. Holman, 1986; Hughes, 2004).

The elevation values obtained for R_{HIGH} can then be compared to the dune base and dune crest topography. In order to assess the vulnerability to collision, overwash and inundation regimes, maps can be designed setting the topographic layers relative to the dune base (D_{LOW}) and crest (D_{HIGH}) and comparing them to previously calculated R_{HIGH} and R_{LOW} values. If a given point where the dune base/crest topography is lower than the calculated storm-induced run-up, the area is categorised as vulnerable to collision/overwash and if lower than the sea level (surge + tide) becomes prone to inundation.

Setting the correct ocean forcing is crucial when analysing potential areas of overwash and inundation. Wave parameters (significant wave height, wave period and wave length), and astronomical and meteorological tides are key factors driving overwash occurrences. By choosing the best method for calculating the total water level (including run-up and storm surge effects) and matching it to dune elevation, the overwash and inundation potential along dune ridges can be defined. However, storm impacts are not equally identified along a barrier island due to the combined effects of alongshore variable morphology of the beach and dune relative to the intensity of the ocean's forcing. Foreshore factors governing overwash must also be taken into consideration in more complex methodologies and modelling, since alongshore changes on overwash and inundation depend on foreshore morphology and on its dynamic. Therefore, the simple model provided by Sallenger (2000) may face some difficulties in complex areas with tridimensional morphological variability (Houser, 2013).

The impact scale of Sallenger (2000) is used extensively because it is a rapid and simple way of determining the likely effects of storms on dune systems and barrier islands. The required variables are easily computed and the scale of the impact is determined by comparing the dune crest/base and the water elevation during storms. The dune crest/base elevations are the simplest morphological characteristics that can be extracted from *in situ* and/or remotely-sensed beach surveys. However other and more important coastal information should be considered if a more accurate prediction of storm impacts is to be carried out.

An important parameter to consider is the storm duration that was instead taken into account by Kriebel *et al.* (1997) in their Intensity Index. This index includes dynamic information, such as storm surge (tide + wave set-up), breaking wave height and storm duration, but does not include beach morphological characteristics. A previous parameter for potential dune face retreat was defined by Hallermeier and Rhodes (1988) and was used extensively in the US for risk assessment exercises (FEMA, 2003, 2005, 2011). This parameter, called the Median Dune Erosion, relates the cross-sectional area of erosion to the return period of flood elevation. Both the Kriebel *et al.* (1997) and Hallermeier and Rhodes (1988) parameters do not consider beach morphodynamic characteristics, such as beach slope and the consequent run-up elevation.

To quantify the risk of coastal flooding, Gares (1990) computed the maximum water level return periods, as the combination of surge + run-up, and compared the results to the dune height. Two additional indicators, named the *cross-section centroid* and *erosion resistance*, were defined by Judge, Overton and Fisher (2003) using both the distance between the dune mound and the shoreline, that is a descriptor of the beach width in front of a dune system, and the cross-sectional area of the dune. The cross-section centroid is defined using the centre of mass of the dune cross-shore profile (*x*,*y* centroid coordinates with: *x*, cross-shore, *z*, elevation) that is compared to the cross-shore and elevation coordinates of R_{LOW} as defined by Sallenger (2000), thus including the beach slope. The erosion resistance (ER), on the other hand, incorporates in its formulation the cross-sectional area of the dune. In the paper of Judge, Overton and Fisher (2003) series of pre- and post-storm beach surveys were compared to classify dune survival or failure, and each of the above mentioned indicators were tested. The authors found that the ER as well as the cross-section centroid were able to describe adequately the observed effects of the storm on the dune system. They also found that the use of the dune crest elevation only is the least suitable method for storm prediction purposes.

Recently, Armaroli *et al.* (2012) developed an indicator known as the *dune stability factor* (DSF, Figure 3.4.3). This factor includes both the sum of several "dynamic" parameters (run-up, set-up, surge and tide) to compute the maximum water level of a storm (measured or predicted) and the cross-sectional area of the dune. The area between the dune foot and the dune crest is compared to the residual area between the maximum water level (measured or predicted) and the dune crest. The comparison is then used to define three impact levels: safety condition, dune frontal erosion, dune removal (Figure 3.4.3). The main limitation of the indicator is that it has only been applied at one field site and thus may need further testing to check its predictive capability. Furthermore, it does not take into account storm duration. Another recent

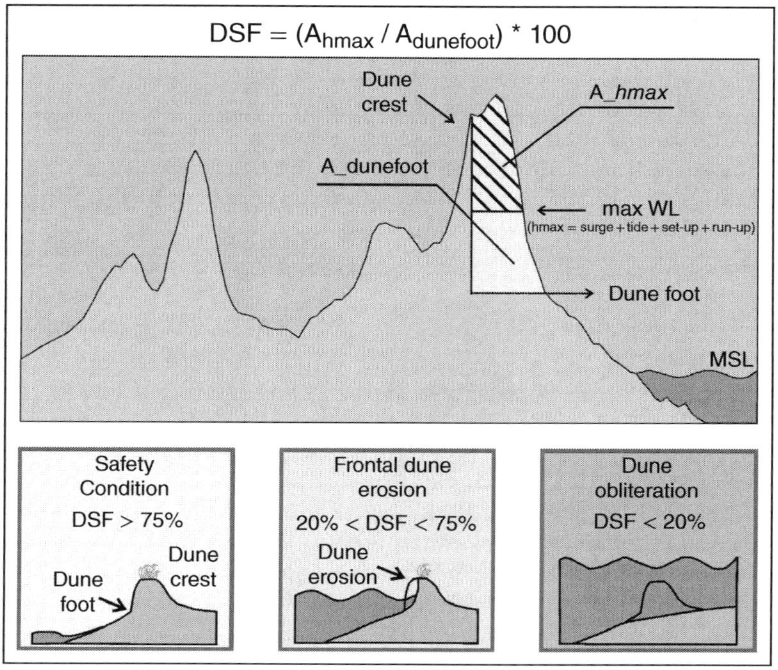

Figure 3.4.3 Sketch describing variables used in scaling the impact of storms on a dune system, the Dune Safety Factor of *Source:* Adapted from Armaroli *et al.* (2012). See plate section for colour representation of this figure.

index is that of Sancho, Oliveira and Freire (2012) who defined an *Overwash Dune Index* using the comparison between the dune crest and the maximum water level of a storm event, and an *Erosion Index* defined by comparing the dune/beach volume after a 1-in-10 year return period storm (beach and dune profile variation modelled with XBeach) and the pre-storm dune volume. The authors state that further testing of these indexes is needed to define their potential.

One of the main problems with the application of the Sallenger (as well as other author's) classification is that accurate estimates of run-up are needed and most formulas are calibrated with only limited field datasets. Presently, only the Holman (1986) and Stockdon *et al.* (2006) formulas rely on a substantial dataset from field observations, albeit entirely based on US coastlines. The other problem is that run-up formulas are based predominantly on beach slope and do not consider beach morphology, such as the presence of a well-developed and wide berm on the upper beach. Care must be taken when considering the run-up estimation in these cases, since these formulas do not consider processes such as infiltration and shear stresses that help to reduce run-up velocities across the berm. This could ultimately prevent the waves from reaching the dune foot of very wide berms.

Another critical point that deserves further attention in conceptual and numerical modelling is the role of vegetation in controlling frontal dune erosion. This role is twofold. First, the vegetation will decrease run-up velocities due to increased friction over plant surfaces. This effect could be modelled numerically simply as an increase in roughness using a high Manning coefficient. However, the roughness will change spatially due to variation in the density and characteristics of plants. Secondly, roots will completely control the slope of the foredune. It is argued (Rosati *et al.*, 1993) that models like S-beach are not sensitive to variations in the values of maximum slope prior to avalanching (by default set to 45°). Holding the sediment together by adhesion to the roots, an unnaturally steep profile may be reached, especially under wet conditions. Drying of the sediment or removal of material at this "cliffed" foredune may cause slumping even under low energy conditions, as observed during a three-year field monitoring of a dune system in the northern Adriatic by Armaroli *et al.* (2013).

3.4.5 Modelling of storm impacts

In an exhaustive review of models which can predict general beach profile changes, Roelvink and Brøker (1993) identify three different types of models, which are relevant for the prediction of beach–dune interaction:

- Descriptive conceptual models (e.g. Sallenger, 2000);
- Models based on the equilibrium profile theory (Bruun, 1962)
- Process based models (see Roelvink and Brøker, 1993).

Descriptive conceptual models (i.e. the storm impact scales of Sallenger discussed in the previous section) describe and classify the response during a storm based on our physical understanding of this response and are not discussed here further. Instead the focus will be on quantitative models, which are classified as empirically-derived equilibrium models and process-based models.

3.4.5.1 Models based on the equilibrium profile theory

The philosophy of equilibrium profile models for storm impacts is that the coastal profile during a storm develops towards a new equilibrium shape that better fits the hydraulic storm conditions. The response to a storm is considered as a cross-shore (1D) process and sediment exchange in the alongshore direction is usually not accounted for. Equilibrium models have usually been developed from measured profile response in the laboratory and field, without necessarily analysing and understanding the physical processes causing this response.

Most of the equilibrium models (Edelman, 1968, 1972; Dean, 1977; Van de Graaff, 1977; Vellinga, 1986; Van Gent *et al.*, 2008) are time independent and predict the post-storm profile independently from the pre-storm profile, where the post-storm profile shape depends on wave properties and sediment properties. The applicability of these models is restricted to dune erosion only. More advanced storm impact models based on equilibrium theory use the initial profile as a starting point and simulate the temporal evolution of the pre-storm profile towards a post-storm equilibrium configuration (Kriebel, 1982; Kriebel and Dean, 1993; Madsen and Plant, 2001). Below we discuss some examples of both model variants.

The dune erosion model DUROS described in Vellinga (1986) is based on the concepts of Edelman (1968, 1972) and Van de Graaff (1997) and extended by van Gent *et al.* (2008). The model has been routinely used in the safety assessment of the Dutch dune coast and predicts the post-storm coastal profile independently from the pre-storm profile. The post-storm profile shape is described between the dune foot (set equal to the maximum storm surge level) and a transition point to the pre-storm profile and is a function of the wave height, wave period and grain diameter (see Van Gent *et al.*, 2008 for more details). The post-storm profile shape has been derived from large-scale dune erosion experiments in flumes and is not necessarily an equilibrium shape but rather a shape that corresponds to representative storm duration for the North Sea (see Vellinga, 1986 for more details). The amount of dune erosion and dune crest retreat results from a horizontal cross-shore shift of the post-storm coastal profile such that continuity of sediment is guaranteed (the erosion volume above maximum storm surge level should balance the deposition below it).

One of the most commonly used examples of a model that predicts the morphological evolution towards an equilibrium profile is the Convolution Model, developed by Kriebel and Dean (1993). This model consists of simple analytical solutions to predict the one-dimensional, time-dependent beach and dune profile response to variations in water level and breaking wave height as they occur during a storm. Beach and dune erosion results solely from cross-shore transport caused by breaking of short-period waves. Wave transformation in the nearshore is approximated using the assumptions of straight and parallel bathymetric contours and linear wave theory.

The model assumes that beaches subjected to steady-state erosion forcing conditions respond towards a stable or equilibrium form in an exponential manner, according to the equilibrium profile theory (Kriebel and Dean, 1993). The breaking wave height is assumed to remain constant over the duration of the storm. The main forcing mechanisms that are considered are storm-induced variations in water level and the magnitude of erosion response is determined by two parameters: the maximum potential response (R_∞), that would occur if the beaches were allowed to have enough time to develop around a new equilibrium; the characteristic erosion timescale (T) that governs the exponential rate at which the profile moves towards this new equilibrium.

The occurrence of high water levels due to storm surge phenomena, if introduced into the simulation, can become one of the main agents controlling profile evolution (Dally and Dean, 1984), as proved by several authors using the convolution method (e.g. Ferreira *et al.*, 1995; Ferreira and Dias, 2000).This gives much importance to the predictive capability of the method in the form of the maximum retreat ($R_{\max}$) of the profile, in order to achieve the mass balance necessary to provide a numerical solution to the model. Ferreira and Dias (2000) have used this concept to obtain the maximum potential retreat of the dune crests along the coast of northern Portugal. Ferreira (2005, 2006) and Callaghan *et al.* (2008) used it to evaluate the impact of storms and storm groups on exposed sandy shores, while recent applications for assessing storm impacts in a GIS environment are provided by Almeida, Ferriera and Taborda (2011). Equilibrium models typically fail in complex situations and are usually only applicable to the collision regime.

3.4.5.2 Process-based models

The philosophy of process-based models is that the morphological evolution during a storm can be computed from the interaction between the relevant physical processes. Process-based models are usually far more complex and computationally more intensive than equilibrium-type models. They require that the modeller has a good understanding of the system that is simulated and the physical processes that drive it. When this requirement is fulfilled process-based models can be rather generic (i.e depending on the implemented physics they can simulate the transition from the collision regime towards the overwash regime) and can be applied to complex situations (i.e. with hard structures or alongshore interactions). In a recent application to a sandy beach dune system at Narrabeen, Australia, Callaghan *et al.* (2013) compares equilibrium models with process-based models. They found that unless properly calibrated, process-based models may provide probabilistic storm erosion volume estimates less reliably than more simple equilibrium models. On the other hand, if process-based models are calibrated, performance improves, becoming optimal for event-based simulations of dune erosion (Harley, Armaroli and Ciavola, 2011; Armaroli *et al.*, 2013).

In daily practice, process-based models simulate only part of the physical processes directly. The remaining physics is represented by parameterizations, either because the physical processes are not yet understood completely or because the processes require too much computational effort. Most process-based models run in profile mode and focus on the underwater cross-shore transport. They estimate the erosion from the dry dune by extrapolating near dune sediment transport to the emerged part of the dune. An alternative is proposed by Overton and Fisher (1988), Nishi and Kraus (1996), Larson, Erikson and Hanson (2004), Palmsten and Holman (2011) who estimate the supply rate of sand from the dunes by relating it to the wave impact force. Both approaches rely on heuristic estimates of run-up, but the advantage of the second approach is that the sediment supply from the dunes does not necessarily need to fulfil the seaward transport capacity of near-dune flows (i.e. the beach in front of the dune is allowed to "breathe").

As mentioned above, most process-based models are limited to profile mode (1D or 2DV). However, depending on the complexity of the coastal system to be considered (i.e. the occurrence of overwash with lateral non-uniformity in dune topography, wave and water level forcing, strong coastal curvature, transition of a seawall into a sandy system, etc.) the profile mode approach will be ineffective and an area model (2DH or 3D) is required. Below we discuss examples of profile model variants and a 2DH model.

Profile models DUROSTA (Steetzel, 1993) is a numerical dune erosion model which is also known as UNIBEST-DE. It is a 2DV cross-shore model, solving wave propagation, flow, sediment transport and bathymetry development for time-varying hydraulic conditions. Although the alongshore dimension is not explicitly taken into account by the model, it has the option to deal with wave obliquity, alongshore current gradients and coastal curvature. DUROSTA simulates the cross-shore profile development in time, based on instantaneous sediment transport rates. The basic assumption is that the net local cross-shore sediment transport rate can be computed as the product of local flow velocity with the distribution of sediment concentration.

The model computes the wave transformation, including refraction, using a random wave breaking dissipation expression as proposed by Battjes and Janssen (1978). The cross-shore flow is described as a time-averaged velocity profile below the mean wave trough level and is based on the vertical distribution of the shear stress. The mass transport towards the coast is concentrated in a narrow zone above the mean wave-trough level and is compensated by a return current in the lower part of the water column. The alongshore flow model describes the cross-variation as well as the vertical distribution of the alongshore flow. DUROSTA only includes suspended sediment transport and grain suspension in the vertical results from the turbulence produced by breaking waves. Bed-load transport is neglected, which seems a reasonable assumption for fine and medium sandy beaches, since the majority of the sediment transport takes place as suspended transport. Morphological changes are computed from cross-shore transport gradients and longshore transport gradients, where the latter are mimicked by imposing a coastal curvature. Dune erosion is simulated by extrapolating near dune sediment transports over the dry dune face using a heuristic estimate for the wave run-up.

SBEACH calculates dune and beach erosion produced by storm waves and water levels; bar formation and movement produced by breaking waves are also simulated. The model is empirically based and was originally developed from a large dataset of net cross-shore sand transport rates and beach profile change observed in large tanks. The empirical formulation, model sensitivity tests, and a field validation case are described in Larson and Kraus (1989) and Larson, Kraus and Byrnes (1990) and Larson, Wise and Kraus (2005). SBEACH is a 1DV semi-empirical time-dependent dune erosion model on the basis of time-averaged process descriptions of wave transformation and sediment transports, that is the hydrodynamics are stationary and the morphodynamics are non-stationary. At every morphological time-step the wave transformation is calculated on the basis of the conservation of energy flux. From the wave transformation the sediment transports and the bed level changes are computed. The sediment transport is calculated on the basis of the empirical formulations. The supply of sediment from the dry dune is based on a linear relation between wave impact and the weight of the sand that is eroding (Overton and Fisher, 1988).

2DH models The above models assume alongshore uniform conditions, both in the hydrodynamic forcing and in the coastal response, and have been applied successfully along relatively undisturbed coasts. However, they are not adequate to assess the more complex situation where the coast has significant alongshore variability. This variability may result from anthropogenic causes, such as the presence of artificial inlets, sea walls, and revetments, but also from natural causes, such as the variation in dune height along the coast or the presence of rip channels and shoals on the shoreface. Therefore, the assessment of storm impact in these more complex situations requires a two-dimensional process-based prediction tool.

XBeach (Roelvink *et al.*, 2009) is a model that was developed to simulate the impact of storms and hurricanes on complex sandy barrier islands, with an initial application of hurricane impact on the sandy coast of Florida, USA. The model formulations do not assume longshore uniformity and are capable of modelling all three regimes of collision, overwash and inundation (Sallenger, 2000).

The model includes all hydrodynamic and morphodynamic processes relevant for the nearshore zone and for storm durations from hours to days. In the model the hydrodynamic processes are separated into motions at the timescale of the short waves and motions at longer timescales, such as currents and long waves. The phase of the short wave motion is not simulated but rather the model solves the wave action equation for the variation of the short-waves envelope (wave height) on the scale of wave groups (Phillips, 1977). It employs a dissipation model for use with wave groups (Roelvink, 1993; Daly *et al.*, 2012) and a roller model (Svendsen, 1984; Nairn, Roelvink and Southgate, 1990; Stive and de Vriend, 1994) to represent momentum stored at the surface after breaking. These variations, through radiation stress gradients (Longuet-Higgins and Stewart 1962, 1964) exert a force on the water column and drive longer period waves (infragravity waves) and unsteady currents, which are solved by the nonlinear shallow water equations. Thus, wave-driven currents (longshore current, rip currents and undertow), and wind-driven currents (stationary and uniform) for local wind set-up, long (infragravity) waves, and run-up and run-down of long waves (swash) are included. The hydrodynamics drive sediment transport under wave and flow conditions, including overwash (sheet flow) conditions and avalanching (slumping) of the dune front. On the basis of transport gradients the bathymetric update is computed using a so-called "morphological factor" for computational speed-up. We refer to Roelvink *et al.* (2009) for a full description of the model.

The model is applicable on spatial scales of up to 15 by 15 kilometres, including the wave shoaling zone, barrier islands and the back-barrier lagoon system (if present). It allows modelling of "hard" structures such as seawalls and buildings. The model is boundary curve fitting (curvilinear) and can be driven with measured or modelled boundary conditions, obtained from larger area models. The model has been validated with a series of analytical, laboratory and field test cases (Roelvink *et al.*, 2009; van Thiel de Vries, 2009; Van Dongeren *et al.*, 2009; McCall *et al.*, 2010), and has until now been used in and adapted for many coastal applications, ranging from sandy to gravel beaches (Williams *et al.*, 2011), beaches with coastal structures (Harley and Ciavola, 2013), coral (Van Dongeren *et al.*, 2012) and potentially to mangrove coasts.

3.4.6 Storm impact indicators, early warning systems and disaster risk reduction

The most rapid and dramatic changes in beach morphology occur during storms. Extreme storms are one of the most important agents causing beach erosion and, therefore, responsible for the destruction of coastal properties on urbanised areas. The main storm variables are wave parameters (namely height), storm duration, sea-level and storm surge. When one or more of these variables reaches a critical level (threshold) major changes are expected at coastal systems. Thresholds can be used as indicators to determine coastal erosion or damage associated with important morphological changes, linking wave parameters with beach profile changes or coastal damage.

Indicators for coastal erosion can be determined by establishing significant wave height (H_s) thresholds responsible for important morphological changes (vertical variability) across the beach profile (eg., Almeida *et al.*, 2012, Del Río *et al.*, 2012). The definition of thresholds for coastal damage can be made by the analysis of the hydrodynamic conditions associated with historical storm impacts with relevant consequences (i.e. significant damage to infrastructure, such as houses, roads and walls). For this purpose quantitative and qualitative information is gathered, namely from regional newspapers, unpublished reports, beach surveys and monitoring, and time-series of aerial photographs. This information is used to make a qualitative assessment of storm impacts on the coast in the last decades and to identify the associated major storm events (Almeida *et al.*, 2012; Del Río *et al.*, 2012). Hydrodynamic conditions for each defined extreme event are described using the following parameters: number of storms (from single storm to groups of storms), duration, power, maximum H_s, direction, maximum tide and storm surge levels. Where relevant, storms can be aggregated into storm groups if the maximum H_s of consecutive storms occurs within a short period (hours to days normally). Thresholds for impacts on infrastructures are defined for the main wave directions, storm groups and individual storms by comparison with historical records (decades) of damage and associated wave conditions, namely H_s and storm duration.

In a less empirical approach, the above-mentioned models can be used to estimate the expected level of hazard for each hydrodynamic condition at a given coastal area. The model output (e.g. run-up level, dune erosion volume, berm retreat, dune height reduction, dune breaching, flood level) can be used directly or against thresholds to identify areas prone to a given hazard. To assist in this, a system of storm impact indicators (SIIs) can be created. which help to translate the quantitative model results, in the form of physical parameters, to useful information for civil protection and coastal management agencies using the Frame of Reference approach (Van Koningsveld *et al.*, 2005, 2007).

The Frame of Reference (FoR) method (Figure 3.4.4) aims to structure the end user–specialist interaction, optimizing the application of the developed knowledge. Applying the method increases the probability that specialist research produces results that are applicable in policy development or practical applications. SIIs, used within the FoR, are, therefore, a quantification of the impact of marine events on the coastal system in a form suitable for decision making. Linked to these SIIs are predefined threshold levels of impact, susceptibility, vulnerability or risk that trigger various degrees of action by authorities. A SII integrates a Strategic Objective, an

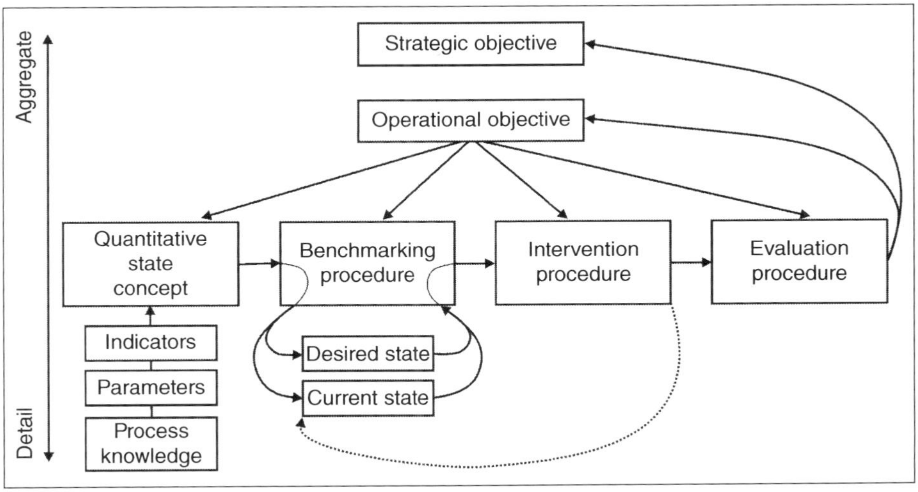

Figure 3.4.4 The Frame of Reference approach of *Source:* Adapted from Van Koningsveld *et al.* (2005, 2007).

Operational Objective, a Quantitative State Concept, a Benchmarking Desired State, a Benchmarking Current State, an Intervention Procedure and an Evaluation Procedure (see Ciavola *et al.*, 2011b and Haerens *et al.*, 2012 for examples). These indicators are intended to be used by competent civil protection and/or coastal management agencies for organising evacuation of people, sending staff to monitor dyke failure at vulnerable locations, locating emergency measures like sand bags or temporary dykes, designating safe areas to which people can move in case of overwash events and so on (Ciavola *et al.*, 2011a). The SIIs are the basis for the early warning systems (EWSs) as the thresholds of these indicators control if, and at what level, a warning should be issued. Examples of management approaches that can be optimized by using SIIs and EWS include dyke and dune safety, protection of beach property, conservation of natural areas or swimmer safety (see Ciavola *et al.*, 2011a, b for a more detailed explanation).

Morton (2002) in a review of the most severe storms that hit the US Atlantic Coast and the Gulf provides an interesting review of the physical, biological and anthropogenic parameters that controlled the extent of observed damage. He concluded his work by advocating an operational approach to the problem, which at the time (i.e. over 10 years ago) was difficult to implement because of computational power as well as availability of numerical tools for the prediction of dune erosion. Nowadays, such an operational approach can be implemented by using EWSs.

An EWS based on indicators should integrate several modules (illustrated in Figure 3.4.5), namely:

- An observation module, including weather, wave, surge measurements;
- A forecast module, including weather, wave, surge and morphological forecasts (XBeach or other modelling tool or analytical approach);
- A decision support module, including SIIs and hazard maps;
- A warning module, including warning at different levels which are site-specific;
- A visualisation module, which can include on-line GIS-based maps.

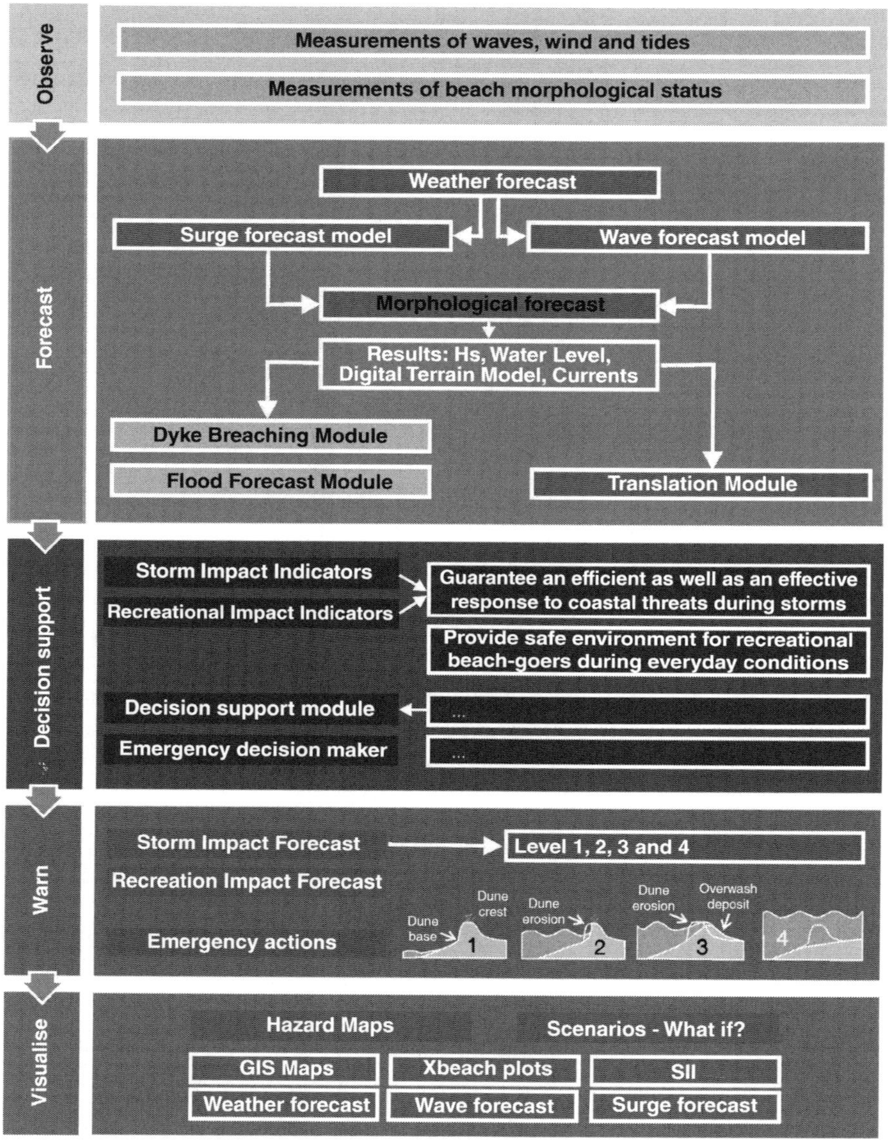

Figure 3.4.5 Methodology suggested during the MICORE Project (www.micore.eu) for the construction of an EWS for the prediction of storm impacts. Further details can be found in Ciavola *et al.* (2011a, b) and Haerens *et al.* (2012).

The successful application of early warnings is one of the most cost-effective, practical and efficient measures for disaster risk reduction and the saving of lives and is regarded as one of the greatest challenges for the future with respect to the interaction between coastal researchers and managers.

SIIs and EWSs will only be effective if followed by actions tending to minimize the impact of storms at coastal areas. In an international context, as an effort to drive the agenda on disaster risk reduction (DRR), the United Nations Office for Disaster Risk Reduction (UNISDR) formulated the disaster reduction goals in the Hyogo Framework for Action (HFA).[1] Concurrently, the EU has issued the Floods Directive[2], focused on the hazard of flooding, which requires Member States to implement flood risk management plans by 2015. In the United States the matter is being dealt by FEMA at a federal level, with specific advice for coastal flooding issues (FEMA, 2003).

By their nature, neither the HFA nor the EU Floods Directive are specific about the methods to be used to assess coastal risks, particularly those risks resulting from multiple, synergistic hazards, such as overtopping, breaching and erosion. On coasts, river flood risk assessment methods (GIS-based flood mapping) will misrepresent the risk because of the non-stationarity of surge and flash flood events. Also, the morphodynamic response of the coast has a strong effect on the flooding of the hinterland and this is where the improved understanding and predictive capability of coastal response models fits in. Finally, effective preparedness measures, specifically coastal EWS, do not as yet play an integral role in the European approach, even though they are encouraged in the EU Floods Directive and have been identified as the most effective method for reducing the risks of loss of life and economic value[3]. Here again, morphodynamic predictive models are essential in order to accurately predict coastal risks. The European Union is still supporting the development and application of coastal morphodynamic EWS in the conclusive part of the FP7-Climate Change research effort, through a contemporary project, RISC-KIT (Van Dongeren *et al.*, 2014) which is currently underway and will contribute to a better definition of DRR plans to coastal areas.

3.4.7 Conclusions

The methods described above (storm impact scale, modelling, storm impact indicators) allow the definition of setback lines and the establishment of hazard and vulnerability maps, which provide an objective identification of areas prone to overwash and inundation. These methods constitute an effective tool for coastal management and planning, since from a coastal management viewpoint overwash and inundation are significant phenomena given the range and extent of the consequences. The methods discussed can be used for both long-term management plans (decadal to century), using different return periods or as a support for EWSs, especially when coupled with SIIs (see examples in Ciavola *et al.*, 2011; Haerens *et al.*, 2012; Vousdoukas, Almeida and Ferreira, 2012). EWSs improve the predictive capability required to prevent consequences from overwash events and inundation in the short-term (hours to days) and should be tied into Disaster Risk Reduction Plans at coastal areas. Recent examples of EWS applications to improve risk reduction are the optimal design of temporary flood protection schemes (Harley and Ciavola, 2013) and operational forecasting of dune erosion and beach width (Harley, Armaroli and Ciavola, 2012).

[1] UNISDR Hyogo Framework for Action 2005-2015: Building the Resilience of Nations and Communities to Disasters.

[2] Directive 2007/60/EC on the assessment and management of flood risks.

[3] UNISDR (2002) Guidelines for reducing flood losses.

Acknowledgements

This chapter is a contribution to the MICORE Project (EU contract 202798) and the RISC-KIT project (EU contract 603458), further details can be found on www.micore.eu and www.risckit.eu.

References

Almeida L.P., Ferreira Ó. and Taborda R. (2011) Geoprocessing tool to model beach erosion due to storms: application to Faro beach (Portugal). *Journal of Coastal Research*, **SI64**, 1830–1834.

Almeida L.P., Vousdoukas M.V., Ferreira Ó. *et al.* (2012) Thresholds for storm impacts on an exposed sandy coastal area in southern Portugal. *Geomorphology*, **143–144**, 3–12.

Armaroli C., Ciavola P., Perini L. *et al.* (2012) Critical storm thresholds for significant morphological changes and damage along the Emilia-Romagna coastline, Italy. *Geomorphology*, **143–144**, 34–51.

Armaroli C., Grottoli E., Harley M.D. and Ciavola P. (2013) Beach morphodynamics and types of foredune erosion generated by storms along the Emilia-Romagna coastline, Italy. *Geomorphology*, **199**, 22–35.

Barnard P.L., Allan J., Hansen J.E. *et al.* (2011) The impact of the 2009-10 El Nino Modoki on U.S. West Coast beaches. *Geophysics Research Letters*, **38**(13).

Battjes J.A and Janssen J.P.F.M. (1978) Energy loss and set-up due to breaking of random waves, *Proceedings of the 16th International Conference on Coastal Engineering*, ASCE, pp. 569–587.

Birkemeier WA. (1979) The effects of the 19 December 1977 coastal storm on beaches in North Carolina and New Jersey. *Shore and Beach*, **47**, 7–15.

Bruun P. (1962) Sea level rise as a cause of shore erosion, *Journal of the Waterways and Harbors Division*, **88**, 117–130.

Callaghan D.P., Nielsen P., Short A., Ranasinghe R. (2008) Statistical simulation of wave climate and extreme beach erosion. *Coastal Engineering*, **55**(5), 375–390.

Callaghan, D. P., Ranasinghe, R. and Roelvink, D. (2013). Probabilistic estimation of storm erosion using analytical, semi-empirical, and process based storm erosion models. *Coastal Engineering*, **82**, 64–75.

Ciavola P., Armaroli C., Chiggiato J. *et al.* (2007) Impact of storms along the coastline of Emilia-Romagna: the morphological signature on the Ravenna coastline (Italy). *Journal of Coastal Research*, **SI50**, 540–544.

Ciavola P., Ferreira O., Haerens P. *et al.* (2011a) Storm impacts along European coastlines. Part 1: The joint effort of the MICORE and ConHaz Projects. *Environmental Science Policy*, **14**(7), 912–923.

Ciavola P., Ferreira O., Haerens P. *et al.* (2011b) Storm impacts along European coastlines. Part 2. Lessons learned from the MICORE Project. *Environmental Science Policy*, **14**(7), 924–933.

Coco G., Senechal N., Rejas A. *et al.* (2014). Beach response to a sequence of extreme storms, *Geomorphology*, **204**, 493–501, doi: 10.1016/j.geomorph.2013.08.028.

Corbella S. and Stretch D.D. (2012) Multivariate return periods of sea storms for coastal erosion risk assessment. *Natural Hazards and Earth System Sciences*, **12**(8), 2699–2708. doi:10.5194/nhess-12-2699-2012.

Daly C., J.A. Roelvink, A.R. van Dongeren *et al.* (2012) Validation of an advective-deterministic approach to short wave breaking in a surf-beat model, *Coastal Engineering*, **60**, 69–83, 10.1016/j.coastaleng.2011.08.001.

Dally W.R. and Dean R.G. (1984) Suspended sediment transport and beach profile evolution. *Journal of Waterway, Port, Coastal and Ocean Engineering*, **110**(1), 15–33.

Dean R.G. (1977) *Equilibrium beach profiles*: U.S. Atlantic and the Gulf Coasts. Department of Civil Engineering, Ocean Engineering Report No. 12, University of Delaware, Newark, DE.

Dette H.H., Larson M., Murphy J. *et al.* (2002) Application of prototype flume tests for beach nourishment assessment, *Coastal Engineering*, **47**(2), 137–177, 10.1016/S0378-3839(02)00124-2.

Del Río L., Plomaritis T.A., Benavente J. *et al.* (2012) Establishing storm thresholds for the Spanish Gulf of Cádiz coast. *Geomorphology*, **143–144**, 13–23, 10.1016/j.geomorph.2011.04.048.

Dolan R. and Davis R.E. (1992) An Intensity Scale for Atlantic Coast Northeast Storms. *Journal of Coastal Research*, **8**(4), 840–853.

Donnelly C., Kraus N. and Larson M. (2006) State of knowledge on measurement and modeling of coastal overwash. *Journal of Coastal Research*, **22**(4), 965–991.

Donnelly C. and Sallenger A.H. (2007) Characterisation and modeling of washover fans. *Proceedings of Coastal Sediments '07*, USACE, New Orleans, USA, pp. 2061–2073.

Edelman T. (1968) Dune erosion during storm conditions. *Proceedings of 11th International Conference on Coastal Engineering*, ASCE, London, UK, pp. 719–722.

Edelman T. (1972) Dune erosion during storm conditions. *Proceedings of the 13th International Conference on Coastal Engineering*, ASCE, Vancouver, Canada, pp. 1305–1311.

Erikson L., Larson M. and Hanson H. (2005) Prediction of swash motion and run-up including the effects of swash interaction, *Coastal Engineering*, **52**, 285–302.

FEMA (Federal Emergency Management Agency) (2003) *Guidelines and Specifications for Flood Hazard Mapping Partners*. Appendix D: Guidance for Coastal Flooding Analyses and Mapping, April 2003.

FEMA (Federal Emergency Management Agency) (2005) *National Flood Insurance Program*. Flood Insurance Manual, May 2005.

FEMA (Federal Emergency Management Agency) (2011) *Coastal Construction Manual*. Principles and Practices of Planning, Sitting, Designing, Constructing, and Maintaining Residential Buildings in Coastal Areas (Fourth Edition). FEMA P-55, Volume I, August 2011.

Fenster M.S., Dolan R. and Morton R.A. (2001) Coastal storms and shoreline change: signal or noise? *Journal of Coastal Research*, **17**(3), 714–720.

Ferreira O. (2005) Storm groups versus extreme single storms: predicted erosion and management consequences. *Journal of Coastal Research*, **SI42**, 221–227.

Ferreira Ó. (2006) The role of storm groups in the erosion of sandy coasts. *Earth Surface Processes and Landforms*, **31**, 1058–1060.

Ferreira O., Dias J.A., Gama C. and Taborda R. (1995) Quantification of beach erosion caused by storms on the Portuguese coast. in *Directions in European Coastal Management*, (eds. M.G. Healy and J.P. Doody), Samara Publishing Limited, Wales, UK, pp. 267–273.

Ferreira O. and Dias J.A. (2000) Prediction of storm impacts and shoreline retreat induced by hypothetical storms on open coastlines. *Proceedings of 3rd International Conference on Environmental Problems in Coastal Regions*, (eds G.R. Rodriguez, C.A. Brebbia and E. Perez-Martell), WIT Press, Southampton, UK, pp. 137–147.

Fisher J.S., Leatherman S.P. and Perry F.C. (1974) Overwash processes on Assateague Island. *Proceedings of 14th International Conference on Coastal Engineering*, ASCE, Copenhagen, Denmark, pp. 1194–1211.

Fletcher C.H., Richmond B.M., Barnes G.M., Schroeder T.A. (1995) Marine flooding on the coast of Kauai during Hurricane Iniki: hindcasting inundation components and delineating washover. *Journal of Coastal Research*, **11**(1), 188–204.

Fritz H.M., Blount C., Sokoloski R. *et al.* (2007) Hurricane Katrina storm surge distribution and field observations on the Mississippi Barrier Islands, *Estuarine Coastal and Shelf Science*, **74**, 12–20.

Gares P.A. (1990) Predicting flooding probability for beach dune systems. *Environmental Management*, **14**, 115–123.

Gencarelli R., Tomasicchio G.R., Kobayashi N. and Johnson B.D. (2008) Beach profile evolution and dune erosion due to the impact of Hurricane Isabel. *Proceedings of 31st International Conference on Coastal Engineering*, ASCE, Hamburg, Germany, pp. 1697–1709.

Haerens P., Ciavola P., Ferreira O. *et al.* (2012) Online operational early warning system prototypes to forecast coastal storm impacts (CEWS). *Coastal Engineering Proceedings*, **1**(33), doi:10.9753/icce.v33.management.45.

Hall M.J., Young R.S., Thieler E.R. *et al.* (1990) Shoreline response to Hurricane Hugo. *Journal of Coastal Research*, **6**(1), 211–221.

Hallermeier R.J. and Rhodes P.E. (1988) *Generic treatment of dune erosion for 100-year event. Proceedings 21st International Conference on Coastal Engineering*, ASCE, New York, pp. 1197–1211.

Harley M., Armaroli C. and Ciavola P. (2011) Evaluation of XBeach predictions for a real-time warning system in Emilia-Romagna, Northern Italy. *Journal of Coastal Research*, **SI64**, 1861–1865.

Harley M.D., Armaroli C. and Ciavola P. (2012) An Early Warning System for the on-line prediction of coastal storm risk on the Italian coastline. *Coastal Engineering Proceedings*, **1**(33), doi:10.9753/icce.v33.management.77.

Harley M.D. and Ciavola P. (2013) Managing local coastal inundation risk using real-time forecasts and artificial dune placements. *Coastal Engineering*, **77**, 77–90.

Harley M.D., Turner I.L., Short A.D. and Ranasinghe R. (2010) Interannual variability and controls of the Sydney wave climate. *International Journal of Climatology*, **30**(9), 1322–1335.

Hayden B. and Dolan R. (1977) Storm-surge overwash along the US mid-Atlantic coast. *Bulletin of the American Meteorological Society*, **58**(10), 1128.

Holman R.A. (1986) Extreme value statistics for wave run-up on a natural beach. *Coastal Engineering*, **9**, 527–544.

Houser, C. (2013). Alongshore variation in the morphology of coastal dunes: Implications for storm response. *Geomorphology*, **199**, 48–61.

Houser C. and Hamilton S. (2009) Sensitivity of post-hurricane beach and dune recovery to event frequency. *Earth Surface Processes and Landform*, **34**, 613–628.

Hughes S.A. (2004) Estimation of wave run-up on smooth, impermeable slopes using the wave momentum flux parameter. *Coastal Engineering*, **51**, 1085–1104.

Huntley D.A., Guza R.T. and Thornton E.B. (1981) Field observations of surf beats, 1, progressive edge waves, *Journal of Geophysical Research*, **86**, 6451–6466.

Irish J.L., Lynett P.J., Weiss R. *et al.* (2013) Buried relic seawall mitigates Hurricane Sandy's impacts, *Coastal Engineering*, **80**, 79–82, doi: dx.doi.org/10.1016/j.coastaleng.2013.06.001.

Judge K.E., Overton M.F. and Fisher J.S. (2003) Vulnerability indicators for coastal dunes. *Journal of Waterway, Port, Coastal and Ocean Engineering*, **129**, 270–278.

Karunarathna, H., Pender, D., Ranasinghe, R. *et al.* (2014) The effects of storm clustering on beach profile variability. *Marine Geology*, **348**, 103–112.

Kriebel D.L. and Dean R.G. (1985) Numerical simulation of time-dependent beach and dune erosion. *Coastal Engineering*, **9**, 221–245.

Kriebel D.L. and Dean R.G. (1993) Convolution method for time-dependent beach-profile response. *Journal of Waterways, Port, Coastal and Ocean Engineering*, **119**(2), 204–226.

Kriebel D.L. (1982) *Beach and dune response to hurricanes*. M.S. Thesis, Department of Civil Engineering, University of Delaware, Newark, NJ, USA.

Kriebel D. L., Dalrymple R., Pratt A. and Sakovich V. (1997) *A shoreline risk index for northeasters*. *Proceedings International Conference on Natural Disaster Reduction*, ASCE, New York, pp. 251–252.

Larson M., Erikson L. and Hanson H. (2004) An analytical model to predict dune erosion due to wave impact. *Coastal Engineering*, **51**(8-9), 675–696.

Larson M. and Kraus C.N. (1989) *SBEACH: Numerical model for simulating storm-induced beach change; Report 1, Empirical foundation and model development*. Technical Report CERC-89-9, Coast Eng. Research Center, U.S. Army Engineer Waterways Experiment Station, Vicksburg, Mississippi.

Larson M., Kraus C.N. and Byrnes M.R. (1990) *SBEACH: Numerical model for simulating storm-induced beach change; Report 2, Numerical formulation and model tests*. Technical Report CERC-89-9, Coastal Engineering Research Center, U.S. Army Engineer Waterways Experiment Station, Vicksburg, Mississippi.

Larson M., Wise R.A. and Kraus N.C. (2005) Modeling dune response due to overwash transport. *Proceedings 29th International Conference on Coastal Engineering*, Lisbon, Portugal, pp. 2133–2145.

Leatherman S.P. (1976). *Quantification of overwash processes*. Ph.D. Thesis, University of Virginia, USA.

Lee G., Nicholls R.J. and Birkemeier W.A. (1998) Storm-induced profile variability of the beach-nearshore profile at Duck, North Carolina, U.S.A., 1981–1991. *Marine Geology*, **148**, 163–177.

List J.H. (1992) A model for the generation of two-dimensional surf beat. *Journal of Geophysical Research*, **97**(C4), 56235635, doi:10.1029/91JC03147.

Longuet-Higgins M.S. and Stewart R.W. (1962) Radiation stress and mass transport in gravity waves with application to "surf-beats". *Journal of Fluid Mechanics*, **8**, 565– 583.

Longuet-Higgins M.S. and Stewart R.W. (1964) Radiation stress in waterwaves, a physical discussion with applications. *Deep Sea Research*, **11**, 529–563.

Madsen A.J. and Plant N.G. (2001) Intertidal beach slope predictions compared to field data. *Marine Geology*, **173**, 121–139.

May S.M., Engel M., Brill D. *et al.* (2013) Coastal hazards from tropical cyclones and extratropical winter storms based on Holocene storm chronologies. in *Coastal Hazards* (ed. C. W. Finkl) Springer, pp. 557–585.

Matias A., Ferreira O., Vila-Concejo A. *et al.* (2009) Foreshore and hydrodynamic factors governing overwash. *Journal of Coastal Research*, **SI56**, 636–640.

Matias A., Ferreira O., Vila-Concejo A. *et al.* (2010) Short-term morphodynamics of non-storm overwash. *Marine Geology*, **274**(1–4), 69–84.

McCall R.T., Van Thiel de Vries J.S.M., Plant N.G. *et al.* (2010) Two-dimensional time dependent hurricane overwash and erosion modeling at Santa Rosa Island. *Coastal Engineering*, **57**(7), 668–683.

Mendez F.J., Menendez M., Luceno A. and Losada I.J. (2006) Estimation of the long-term variability of extreme significant wave height using a time-dependent Peak Over Threshold (POT) model. *Journal of Geophysical Research*, **111**(C07024), doi: 10.1029/2005jc003344.

Mendoza E.T., Jimenez J.A. and Mateo J. (2011) A coastal storms intensity scale for the Catalan Sea (NW Mediterranean). *Natural Hazards and Earth Systems Science*, **11**, 2453–2462.

Morton R.A. (2002) Factors controlling storm impacts on coastal barriers and beaches – a preliminary basis for near real-time forecasting. *Journal of Coastal Research*, **18**(3), 486–501.

Morton R.A. and Sallenger A.H. (2003) Morphological impacts of extreme storms on sandy beaches and barriers. *Journal of Coastal Research*, **19**(3), 560–573.

Nairn, R., Roelvink, J. and Southgate, H. (1990) Transition zone width and implications for modeling surfzone hydrodynamics. *Coastal Engineering Proceedings*, **1**(22). doi:10.9753/icce.v22.

Nishi, R. and Kraus, N. (2001) Mechanism and calculation of sand dune erosion by storms. *Coastal Engineering Proceedings*, **1**(25). doi:10.9753/icce.v25.

Nordstrom K.F. and Jackson N.L. (1995) Temporal scales of landscape change following storms on a human-altered coast, New Jersey, USA. *Journal of Coastal Conservation*, **1**(1), 51–62.

Overton, M.F. and Fisher, J.S. (1988) Laboratory investigation of dune erosion. *Journal Waterway, Port, Coastal and Ocean Engineering*, **114**(3), 367–373.

Palmsten M. and Holman R.A. (2011) Infiltration and instability in dune erosion. *Journal of Geophysical Research*, **116**, C10.

Phillips, O.M. (1977) *The Dynamics of the Upper Ocean*, 2nd edn., Cambridge University Press, New York, 336 pp.

Raubenheimer B. and Guza R.T. (1996) Observations and predictions of run-up. *Journal of Geophysical Research*, **101**(C11), 25575–25587, doi:10.1029/96JC02432.

Reniers A.J.H.M., Roelvink D.J.A. and Thornton E.B. (2004) Morphodynamic modeling of an embayed beach under wave group forcing. *Journal of Geophysical Research*, **109**(C01030), doi:10.1029/2002JC001586.

Reniers A.J.H.M., MacMahan J., Thornton E.B. and Stanton T.P. (2006) Modelling infragravity motions on a rip-channel beach. *Coastal Engineering*, **53**, 209–222.

Ritchie W. and Penland S. (1988) Rapid dune changes associated with overwash processes on the deltaic coast of South Louisiana. *Marine Geology*, **81**, 97–122.

Roelvink J.A. (1993) *Surf beat and its effect on cross-shore profiles*. Ph.D. thesis, Delft Univ. of Technology, Delft, Netherlands, 150 pp.

Roelvink J.A. and Brøker I. (1993) Cross-shore profile models. *Coastal Engineering*, **21**, 163–191.

Roelvink D., Reniers A., van Dongeren A. *et al.* (2009) Modelling storm impacts on beaches, dunes and barrier islands. *Coastal Engineering*, **56**, 1133–1152.

Rosati J.D., Randall A.W., Kraus N.C. and Larson M. (1993) *SBEACH: Numerical model for simulating storm-induced beach change; Report 3: User's Manual*. Instruction Report CERC-93-2, U.S. Army Engineer Waterways Experiment Station, Vicksburg, Mississippi.

Ruggiero P., Komar P.D., McDougal W.G. and Beach R.A. (1996) Extreme water levels, wave runup and coastal erosion. in *Proceedings 25th International Conference on Coastal Engineering*, Orlando, Florida, ASCE, 2793–2805.

Ruggiero P., Komar P.D., McDougal W.G. *et al.* (2001) Wave Runup, Extreme Water Levels and the Erosion of Properties Backing Beaches. *Journal of Coastal Research*, **17**(2), 407–419.

Russell P. (1993) Mechanisms for beach erosion during storms. *Continental Shelf Research*, **13**(11), 1243–1265.

Sallenger A.H. (2000) Storm Impact Scale for Barrier Islands. *Journal of Coastal Research*, **16**(3), 890–895.

Sancho F., Oliveira F.S.B.F. and Freire P. (2012) Coastal dunes vulnerability indexes: A new proposal *Coastal Engineering Proceedings*, **1**(33), doi:10.9753/icce.v33.management.68.

Schaeffer H.A. (1994) Edge waves forced by short-wave groups. *Journal of Fluid Mechanics*, **259**, 125–148.

Steetzel H.J. (1993) *Cross-shore Transport during Storm Surges*. Ph.D. thesis, Delft University of Technology.

Stive, M. and De Vriend, H. (2011) Shear stresses and mean flow in shoaling and breaking waves. *Coastal Engineering Proceedings*, **1**(24). doi:10.9753/icce.v24.

Stockdon H.F., Holman R.A., Howd P.A. and Sallenger Jr., A.H. (2006) Empirical parameterization of setup, swash, and runup. *Coastal Engineering*, **53**, 573–588.

Stockdon H.F., Sallenger A.H., Holman R.A. and Howd P.A. (2007) A simple model for the spatially-variable coastal response to hurricanes. *Marine Geology*, **238**, 1–20.

Splinter K.D. and Palmsten M.L. (2012) Modeling dune response to an East Coast Low. *Marine Geology*, **329–331**, 46–57.

Stone G.W. and Muller R.A. (2005) Meteorological effects on coasts, in *Encyclopedia of Coastal Science* (ed. M.L. Schwartz), Springer, pp. 636–637.

Svendsen, I. A. (1984). Mass flux and undertow in a surf zone. *Coastal Engineering*, **8**(4), 347–365.

Terwindt J.H.J., Hulsbergen C.H. and Kohsiek L.H.M. (1984) Structures in deposits from beach recovery, after erosion by swell waves around the southwestern coast of Aruba (Netherlands Antilles). *Marine Geology*, **60**(1–4), 283–311.

Thom B.G. and Hall W. (1991) Behaviour of beach profiles during accretion and erosion dominated periods. *Earth Surface Processes and Landforms*, **16**, 113–127.

Tucker M.J. (1954) Surfbeats: sea waves of 1 to 5 minutes' period. *Proceedings of the Royal Society London, Series A*, **202**, 565–573.

USACE (2004) *Coastal overwash. Part 1: Overview of Processes*. Regional Sediment Management (RSM) Demonstration Program Technical Note, 36 pp.

Van de Graaff J. (1977) Dune erosion during a storm surge. *Coastal Engineering*, **1**, 99–134.

Van Dongeren A., Reniers A., Battjes J. and Svendsen I. (2003). Numerical modeling of infragravity wave response during DELILAH. *Journal of Geophysical Research*, **108**(C9), doi: 10.1029/2002JC001332.

Van Dongeren A., Bolle A., Vousdoukas M.I. *et al.* (2009) *MICORE: dune erosion and overwash model validation with data from nine European field sites*. *Proceedings Coastal Dynamics* 2009, Tokyo, Japan, Paper No. 82.

Van Dongeren A.R., Lowe R., Pomeroy A. *et al.* (2012) Numerical modeling of low-frequency wave dynamics over a fringing coral reef. *Coastal Engineering*, **173**, 178–190, doi: 10.1016/j.coastaleng.2012.11.004.

Van Dongeren A.R., Ciavola P., Viavattene C. *et al.* (2014) RISC-KIT: Resilience-Increasing Strategies for Coasts – toolkit. in Proceedings 13th International Coastal Symposium (Durban, South Africa), *Journal of Coastal Research*, SI70 (eds A.N. Green and J.A.G. Cooper) 366–371.

Van Gent M.R.A., van Thiel de Vries J.S.M., Coeveld E.M. *et al.* (2008) Large-scale dune erosion tests to study the influence of wave periods. *Coastal Engineering*, **55**, 1041–1051.

Van Koningsveld M., Davidson M.A. and Huntley D.A. (2005) Matching science with coastal management needs; the search for appropriate coastal state indicators. *Journal of Coastal Research*, **21**, 399–411.

Van Koningsveld M., Davidson M.A., Huntley D.A. *et al.* (2007) A critical review of the CoastView project: recent and future developments in coastal management video systems. *Coastal Engineering*, **54**, 567–576.

Van Thiel de Vries J.S.M., van Gent M.R.A., Walstra D.J.R. and Reniers A.J.H.M. (2008) Analysis of dune erosion processes in large-scale flume experiments. *Coastal Engineering*, **55**, 1028–1040.

Van Thiel de Vries J.S.M. (2009) *Dune erosion during storm surges*. PhD Thesis, Delft University of Technology, Delft.

Vellinga P. (1986) *Beach and dune erosion during storm surges*. Ph.D. thesis. Delft University of Technology. Delft Hydraulics Communication No. 372.

Vitousek S. and Fletcher C.H. (2008) Maximum annually recurring wave heights in Hawai'i, *Pacific Science*, **62**(4), 541–553.

Vousdoukas M.I., Almeida L.P.M. and Ferreira O. (2012a) Beach erosion and recovery during consecutive storms at a steep-sloping, meso-tidal beach. *Earth Surface Processes and Landform*, **37**, 583–593.

Vousdoukas M., Ferreira Ó., Almeida L. and Pacheco A. (2012b) Toward reliable storm hazard forecasts: XBeach calibration and its potential application in an operational early-warning system. *Ocean Dynamics*, **62**, 1001–1015.

Williams J.J., Ruiz de Alegría-Arzaburu A., McCall R. T. and Van Dongeren A. (2011) Modelling gravel barrier profile response to combined waves and tides using XBeach: laboratory and field results. *Coastal Engineering*, **63**, 62–80.

Wright L.D. and Short A.D. (1984) Morphodynamic variability of surf zones and beaches. *Marine Geology*, **56**, 93–118.

Wright L.D., Short A.D. and Green M.O. (1985) Short-term changes in the morphodynamic states of beaches and surf zones: an empirical. *Marine Geology*, **62**, 339–364.

Yates M.L., Guza R.T. and O'Reilly W.C. (2009) Equilibrium shoreline response: observations and modeling. *Journal of Geophysical Research*, **114** (C09014), doi: 10.1029/2009JC005359.

Part Four

Social and Economic Considerations

4.1

Assessing the Costs of Natural Hazards – State of the Art and the Way Forward

Volker Meyer[1], Reimund Schwarze[1], Nina Becker[1,2], Vasileios Markantonis[3], Jeroen C.J.M. van den Bergh[4,5,6,7], Laurens M. Bouwer[8], Philip Bubeck[9], Paolo Ciavola[10], Elisabetta Genovese[11], Colin Green[12], Stephane Hallegatte[11,13,19], Heidi Kreibich[14], Quentin Lequeux[10], Ivana Logar[15], Elissaios Papyrakis[7,16,20], Clemens Pfurtscheller[17], Jennifer Poussin[7], Valentin Przyluski[11], Annegret H. Thieken[18] and Christophe Viavattene[12]

[1]*Department of Economics, Helmholtz Centre for Environmental Research, Leipzig, Germany*
[2]*Overseas Development Institute, London, UK*
[3]*Faculty of Economics and Business Administration, Chemitz University of Technology, Chemitz, Germany*
[4]*ICREA, Barcelona, Spain*
[5]*Institute of Environmental Science and Technology, Universitat Autònoma de Barcelona, Barcelona, Spain*
[6]*Faculty of Economics and Business Administration, VU University Amsterdam, Amsterdam, The Netherlands*
[7]*Institute for Environmental Studies, VU University Amsterdam, Amsterdam, The Netherlands*
[8]*Deltares, Delft, The Netherlands*
[9]*adelphi, Berlin, Germany*
[10]*Dipartimento di Fisica e Scienze della Terra, University of Ferrara, Ferrara, Italy*
[11]*Centre International de Recherche sur l'Environnement et le Développement (CIRED), Nogent-sur-Marne, France*
[12]*Flood Hazard Research Centre, Middlesex University, Hendon, UK*
[13]*Ecole Nationale de Météorologie, Toulouse, France*
[14]*Helmholtz Centre Potsdam – GFZ German Research Centre for Geosciences, Potsdam, Germany*
[15]*Swiss Federal Institute of Aquatic Science and Technology (EAWAG), Dübendorf, Switzerland*
[16]*School of International Development, University of East Anglia, Norwich, UK*
[17]*Institute of Interdisciplinary Mountain Research, Austrian Academy of Sciences, Innsbruck, Austria*
[18]*Institute of Earth and Environmental Science, University of Potsdam, Potsdam, Germany*
[19]*The World Bank, Sustainable Development Network, Washington, DC, USA*
[20]*International Institute of Social Studies, Erasmus University Rotterdam, The Netherlands*

Hydrometeorological Hazards: Interfacing Science and Policy, First Edition. Edited by Philippe Quevauviller.

4.1.1 Introduction

The costs of natural hazards around the world are rising. While the number of events reported in the last 10 years has decreased slightly, the number of affected people and their related costs is increasing, indicating also an augmented exposure and vulnerability of humankind to natural hazards (see Figures 4.1.1 and 4.1.2, EM-DAT 2013, IPCC 2012). Some outstanding events in the last decade in terms of their economic damage were, for example, the Tsunami in Japan in 2011 with damages of around US$ 210 billion, Hurricane Katrina in 2005 (US$125 billion), the Wenchuan earthquake in China in 2008 (US$85 billion), the floods in Thailand in 2011 (US$40 billion) and the drought in the US in 2012 which caused damage of more than US$20 billion (EM-DAT 2013, Munich Re, 2006, Munich Re, 2012, Munich Re, 2013).

In times of such increasing disaster losses the reduction (or mitigation) of natural hazard risks needs to be effective and efficient (Laframboise and Loko, 2012). An in-depth understanding not only of the effects of disasters, but also of the suitable tools to measure the economic impact of natural hazards (Wahlstrom, 2013), is required in order to develop forms of sustainable risk management as well as risk mitigation and adaptation[1] strategies (WB and UN, 2010; IPCC, 2012). This is especially true given the limited

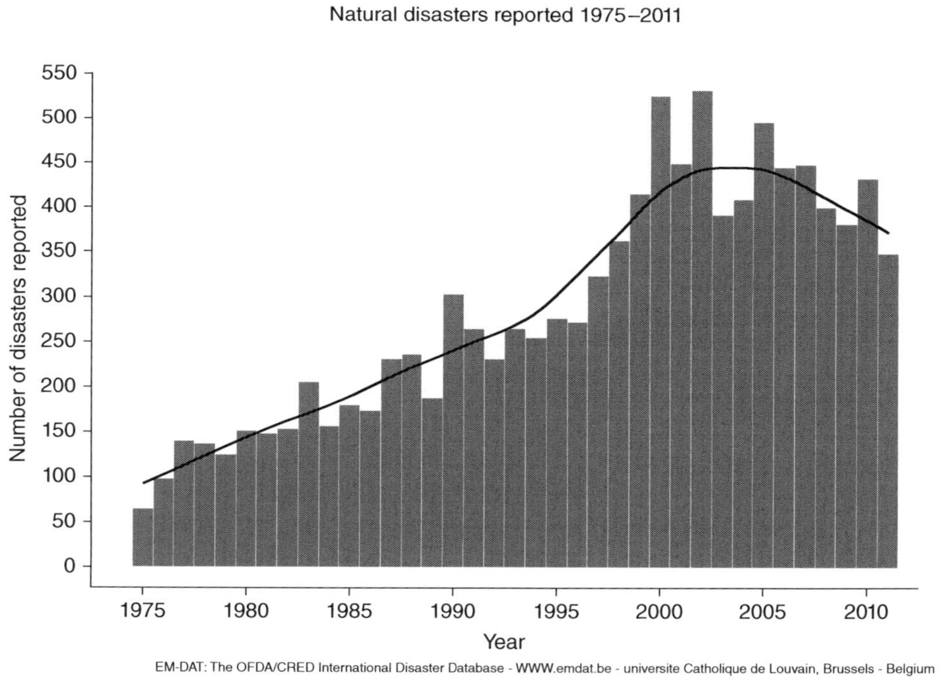

Figure 4.1.1 Natural disasters reported 1975–2011 and trend (linear interpolated smoothed black line). *Source:* EM-DAT, 2013.

[1] In this chapter, mitigation refers to the reduction of natural hazard risks, while in other contexts it usually refers to the reduction of greenhouse gas emissions or the enhancement of carbon sinks. Throughout the paper, mitigation is

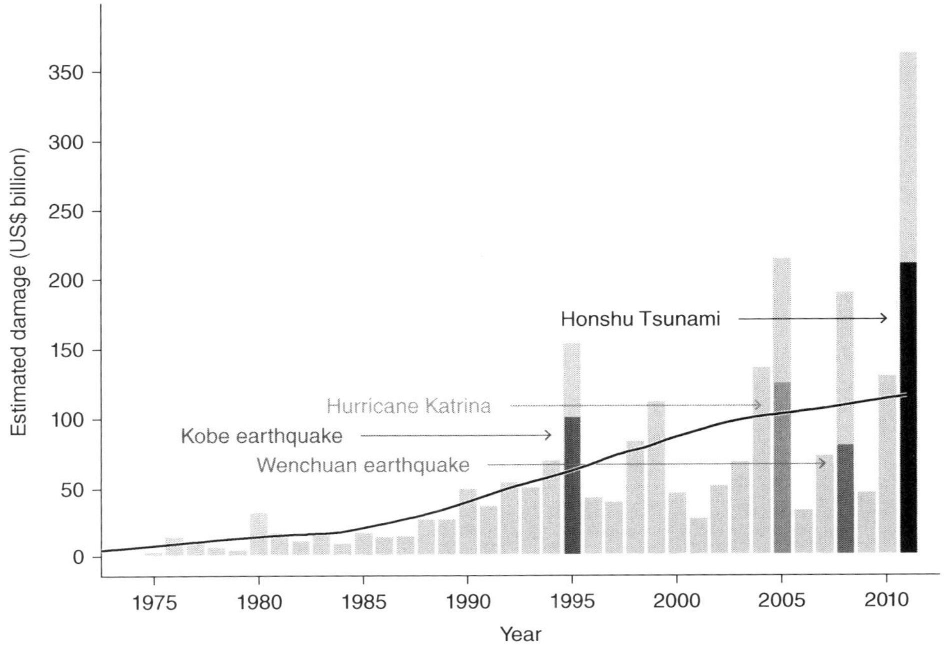

Figure 4.1.2 Estimated damage (US$ billion) caused by reported natural disasters 1975–2011 and trend (linear interpolated smoothed black line) *Source:* EM-DAT, 2013.

financial resources available. In this respect, reliable and comprehensive estimates of the costs and benefits of natural hazards are crucial for informed decision making and, more specifically, for the development of policies, strategies and measures to prevent or reduce the impact of natural hazards on societies and to improve their coping and adaptive capacities.

Cost assessments may be carried out in the aftermath of an event (*ex post*) or with regard to potential future events (*ex ante*). The aim of *ex post* assessments is to inform local or national governments of the overall amount of induced damage and to provide a basis for calculating levels of compensation and recovery support (e.g. Sächsische Staatskanzlei, 2003; McCarty and Smith, 2005; Karunasena and Rameezdeen, 2010). *Ex ante* cost assessments are conducted in order to support governmental decision-making relating to alternative risk mitigation options (e.g. MAFF, 1999; BMLFUW, 2008a; BAFU, 2010). Decision makers in natural hazard risk management will want to consider, as far as possible, all the benefits and costs of alternative courses of action in the course of identifying the best available option. The insurance industry also has an interest in obtaining reliable cost figures so that it can calculate insurance premiums and financial worst case scenarios, that is the loss of the

used as a synonym for risk reduction, including adaptation measures in response to changes in the hazard situation. Adaptation in this respect and throughout the paper is understood as adaptation to anticipate changing risks as a result of climate change. As such, adaptation constitutes an integral component of risk mitigation.

200-year event that is required by the directive Solvency II or the Probable Maximum Loss (PML) e.g. Kleindorfer and Kunreuther, 1999; Woo, 2002a cited in Woo, 2002b; Smolka, 2006; Paudel, Botzen and Aerts, 2013).

Current methods for assessing the costs of natural hazards employ a diversity of terminologies and methodological approaches relating to different hazards and sectors affected. For example, with regard to estimates of the aggregated and sector-based effects of disasters in the long term, the World Bank and United Nations (2010) find that variations in these estimates result from the use of different data and techniques and from the inclusion of different kinds of disasters. This poses a considerable obstacle to arriving at reliable, comprehensive and comparable cost figures.

This contribution aims to synthesise current knowledge about cost assessment methods for various hazards and cost types. It is based on research carried out in the EU project 'Costs of Natural Hazards' (CONHAZ) which ran from February 2010 to January 2012. The project adopted a comprehensive approach, considering natural hazards ranging from droughts, floods, storms and coastal hazards to Alpine hazards.

In addition, different cost types were considered in the project. Terminology relating to cost types sometimes differs in the literature and among hazard communities (see e.g. Parker, Green and Thompson, 1987; Smith and Ward, 1998; H. John Heinz III Center, 2000; Wilhite, 2000; Thieken, Seifert and Merz 2010). For this reason, a working terminology for cost categories was developed within the CONHAZ project which takes these different definitions into account. It is largely based on the classification of direct/indirect and tangible/intangible costs, introduced in relation to flood damages by Parker, Green and Thompson (1987). However, a novelty within the CONHAZ project was to explicitly include risk mitigation costs as well, and to consider business interruption costs as a separate sub-category. This results in five cost categories, namely: 1. direct costs, 2. business interruption costs, 3. indirect costs, 4. intangible costs, and 5. risk mitigation costs (see also Figure 4.1.3). The main rationale behind distinguishing these five cost categories was that they are likely to require different cost assessment methods. These cost categories are defined as follows:

1. *Direct costs* are damages to property due to direct physical contact with the hazard, that isthe physical destruction of buildings, inventories, stocks, infrastructure or other assets at risk (Smith and Ward, 1998). The related losses are therefore related predominantly to the assets' susceptibility to the hazard characteristics.
2. *Business interruption costs* occur in areas directly affected by the hazard. Business interruptions take place, for example, if people are not able to carry out their work because their workplace is either destroyed or not accessible due to a hazard. They also occur if industrial or agricultural production is reduced due to water scarcity. In the literature, such losses are sometimes referred to as *direct damages*, as they occur due to the immediate impact of the hazard (see e.g. Wilhite, Svoboda and Hayes, 2007; Kok *et al.*, 2004). They are often also referred to as *primary indirect damages* because the losses do not result from physical damage to property but from the interruption of economic processes (e.g. Smith and Ward, 1998). However, the methods to evaluate losses due to business interruption are quite different from those used for direct or indirect damages. For this reason, and in order to avoid terminological misunderstandings, 'business interruption costs' is used as a separate cost category for the purpose of this article.
3. *Indirect costs* are losses induced by either direct damages or business interruption costs (Przyluski and Hallegatte, 2011). They can occur inside or outside of the

		Tangible costs	Intangible (non-market) costs
Damage costs	Direct	• Physical damage to assets: – buildings – contents – infrastructure	• Loss of life • Health effects • Loss of environmental goods
	Business interruption	• Production interruption because of destroyed machinery	• Ecosystem services interrupted
	Indirect	• Induced production losses of suppliers and customers of companies directly affected by the hazard	• Inconvenience of post-flood recovery • Increased vulnerability of survivors
Risk mitigation costs	Direct	• Investment costs • Operation & maintenance costs	• Environmental damage - due to the development of mitigative infrastructure - or due to a change in agricultural practices
	Indirect	• Induced costs in other sectors	

Cost categories applied in this article:

Direct costs | Business interruption costs | Indirect costs | Intangible (non-market) costs | Risk mitigation costs

Figure 4.1.3 Cost categorisation applied in this article with examples. *Source:* Adapted from and extended based on Penning-Rowsell *et al.*, 2003; Smith and Ward, 1998. See plate section for colour representation of this figure.

hazard area and often with a time lag. These losses include, for example, induced production losses of suppliers and customers of companies directly affected by the hazard. In most cases, indirect costs are measured by loss of gross domestic product.

4. *Intangible costs* refer to damages to goods and services which are not measurable (or at least not easily measurable) in monetary terms because they are not traded on a market. Hence, they are also referred to as non-market values or costs (Smith and Ward, 1998). This contrasts with tangible costs, which imply an existing market for tangible assets or the goods and services derived thereof; thus tangible costs are easily expressed in monetary terms. Intangible effects include, for instance, environmental impacts, health impacts and impacts on cultural heritage.
5. *Risk mitigation costs*, that is risk reduction, can be regarded as part of the total costs of natural hazards and are thus considered an essential cost category (Bouwer *et al.*, 2011). The costs of risk mitigation can also be classified according to the cost categories introduced above, that is direct costs, indirect costs and intangible costs. The direct costs of risk mitigation refer to any costs attributed to research and design, the set-up, operation and maintenance of infrastructure, or other measures for the purposes of mitigating (or adapting to) natural hazards. The indirect costs of

risk mitigation relate to any secondary costs (externalities) occurring in economic activities/sectors (or localities) that are not directly linked to such infrastructure investment. The intangible costs refer to any non-market health or environmental impacts of risk mitigation measures, such as environmental damage due to the development of mitigating infrastructure or a change in agricultural practices.

As this last example shows and Figure 4.1.3 illustrates, this cost categorisation includes some overlaps. For example, intangible costs can be direct (e.g. injuries) or indirect damage (e.g. social disruption), or induced by risk mitigation investments (e.g. wetland degradation induced by the construction of dykes). In this contribution we deal with these overlaps in the following way. The tangible cost categories are discussed in Sections 4.1.2.1 (direct costs), 4.1.2.2 (business interruption costs), 4.1.2.3 (indirect costs) and 4.1.2.5 (all tangible risk mitigation costs), while all intangible categories are presented in one section (4.1.2.4) as they require similar valuation techniques.

A review of existing methods for cost assessment was carried out for each natural hazard (floods, droughts, coastal hazards and Alpine hazards) and cost category (direct costs, business interruption costs, indirect costs, intangible costs and risk mitigation costs) considered. This was done based on literature reviews, four expert workshops (one in each hazard community), and a final conference to synthesise findings.

The results of these reviews are presented in four reports for each different cost type, that is for direct costs and costs due to business interruption (Bubeck and Kreibich, 2011), for indirect costs (Przyluski and Hallegatte, 2011), for costs due to intangible, non-market effects (Markantonis, Meyer and Schwarze, 2011, 2012), and for costs of risk mitigation (Bouwer *et al.*, 2011). Four other reports present the results for the different hazard types, that is droughts (Logar and van den Bergh, 2011, 2013), floods (Green, Viavattene and Thompson, 2011), coastal hazards (Lequeux and Ciavola, 2011) and Alpine hazards (Pfurtscheller, Lochner and Thieken, 2011).

The objective of this contribution is to synthesise the overall findings and conclusions based on these reviews. The remainder of this chapter is organised as follows. Section 4.1.2 presents the main findings concerning current best practices, based on the literature review and expert workshops. Section 4.1.3 concludes on the state of the art by presenting overall knowledge gaps and recommendations for practice and research. The remainder of Section 4.1.3 addresses the way forward by depicting a vision on integrated cost assessment and risk management.

4.1.2 State of the art of cost assessment for natural hazards – An overview

The following overview examines each of the five distinguished cost types – direct costs, business interruption costs, indirect costs, intangible (non-market) costs, and risk mitigation costs. For each cost category, the main methods for cost estimates are reviewed and enriched with examples from different hazard communities where possible. In addition, within each cost type category, one prominent method is selected and illustrated with examples to further guide the reader how to apply specific methods.

4.1.2.1 Direct costs

Direct tangible costs, such as the destruction of houses and infrastructure, are among the most visible impacts of natural hazards. Table 4.1.1 presents an overview of methods for estimating direct costs, including examples of applications for different hazards.

The most frequently applied approach for assessing the direct costs for most hazards is the use of susceptibility functions (or damage functions) (see Box 4.1.1). What all susceptibility functions have in common is that they describe the relation between one or more hazard parameters (single or multi-parameter models), with a resulting monetary damage for a certain type or use of object at risk.

Table 4.1.1 Direct costs: methods, applications and examples.

General method	Specific method	Application and/or examples
Susceptibility function	Single-parameter models (based on single hazard impact parameter)	Floods: Model of ICPR (2001); Model of MURL (2000), adopted by Glade (2003); Model of Hydroteec (Emschergenossenschaft and Hydrotec 2004) Coastal Hazards: Reese, Markau and Sterr (2003) Droughts: Corti *et al.* (2009) Alpine hazards: Fuchs, Heiss and Hübl (2007b), Huttenlau (2010), Totschnig, Sedlacek and Fuchs (2011)
	Multi-parameter models (based on several hazard impact and /or resistance parameters)	Floods: HAZUS-MH (FEMA, 2011; Scawthorn *et al.*, 2006); FLEMOps and FLEMOcs models (Apel *et al.*, 2009; Elmer *et al.*, 2010; Kreibich *et al.*, 2010; Thieken *et al.*, 2008); Model of Multicoloured Manual (Penning-Rowsell *et al.*, 2003); HIS-SSM (Kok *et al.*, 2004); Model of Maiwald and Schwarz (2010) Coastal hazards: FEMA (2011), HIS-SSM (Kok *et al.*, 2004), Nadal *et al.* (2010) Alpine hazards: BUWAL (1999), Keiler *et al.* (2006), Holub, Suda and Fuchs (2012)
Market valuation techniques	Market price method	Drought: Grafton and Ward (2008)
Integrated Assessment Analysis	Biophysical-Agroeconomic Models	Droughts: Holden and Shiferaw (2004),Fischer *et al.* (2005)
	Hydrological-Economic Models	Droughts: Booker, Michelsen and Ward (2005), Ward, Booker and Michelsen (2006), Grossmann *et al.* (2011)
Computable General Equilibrium Analysis	Computable General Equilibrium Models	Droughts: Horridge, Madden and Wittwer (2005)

Box 4.1.1 Susceptibility analysis to evaluate direct flood damage

A susceptibility analysis relates the damage of a certain element at risk to the flood impact.

For physical elements at risk, such as buildings (contents and structure), two types of susceptibility or damage functions are commonly used: relative or absolute damage functions (Figure 4.1.4a and b).

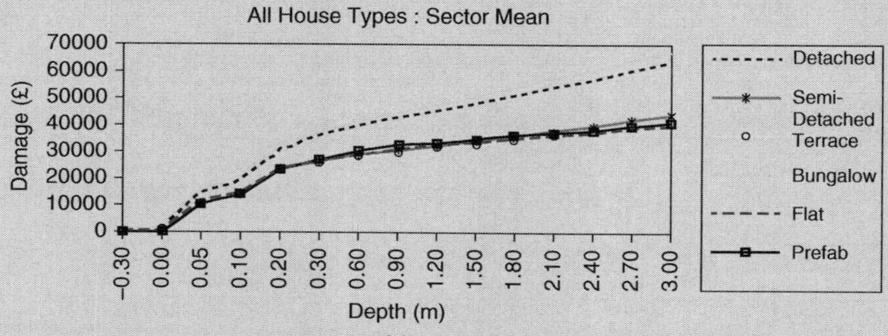

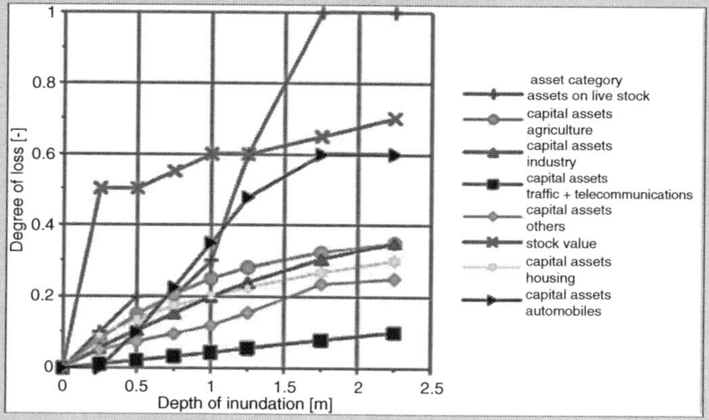

Figure 4.1.4 (a) Absolute damage function (England and Wales). *Source:* Penning-Rowsell *et al.* 2003. (b) Relative depth damage function. *Source:* Translated from Meyer and Mai 2003. See plate section for colour representation of this figure.

The absolute function consists in establishing the damage function for a particular asset in monetary terms, either in relation to the building or per unit area. The relative function provides the susceptibility expressed as a percentage of the total value of the assets. In this case, the total value of the asset has to be estimated, for example based on statistics or object-specific appraisals. In each case the function can be established with an empirical or a synthetic approach. The empirical approach is based on *ex post* damages values. The synthetic approach is based on hypothetical damage estimates by expert assessors (what if analysis) (e.g. Penning-Rowsell *et al.*, 2003).

Assets are very different in shape, in structure, in nature, in lifetime, in numbers and in economic value and thus the variability in terms of response to a flood and of associated loss are very high. Different damage curves are used to reflect it. A first approach consists of, and often is limited to, a standard classification and knowledge on socio-economic or land use categories (residential, commercial, industrial, public, agriculture) (Figure 4.1.4b).

A more refined approach will consider the sub-type within these categories, for example for residential use detached and semi-detached houses, flats, bungalows and so on (see e.g. Figure 4.1.4a).

These functions have a particularly long tradition in the context of flood damage evaluation (e.g. Penning-Rowsell and Chatterton, 1977; Parker, Green and Thompson, 1987; Klaus *et al.*, 1994; IWK, 1999) but have also been developed, for instance, for coastal flooding, droughts, and Alpine hazards. Hazard parameters can include, for example, avalanche pressure, water depth or drought-induced soil subsidence. In addition to these hazard parameters, some damage functions also take into account resistance parameters, such as differences in building structures or the standard of risk mitigation measures undertaken (e.g. BUWAL, 1999; Keiler *et al.*, 2005, 2006; BAFU, 2010).

Over the years a number of multiparameter models have also been developed, such as those by Nicholas, Holt and Proverbs (2001), Zhai, Fukuzonu and Ikeda (2005), Thieken *et al.* (2008) and Kreibich *et al.* (2010) (see Box 4.1.2 as an example). Studies have also shown that the application of multivariate models improves the reliability of flood damage modelling (Apel *et al.*, 2009; Wünsch *et al.*, 2009; Elmer *et al.*, 2010; Seifert *et al.*, 2010).

Box 4.1.2 Multi-parameter flood damage modelling taking precautionary behaviour into account: Flood loss estimation model for the private sector (FLEMOps)

Several studies have outlined the large uncertainties associated with flood damage assessments. The uncertainties stem from the fact that very complex damaging processes are usually described using simple depth-damage functions. An example of a multi-parameter flood damage model that takes several damage influencing parameters into account is the FLEMOps series (e.g. Apel *et al.*, 2009; Elmer *et al.*, 2010; Thieken *et al.*, 2008).

The FLEMOps model has been developed based on comprehensive empirical data of up to 2158 private households that were affected by flood events in 2002, 2005 and 2006 in Germany. In addition to details on suffered damages, this data set also provides information on several damage influencing parameters at the object

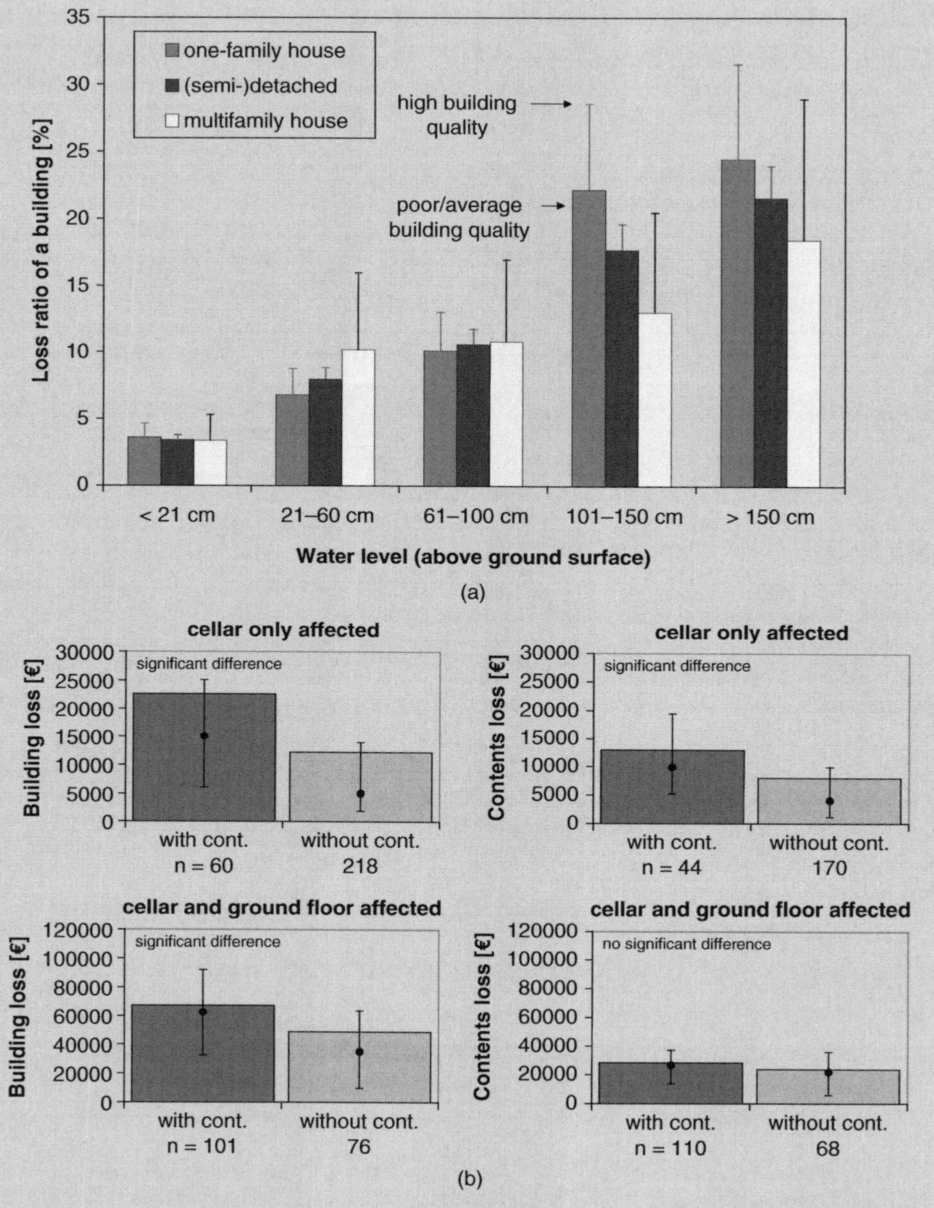

Figure 4.1.5 (a) Loss ratios for different inundation depth and different building types from FLEMOps. (b) Influence of contamination on losses (from FLEMOps+). *Source:* Thieken *et al.* 2008. Reproduced with permission of WIT Press; Kreibich *et al.* (2011). See plate section for colour representation of this figure.

level, such as contamination of flood water, building quality or the level of precautionary measures. Using detailed statistical analysis (Kreibich *et al.*, 2005; Thieken *et al.*, 2005), this information was integrated in a multi-parameter flood loss model. FLEMOps calculates the damage ratio for private households using five different classes of inundation depth, three individual building types and two classes of building quality (Figure 4.1.5). Further model enhancements were made to integrate other damage influencing parameters. In a first additional modelling step, referred to as FLEMOps+, the influence of private precaution and the contamination of flood water can be taken into account, using scaling factors (Büchele *et al.*, 2006). In a second additional modelling step, the influence of flood frequency was included, because average damages were found to be higher for less probable events, independent of water level (Elmer *et al.*, 2010). A procedure to upscale the damage model from the micro-scale to the meso-scale was introduced by Thieken *et al.* (2008). Only recently, FLEMO has been adopted for an Alpine case study in Austria (Cammerer, Thieken and Lammel, 2013).

Market valuation techniques (i.e. market prices, production function, avoided costs, replacement or repair costs), Integrated Assessment analysis and Computable General Equilibrium (CGE) analysis are further methods to assess direct costs, in particular those of droughts. Here market valuation techniques (i.e. market prices, production function, avoided costs, replacement or repair costs) are the most suitable methods for assessing direct tangible costs of droughts. They have the advantage of being easily to apply, covering any economic sector, and delivering fairly precise estimations. CGE-analysis and Input Output analysis are sometimes used to estimate direct costs together with indirect costs (e.g. Horridge, Madden and Witwer, 2005), but they require greater effort in their application. Biophysical and agro-economic models integrate crop models with the economic assessment and are thus specifically focused on agriculture (Kulshreshtha and Klein, 1989; Rosenberg, 1993; Holden and Shiferaw, 2004; and Fischer *et al.*, 2005). Hydrological-economic models link the economic assessment to a hydrological model (see e.g. Booker, 1995; Booker, Michelsen and Ward, 2005; Ward, Booker and Michelsen, 2006). Ricardian hedonic price modelling, as in Easterling and Mendelsohn (2000), links variations in land values across space with variations in climate.

4.1.2.2 Business interruption costs

Three main approaches (see also Table 4.1.2) are applied to assess the costs of natural hazards due to business interruption, namely:

1. Sector-specific reference values, for example for loss of added value, wage losses or relocation expenses per unit affected or sector specific models (see Box 4.1.3 as an example);
2. Event analysis: comparison of production output between hazard and non-hazard years; and
3. Share of direct damage: approaches that calculate production losses using a fixed share of direct damages.

Table 4.1.2 Business interruption costs: methods, applications and examples.

General method	Specific method	Application and/or examples
Sector specific reference values or models	Loss of value added	MURL (2000), Laternser (2000), SLF (2000), BMLFUW (2008a,b)
	Sector specific models	For transport: Department for Transport (2009)
Event analysis	Comparison hazard and non-hazard time periods based on reported cost figures	Benson and Clay (1998), COPA-COGECA (2003), Fink *et al.* (2004), Martin-Ortega, González-Eguino and Markandya (2012), Rijkswaterstaat (2004), Nöthiger (2003)
Share of direct damage	Fixed share of direct damage estimates	ANUFLOOD (NR&M 2002), RAM (NRE 2000)

The latter two approaches involve more uncertainties than the first and are therefore useful only for rapid appraisals (e.g. for emergency planning and budgeting). The first approach is considered more comprehensive. Overall, however, these types of losses are usually assessed in practice using rather simplistic methods, and initial empirical findings support the simplified approach of estimating a fixed share of direct damage in principle (Kreibich *et al.*, 2010).

Box 4.1.3 Estimation of flood losses due to the disruption of production processes in the US

The US model Hazus-MH MR5, which is provided by the US Federal Emergency Management Agency (FEMA, 2011), estimates losses due to the disruption of production processes on the basis of relocation expenses, capital related income losses, wage losses and rental income losses (Figure 4.1.6). Relocation expenses include the cost of shifting and transferring, and the rental of temporary space. These costs are assumed to be incurred once the building reaches a damage threshold of 10%. Cost per day and area factors are specified for various economic sectors in order to derive monetary losses. Capital related income losses, wage losses and rental income losses are estimated depending on the building recovery time. Building recovery time is calculated by summing the time needed for physical restoration of the building, as well as time for clean-up, time required for inspections, permits and the approval process, as well as delays due to contractor availability. All these components are estimated in dependence on water depth and business branch. The thus derived flood and sector specific building recovery time is used to estimate monetary costs per day and area, which are defined for various economic sectors (e.g. wage per square foot per day for the financial sector).

No.	Label	Occupancy Class	Rental Cost (2006)		Disruption Costs (2006)
			($/ft²/month)	($/ft²/day)	($/ft²)
		Residential			
1	RES1	Single-family Dwelling	0.68	0.02	0.82
2	RES2	Mobile Home	0.48	0.02	0.82
3	RES3A	Multi-family Dwelling; Duplex	0.61	0.02	0.82
4	RES3B	Multi-family Dwelling;	0.61	0.02	0.82
5	RES3C	Multi-family Dwelling; 5 - 9 units	0.61	0.02	0.82
6	RES3D	Multi-family Dwelling; 10 - 19 units	0.61	0.02	0.82
7	RES3E	Multi-family Dwelling; 20 - 49 units	0.61	0.02	0.82
8	RES3F	Multi-family Dwelling; 50+ units	0.61	0.02	0.82
9	RES4	Temporary Lodging	2.04	0.07	0.82
10	RES5	Institutional Dormitory	0.41	0.01	0.82
11	RES6	Nursing Home	0.75	0.03	0.82

Figure 4.1.6 Rental and disruption costs for different building types. *Source:* FEMA (2011) Hazus®-mh mr5. Flood model, Federal Emergency Management Agency, Technical manual.

4.1.2.3 Indirect costs

The methods used to assess indirect costs of natural hazards are presented in Table 4.1.3.

Methods include firm- or household-level surveys relating to past events (see event analysis in Table 4.1.3), and more frequently economic models, including:

1. Econometric models at the local or the national level;
2. Input Output models at the regional or national level;
3. CGE models at the regional or national level;
4. Intermediary models between CGE models and Input Output models.

Other less frequent model-based approaches used to estimate indirect costs consider the impact of natural disasters on public finances or else apply idealized models.

The method of collecting data on past events, for example based on firm- or household-level surveys, considers a single event in one location. Econometric approaches, in contrast, rely on several events to derive the main explanatory factors for estimating the costs of future events as well. Input Output models estimate the consequences of a specific impact of one or more economic sectors on other sectors of the economy. This is achieved by applying fixed input–output coefficients, which describe relationships between different economic sectors. Such models assume that there is no technological change and no substitution between production inputs. As a result, economic losses may be overestimated. CGE models allow more flexibility and substitution at different levels, driven by markets and price changes. They may therefore result in lower estimates than Input Output models. However, CGE models are also more difficult to apply than Input Output models and have a

Table 4.1.3 Indirect costs: methods, applications and examples.

General method	Specific method	Application and/or examples
Event analysis	Surveys at firm level	Boarnet (1998), Kroll *et al.* (1991), Tierney (1997)
	Surveys at the household level	McCarty and Smith (2005)
Econometric approaches	Gross regional product effect assessment	Noy and Vu (2009), Strobl (2010)
	National Gross domestic product effect assessment	Albala-Bertrand (1993), Cavallo and Noy (2009), Hochrainer (2009), Jaramillo (2009), Noy (2009), Loayza *et al.* (2009), Noy and Nualsri (2007), Raddatz (2009), Skidmore and Toya (2002)
Input-Output Analysis	Input-Output Models	Hallegatte (2008), HAZUS-E (see also McCarty and Smith, 2005), Haimes *et al.* (2005), Okuyama, Hewings and Sonis (2004), Rose and Liao (2005), Rose and Miernyk (1989)
Computable General Equilibrium Analysis	Computable General Equilibrium Models	Berrittella *et al.* (2007), Boyd and Ibarrarán (2009), Horridge, Madden and Witwer (2005), Pauw, Thurlow and van Seventer (2010), Rose, Oladosu and Liao (2007), Tsuchiya, Tatano and Okada (2007), Wittwer and Griffith (2010)
Intermediate models	Hybrid Input-Output / Computable General Equilibrium Models	Hallegatte (2008),Rose, Oladosu and Liao (2007)
Public Finance Analysis	Analysis of the impact on public finance	Mechler, Linnerooth-Bayer and Peppiatt (2006)
Idealized Models	Modeling interactions of hazard impacts with technical change or business cycles	Hallegatte and Dumas (2008), Hallegatte and Ghil (2008)

major shortcoming in that they assume that markets will function perfectly (even in post-disaster situations).

Intermediate models are either Input Output models with flexibility, as in Hallegatte (2008) see Box 4.1.4, or CGE models with reduced substitution elasticity, as in Rose, Oladosu and Liao (2007).

Box 4.1.4 An adaptive regional input-output model and its application to the assessment of the economic cost of Hurricane Katrina

Using an adaptive regional input–output (ARIO) model, that is a model based on input–output tables, the applied methodology assesses indirect losses resulting from a hurricane, through production and job losses and reconstruction phase (duration and cost). The main particularity of the model is that it considers (i) changes in sector production capacities and both forward and backward propagations in the economic system (i.e. indirect effects resulting, respectively, from modifications in supply and demand capacities); and (ii) adaptive behaviours in the aftermath of the hurricane. Indirect economic losses in Louisiana after Katrina amounted to $42 billion compared to $107 billion of direct economic losses. More generally, this analysis concludes that regional indirect losses increase nonlinearly with direct losses, suggesting the existence of a threshold in the coping capacity of economic systems. In this analysis of Louisiana, indirect losses remain negligible (or even negative) for direct losses below $50 billion, and then increase nonlinearly to reach $200 billion for direct losses of the same amount (see Figure 4.1.7). Also, indirect losses decrease rapidly if it is possible to "import" reconstruction means (workers, equipment, finance) from outside the affected region. This result highlights the importance of taking into account interregional interactions in assessing indirect loss.

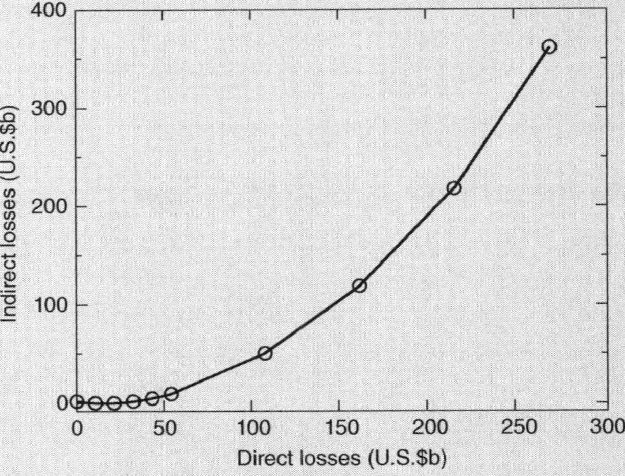

Figure 4.1.7 The direct losses – indirect (output) losses as a function of direct (asset) losses, in Louisiana for Katrina-like disasters. *Source:* Hallegatte (2008). Reproduced with permission of Wiley.

Another approach uses the impact of natural disasters on public finances. Its aim is to assess indirect costs in terms of the government's capacity to cope with large amounts of expenditure due to natural disasters and their subsequent ability to deliver basic services in the aftermath. This approach implicitly considers the impacts on public finances and their capacity to overcome these challenges as a proxy for indirect costs. One example of this is the IIASA CATSIM model developed by Mechler, Linnerooth-Bayer and Peppiat (2006) and applied to Honduras.

Another approach used to assess indirect costs is to develop idealised models aimed at emphasising one or more particular relation(s) or mechanism(s) at play in the economic system after a natural hazard. These include interactions between hazard impacts and technical change (Hallegatte and Dumas, 2008) or business cycles (Hallegatte and Ghil, 2008). Even though their main aim is not to assess the costs of extreme events directly, nonetheless they help to identify important mechanisms and to investigate their role.

Green, Vivattene and Thompson (2011) and Pfurtscheller, Lockner and Thieken (2011) suggest that the usefulness of CGE and Input Output models for decision support on smaller scales in particular may be debatable, as they fail to meet stakeholders' needs. For instance, most stakeholders are interested in assessing the indirect impact of various types of events, both large and small, at micro (cities) or meso (catchment) scale, with or without risk mitigation measures. However, most of the methods discussed can only assess the impacts of an extreme event on the national or sometimes regional scale. Furthermore, Green, Vivattene and Thompson (2011) regard the potential transfer of these methods to practitioners as being quite unrealistic. The reasons for this are that the models require a high degree of skill to run and entail complex mechanisms and uncertainties. Alternative approaches for assessing the indirect costs at a local scale are currently still lacking (Green, Vivattene and Thompson, 2011; Pfurtscheller, Lockner and Thieken, 2011).

4.1.2.4 Intangible (non-market) costs

Intangible (non-market) costs do not necessarily have to be expressed in monetary terms in order to be included in decision support frameworks. In a Multicriteria analysis framework, for example, they can be included as non-monetary decision criteria or, in a Cost-Effectiveness analysis framework, as a non-monetary target measure. However, to be included in a Cost-Benefit analysis framework, intangible costs have to be expressed in monetary terms. Methods for estimating the monetary value of intangible effects of natural hazards consider both use and non-use values that individuals derive from environmental or health goods and services, for instance. While use values relate to the direct, indirect or even optional use, non-use values relate to the value individuals derive from just knowing that a certain environmental good exists or that it is being preserved for future generations. Different valuation methods are proposed according to each type of (non-)use value. These methods can be categorised (as presented in Table 4.1.4) into:

1. revealed preference and
2. stated preference valuation methods.

Table 4.1.4 Intangible effects: methods, applications and examples.

General method	Specific method	Application and/or examples
Revealed preferences methods	Travel Cost method	Hartje, Meyer and Meyerhoff (2001)
	Hedonic Pricing method	Hamilton (2007), Chao, Floyd and Holliday (1998), Cavailhes *et al.* (2009)
	Cost of Illness approach	DEFRA (2007)
	Replacement Cost method	Leschine, Wellman and Green (1997)
	Production Function Approach	n.a.
Stated preferences methods	Contingent Valuation method	Birol, Karousakis and Koundouri (2006), Daun and Clark (2000), DEFRA (2004), Leiter and Pruckner (2007), Pattanayak and Kramer (2001b), Turner, Doktor and Adger (1993), Zhai and Ikeda (2006), Zhongmin *et al.* (2003)
	Choice Modelling method	Brouwer and Schaafsma (2009), Daun and Clark (2000), Hensher, Shore and Train (2006), Olschewski *et al.* (2011)
	Life Satisfaction Analysis	Carroll, Frijters and Shields (2009)
Benefit or Value Transfer methods		Martin-Ortega, González-Egiuno and Markandya (2012)

The accuracy and effectiveness of these valuation methods depend on the availability and quality of the data (including the quality of the survey design), the available resources and the decision made in each case on the most appropriate method for estimating the intangible effects. Revealed preference methods have the advantage of estimating the value of a particular good based on actual market behaviour, that is *ex post*. Information derived from observed behaviour is used to derive an individual's willingness to pay for an environmental improvement or for avoiding environmental deterioration. The main revealed preference methods are the following (see Pearce and Turner, 1990; Hanley and Spash, 1993; Birol, Karousakis and Koundouri, 2006):

- Travel Cost method: recreational or environmental sites are valued by analysing observed travel time and expenditure of visitors.
- Hedonic Pricing method: the value of environmental characteristics is estimated based on actual market prices, in particular from the housing market.
- Cost of Illness approach: costs of health impacts are estimated based on medical costs and lost wages due to illness.
- Replacement Cost method: the value of an ecosystem good or service is estimated based on the costs of replacing that good or service.
- Production Function approach: the value of an environmental good which is used to produce a market good is estimated based on the producer's production function.

In contrast, stated preference methods create a hypothetical or contingent market in a survey. They use willingness to pay, or willingness to accept compensation for relinquishing an environmental deterioration or to forego an environmental improvement. Important stated preferences approaches for estimating the environmental and health goods or services are the following (see Pearce and Turner, 1990; Hanley and Spash, 1993; Birol, Karousakis and Koundouri, 2006):

- Contingent Valuation (CV) method: in order to valuate non-market goods, people are asked in surveys about their willingness to pay to avoid a given decrement of this particular non-market good, or about their willingness to accept its deterioration by receiving a certain amount of compensation (see Box 4.1.5 as an example).
- Choice Modelling (CM) method: willingness to pay is elicited by choice experiments in which people can choose between different bundles of goods with varying characteristics. These can be either market or non-market goods.
- The Life Satisfaction Analysis: welfare estimations of public goods (health, environment) are estimated based on life satisfaction surveys.
- Furthermore, Benefit or Value Transfer methods can be used to transfer the results of previously applied valuation methods to a new case study in order to estimate the intangible costs.

Among the stated preference methods, CV has been the most commonly used method in valuating non-market goods and services for a long time. It has also been applied in the assessment of the intangible costs of natural hazards (see Turner, Doktor and Adger, 1993; Daun and Clark, 2000; DEFRA, 2004; Leiter and Pruckner, 2007).

Box 4.1.5 Contingent valuation method and willingness to pay for drought mitigation from watershed protection in Eastern Indonesia

A household survey was conducted in eastern Indonesia (500 face-to-face interviews with local farmers) to estimate the economic value of an ecosystem service of drought mitigation provided by tropical forest watersheds in Ruteng Park protected area to local agrarian communities in Indonesia. The mean (median) annual stated willingness to pay through an annual fee is \$2.79 (\$1.64) per household, which aggregates to a total annual value of \$27.000. The survey is part of a larger project on the economic analysis of protected areas. The study intends to provide signals to watershed managers and policy makers regarding the economic magnitude and spatial distribution of the local economic value of watershed protection.

Cost types addressed include intangible costs, that is loss of local farmers' welfare (well-being) due to decreased agricultural production. The study focused on agriculture as the impacted sector (services provided by protected watershed primarily contribute as inputs to agricultural production). Effort and resources required for such a survey are high due to the design and administration of a survey and skills needed to undertake econometric analysis.

The precision is expected to be moderate. Applying the CV method to a hardly measurable ecological service in a developing country setting includes a high risk of commodity and context mis-specification despite a good practice in survey design and administration. The authors themselves point out that because of the imprecision in their economic data, indices of ecological attributes and household opinions they do not recommend using the estimates to predict precise values of drought mitigation services.

The main problem in assessing the validity of the willingness to pay estimates is the absence of actual values against which to compare the results. However, validity of the willingness to pay can be tested by comparing the result with those from other valuation studies using other methods, the findings of cross-study analyses (e.g. meta-analyses or benefits transfer exercises), or simulated markets. Such comparisons often showed that CV is likely to slightly overestimate the actual value due to its hypothetical nature. Validity can also be evaluated by examining consistency of CV estimates with theoretical expectations derived from economic theory. For example, when the price of a good increases, consumption of that good declines. Moreover, a positive relationship between stated values and the respondent's disposable income is expected.

Source: Pattanayak and Kramer (2001a)

CM has become more popular in recent years. Both CV and CM can estimate economic values for any environmental resource. Their advantage over other methods is that they can estimate non-use values as well as use values. CM, however, additionally makes it possible to estimate the implicit values of its attributes, their implied ranking and the value of changing more than one attribute at a time (Hanley, Wright and Adamowicz, 1998; Bateman *et al.*, 2003).

Revealed and stated preferences methods are also applied when assessing intangible costs caused by risk mitigation measures. These include the Hedonic Pricing method, applied for aesthetic costs arising from land use planning measures (Hamilton, 2007) and land fragmentation and habitat loss due to infrastructure investments (Cavailhes *et al.*, 2009). In addition, CV and CM have been applied to value the intangible costs of infrastructure measures (Van der Heide *et al.*, 2008), hazard modification measures (Brouwer and Schaafsma, 2009) and emergency response and evacuation (Zhai and Ikeda, 2006).

4.1.2.5 Risk mitigation costs

Measures to mitigate risk can span over the following categories (Bouwer *et al.*, 2011, 2013):

1. risk management planning and adaptation plans
2. hazard modification
3. infrastructure
4. mitigation measures *sensu stricto*
5. communication
6. monitoring and early warning
7. emergency response and evacuation
8. financial incentives
9. risk transfer

Table 4.1.5 Risk mitigation costs: methods, applications and examples.

General method	Specific method	Application and/or examples
Market price	Cost of Implementation	Fuchs *et al.* (2007a), Morton *et al.* (2005), Pfurtscheller and Schwarze (2010) (emergency response and evacuation); Wegmann, Merz and Meierhans Steiner (2007), Pfurtscheller and Thieken (2013) (total costs of mitigation); Pattanayak and Kramer (2001, b), Stoffel (2005) (hazard modification); Fischer *et al.* (2005), Kind (2013) (infrastructure); Kreibich, Christenberger and Schwarze (2011) (mitigation measures – *sensu stricto*); European Commission (2008), Hallegatte (2012) (monitoring and early warning systems); Fisher *et al.* (1995), Woo (1994), Von Ungern-Sternberg (2004) (financial incentives); Michelsen and Young (1993), Schwarze *et al.* (2011) (risk transfer)
	Income Loss from Disruption of Economic Activity	Rogers and Tsirkunov (2010) (monitoring and early warning systems)

Table 4.1.5 presents the main methods to assess the direct costs attributed to risk mitigation.

Cost assessment for risk mitigation measures focuses almost exclusively on estimating direct costs, in particular research and design, and investment costs. Operation and maintenance costs as well as intangible and indirect costs are often neglected. The reason for the focus on direct costs is that they are most easily quantifiable. The direct costs of mitigation are based on readily available market prices that relate to either:

1. the actual cost of implementation
2. or income loss due to the disruption of economic activity (see Table 4.1.5).

The focus is often on direct investments in 'hard' risk mitigation measures (see e.g. Box 4.1.6), that is the categories infrastructure and mitigation measures (*sensu stricto*). Thus, with a few exceptions (see e.g. Wegmann, Merz and Meierhans Stetner, 2007 for Alpine hazards), comprehensive and comparable overviews of total mitigation costs, whether at regional or national level, are rarely available.

Although different approaches exist for estimating indirect and intangible costs of risk mitigation measures (see Sections 4.1.2.3 and 4.1.2.4), less emphasis is given to these costs in studies that focus on the cost assessment of these measures. Such costs can be important and their exclusion can lead to incomplete and biased estimates of the overall costs of risk reduction.

There also tends to be a heavy emphasis on implementation costs. The extent of operation and maintenance costs is not properly included, as these costs are often estimated

Box 4.1.6 Structural mitigation measures: Governmental expenditure for coastal protection in the north sea countries

Estimates have been made of governmental investments on flood protection infrastructure for different countries around the North Sea for the period 2000–2006. It is found from this study that actual government investments on flood protection are quite low, and far below 0.1% of GDP in the countries of Denmark, Germany, the Netherlands, Flanders (Belgium) and England (Figure 4.1.8). However, the report (SAFECOAST Synthesis Report, p.30) notes that "The figures are estimates and there are information gaps, and are therefore not strictly comparable since some of the figures also include flood risk management from river flooding. Hence, the figures may be observed in terms of their order of magnitude", and "Costs related to private flood insurance (e.g. England) or private ownership (e.g. Denmark) are not included in the table". For comparison, Switzerland invests 0.6% of its GDP for protection against natural hazards (including 37% for insurance) (Wegmann, Merz and Meierhans Stetner, 2007).

		Denmark	Germany	Netherlands	Flanders	England
Annual spending on coastal protection 2000-2006 averaged	[Mln € / yr]	11 [a] (coast)	110 [b] (coast)	550 [c] (coast / rivers)	20 [d] (coast)	620 [e] (coast / rivers)
GDP (nominal) Current prices	[Bln €] national	188	2156	468	274	1672
2000-2005 averaged	[Bln €] subnational		353 [b2]		157 [d2]	1434 [e2]

Sources: National and regional policy documents and master plans, National treasuries, Eurostat

a) DCA (2008) – Annual coastal protection spending on 110 km of Danish west coast in 2004-2008 are 86 mln Krone – Euro conversion fixed at 0,134. Dikes at Danish Wadden coast funded locally (no data, but expected to be below € 5 million).
b) Estimate total of average coastal protection spending in the 4 coastal states that border the North Sea.
b2) Cumulative (regional) GDP of 4 North Sea coastal states (sum of the figures as presented in table 2.1)
c) This amount includes total flood defence (rivers and coasts) spending from both national government and the water boards. The part of this amount that is allocated annually for coastal protection alone is estimated at 10-20% (among which annual nourishments of 45 million). Water in beeld, 2007 and Dutch National treasury (2008)
d) Figures by MOW MDK Flanders coastal division (~50% nourishments). Excluding spending related to the Sigma plan (880 million until 2030)
d2) GDP of Flanders region only (Vlaams Gewest).
e) Includes central and local government spending. Pound – Euro conversion fixed at 1,45 (Defra, 2006c)
e2) GDP England (excluding Scotland, Wales and Northern Ireland)

Figure 4.1.8 Estimates of current governmental expenditure and GDP in the North Sea countries. *Source:* SAFECOAST (2008).

by simply assessing them as percentages of construction costs (Bouwer *et al.*, 2011). Any reliable Cost–Benefit analysis of infrastructure investment (for mitigation of or adaptation to natural hazards) requires an accurate estimation of all costs associated with the inception and implementation of the project (i.e. during the asset's entire life cycle). The Whole Life Cycle Costing approach (see e.g. Langdon, 2007; Viavattene and Faulkner, 2012) attempts to provide such a systematic consideration of all present and future costs linked to risk mitigation investment (and assets more broadly). Monitoring and early warning activities next to structural measures have also received considerable attention (Holub and Fuchs, 2009; Holub, Suda and Fuchs, 2012), and there have also been some initial attempts to assess the costs of emergency response and evacuation measures (Fuchs *et al.*, 2007a; Pfurtscheller and Schwarze, 2010, Pfurtscheller and Thieken, 2013) in the case of Alpine hazards (see Box 4.1.7).

Box 4.1.7 Non-structural mitigation measures – An analysis of the costs of emergency services during and after the 2005 floods in Western Austria

Voluntary emergency and relief services, such as fire brigades or the Red Cross, form the backbone of disaster management in Austria and in other European countries. The assessment of the costs of these emergency services is a crucial component of risk and disaster management strategies. These costs can reach economically critical levels for (low-income) municipalities and often exceed public funds. Because empirical evidence on the costs is largely missing, especially for Alpine hazards, a case study was carried out in the aftermath of the 2005-flood event in the Federal State of Tyrol/Austria, to assess the costs of the federal fire departments. This flood event in Tyrol caused an estimated total direct loss of €410 million. Reports of the fire departments data stated that approximately 320 000 mission hours were needed to deal with the emergency, clean-up and evacuation. This translates into up to 13 400 persons at work per day, including fire departments and additionally about 1500 members of the armed forces, police and Red Cross. For the post-disaster spending it is necessary to count only the marginal (extra-occurring) costs. The costs of emergency can

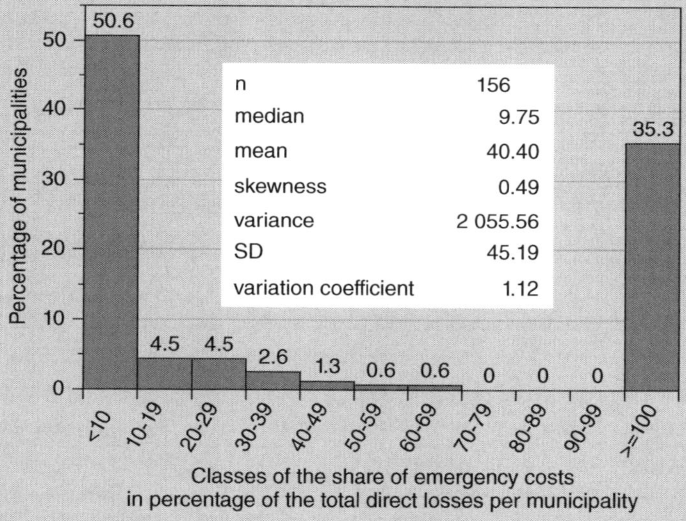

Figure 4.1.9 Histogram of the share of emergency costs as a percentage of of the total direct losses per municipality. *Source:* Pfurtscheller and Thieken 2013.

be computed by summing up the monetary equivalent of service hours, that is service hours multiplied by an average hourly wage of €20. This is the rate charged by the federal fire brigade association in Tyrol. Finally, the total emergency costs of the fire brigades were calculated as €2.4 million. These numbers can then be compared with the total direct losses occurring at the municipalities (€126 million). On average, the costs of emergency per municipality were about €15 390 (median €1 895) and the mean of emergency costs is about 40.4% of the total direct loss (median 9.8). A breakdown of the costs of emergency across municipalities showed that in more than 50% of the municipalities emergency costs were less than 10% of the total direct losses, but also that these costs were often the only listed cost category during the 2005 floods in Tyrol (Figure 4.1.9). More than 35% of the municipalities had to face emergency costs amounting to more than 100% of the total direct losses.

The study also identified triggers of emergency efforts of the fire brigades. Multiple regressions revealed that mainly the flood duration and the occurrence of multiple hazards have both a high impact on the total hours spent and, thus, the costs of emergency services.

Source: Pfurtscheller and Schwarze (2010), Pfurtscheller and Thieken (2013).

4.1.3 Conclusions and the way forward

Concluding from the foregoing state of the art overview of cost assessment methods of natural hazards, a range of knowledge gaps can be identified (see Meyer *et al.*, 2013 for a more detailed discussion of knowledge gaps, see also IPCC, 2012). These include (i) overarching problems related to the incompleteness and uncertainty of cost assessments. Such uncertainties are due to, (ii) the lack of data sources and (iii) shortcomings of existing methods. Further important knowledge gaps relate to (iv) the future dynamics of risk in cost assessments, (v) the distribution of costs within society and (vi) using cost assessments for decision support. Based on this, Meyer *et al.* (2013) extracted some key recommendations for practice and research (see Table 4.1.6).

Based on the compilation and synthesis of currently available and applied methods for cost assessments of natural hazard risks (see Section 4.1.2), and especially the identified knowledge gaps and recommendations described above, a vision for a cost assessment integrated into risk management as a way forward is outlined by the authors. "Vision" here refers to an outline of what cost assessment could look like in Europe in about 10 or 15 years from now considering the recommendations given rather than developing and describing a cost assessment and risk management framework in detail.

The objective of this vision is to outline a cost assessment framework that can be applied by different actors in risk management for their specific aims (context-specific and applicable), which includes all relevant cost types (comprehensive), considers and communicates uncertainties in an appropriate way (transparent) and accounts for changing hazards and risks (considers dynamics). In the following, basic steps of a framework which would integrate these guiding principles are proposed and described (also see Figure 4.1.10).

Table 4.1.6 Key recommendations for practice and research in the field of cost assessment of natural hazards (*Source*: Meyer *et al.* 2013).

Key Recommendations for practice and research

- Cost assessments are often incomplete and biased. In order to obtain a complete picture of the costs of natural hazards, not only direct costs but also costs due to business interruption, indirect and intangible/non-market costs as well as the costs of risk mitigation should be considered.
- Although improvements have been made over the last few decades, considerable uncertainties still exist in all parts of cost assessments. In any appraisal it is therefore important to identify the main sources of uncertainty at an early stage and try to reduce or handle them. Any residual uncertainties in cost estimates should be documented and communicated to decision makers.
- One of the main sources of uncertainty in the estimation of the costs of natural hazards is the lack of sufficient, comparable and reliable data. A framework for supporting data collection should be established at the European level, both for object-specific ex post damage data (event analysis) and risk mitigation costs. Such a framework should ensure sufficiently detailed information and minimum data quality standards to facilitate the development and consistency of European and national databases.
- In general, there is a need for a better understanding of the processes leading to damage so that they can be modeled appropriately. With regard to direct damages, multi-parameter damage models are needed that better capture the variety of damage influencing parameters, including resistance parameters.
- With regard to indirect costs, more research is needed to understand and to model how markets function outside the state of equilibrium and at different scales. This applies particularly to the dynamics of return to equilibrium after a hazardous event, the associated social and institutional interactions and how agent expectations are formed in situations of high uncertainty.
- With regard to intangible costs, further research is needed on the physical impacts of natural hazards on the environment and human health.
- With regard to the costs of risk mitigation, special emphasis should be given to a better estimation of the costs of non-structural measures.
- More research is needed on the impacts of climate and socio-economic change on the future costs of natural hazards and the costs of adaptation to these changes. At the same time, it remains important to determine how such findings can be integrated into cost assessment methods. In this respect, the exchange of knowledge between the natural hazard risk community and the climate change community should be improved.
- There is a need for appropriate tools, guidance and knowledge transfer to support decision makers when integrating cost assessment figures into their decision making process. Such tools or frameworks should communicate and consider uncertainties in cost figures and ensure the transparency of the decision rules.

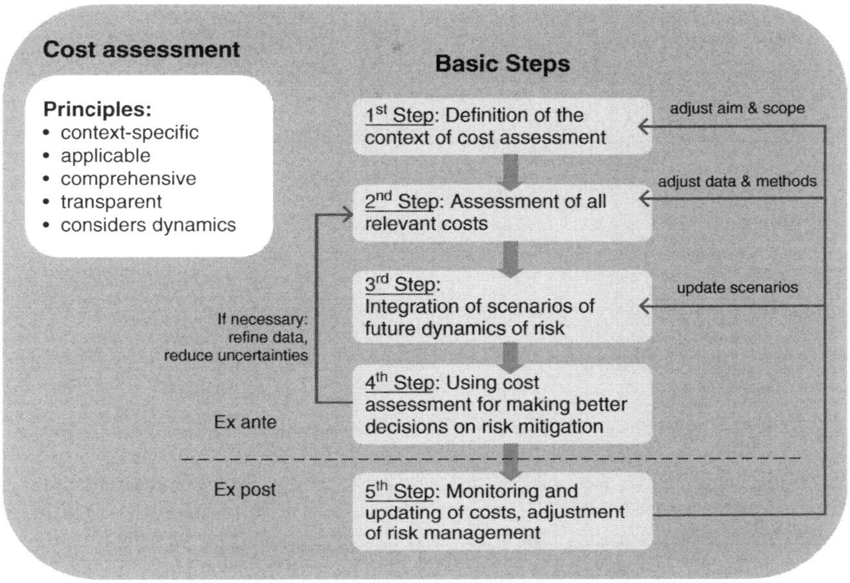

Figure 4.1.10 Cost assessment framework: guiding principles and basic steps.

Step 1: Definition of the context of cost assessment

- Cost assessments are always purpose-related. Consequently, the aim and scope of the assessment are defined first. This includes the system boundaries and relevant hazard(s) to be considered.
- Socio-economic aspects which may influence the system's recovery or response after a hazardous event are also taken into account.
- Based on these aspects the appropriate spatial scale and time horizon are determined.
- Taking preliminary assessments or expert judgements as a starting point, relevant cost categories are defined. "Relevant" means that there is (i) a considerable impact expected for these cost categories and (ii) that these cost categories could make a difference when evaluating alternative risk mitigation options. Relevant cost categories can be of any cost type, that is direct cost, costs due to business interruption, indirect costs, non-market/intangible costs, and also the costs of risk mitigation.
- Together with stakeholders, potential strategies and measures, which may help to reduce natural hazards' risks, are identified and pre-selected. This also includes the discussion about potential target levels of risk or risks which are considered unacceptable (even if costs for mitigating these risks are very high). Defining these constraints is an important social discourse of defining system boundaries for cost assessments. As the level of acceptable risks is changing in time, the risk preferences of stakeholders need to be investigated and risk discourses repeated on a regular basis as part of Step 5.

Step 2: Assessment of all relevant costs

- Costs are assessed for all relevant cost categories identified in Step 1. Appropriate methods can be identified and selected from a toolbox which provides an overview of existing methods for assessing different cost types along with their specific properties and potential fields of application.
- For intangible/non-market costs decision makers have to decide if it is necessary or helpful to include them in monetary terms or if they wish to consider them in a non-monetary or even qualitative way. This decision has to be reflected in the choice of the decision support framework used in Step 4.
- In order to ensure comprehensiveness, it is important not only to evaluate the shock to the system (i.e. the immediate impact of the event on the economy), but also the long-term welfare effects, that are based on the ability and path of the system to recover from and respond to the shock. Considering different scenarios of recovery, their effects on the growth trajectory of the system and hence their opportunity costs are also relevant for estimating total costs.
- The toolbox further provides guidance on how to combine methods (e.g. for different impacted sectors, for different hazards and combinations of hazards, and for different cost types). This also contributes to achieving comprehensiveness of the cost assessment, but also prevents double-counting.
- Additionally, the toolbox contains information on the data requirements of the chosen methods and provides a link to potential databases or to the required data collection approaches where no secondary data sources are available. A meta-database, linked to the toolbox, provides information on existing databases relevant for cost assessment of natural hazards, such as for example *ex post* damage data, mitigation costs, land-use and asset value data, and Input–Output tables.
- For all measures selected in Step 1 (including a baseline scenario) costs are assessed for a range of different hazard scenarios with different intensities and different probabilities of occurrence. Including cost estimates under various scenarios enables a proper estimate of the shape of a loss-probability curve, ranging from frequent, high probability events to extreme, low probability events. Based on this, the potential variety of costs can be outlined and the expected annual average costs can be estimated. Scenarios assuming the failure of risk mitigation measures are also included to consider the risk of failure.
- Natural and epistemic uncertainties in cost estimates are made transparent in the results and are communicated to the decision makers. While natural uncertainty stems from the variability of the underlying stochastic process and cannot be reduced, epistemic uncertainty results from incomplete knowledge about the process under study, for example from aggregated input data or model uncertainties. The latter can be at least partly reduced by improving the database and the models.

Step 3: Integration of scenarios of future dynamics of risk

- Natural hazard risk is essentially dynamic, depending on climate variability and change, as well as on changes in vulnerability patterns. In order to account for this, dynamic scenarios for the future development of major risk drivers are developed and included in the cost assessment, as described in Step 2.

- Stakeholders should be involved in all stages of developing such scenarios to ensure a realistic and deliberately democratic representation of the evolution of risk drivers. Potential or likely changes in the cost estimates based on these scenarios are described and their influence on the evaluation of risk mitigation measures is depicted (Step 4).
- Uncertainties pertinent to the dynamic scenarios should be clearly communicated. If necessary, a sensitivity analysis could study alternative dynamic scenarios and their implications for hazard impacts and mitigation policies.

Step 4: Using cost assessment for making better decisions on risk mitigation

- The aim of cost assessment is to support decision makers in selecting alternative risk mitigation options by providing them with informed and well-reasoned arguments. Cost estimates are therefore integrated in decision support frameworks (as, for example CBA or MCA), which help the decision makers to evaluate the different risk mitigation options.
- It is made transparent to the decision makers how the choice between different decision support frameworks, their associated decision rules or the selection and weighting of evaluation criteria influences the outcome of the evaluation and the ranking of options.
- Uncertainties in the results are explicitly communicated to the decision makers and guidance is provided on how to interpret or use this uncertain information. If decision makers feel that more detailed or precise cost figures are needed for making a decision, more efforts on data collection and modelling are required for the pre-selected options to reduce the most important sources of epistemic uncertainty (see Step 2).
- Other criteria, such as robustness (performance of an option with regard to different risk development scenarios) and flexibility (ability to adjust an option according to future risk developments), are also considered in the evaluation of risk mitigation options to show their ability in dealing with different development scenarios.

Step 5: Monitoring and updating costs and adjusting risk management

- Actual losses caused by natural hazards and real expenditures for risk reduction are frequently monitored. Such *ex post* evaluations of object-specific damages or risk mitigation costs are entered into databases and are utilised to update, improve, validate and adjust cost assessment models (see also recommendations in Section 4.1.3.5) and, hence, cost estimates (see Step 2).
- Furthermore, new or updated information on the expected development of major risk drivers is used to update cost estimates (see Step 3).
- Regular checks confirm if such new insights or other developments lead to necessary adjustments in the decision context of risk management (see Step 1: aim, scope, scale of the assessment, target level of risks).
- Updated cost estimates are used for a new evaluation of risk mitigation options (Step 4). If necessary, decisions are revised or chosen risk mitigation options are adjusted (if possible, see "flexibility" in Step 4).

The presented integrated cost assessment framework relates to a better decision-making in natural hazard risk management. This can be understood as a starting point towards a more long-term perspective that emphasises the integration of cost assessment into a wider sustainable development framework. Furthermore, it is a step towards a more iterative understanding of decision-making which focuses on learning and revision, and at the same time not only allows for but also requires stakeholder engagement.

References

Albala-Bertrand J.M. (1993) *The Political Economy of Large Natural Disasters with Special Reference to Developing Countries*, Oxford, UK: Clarendon Press.

Apel H., Aronica G.T., Kreibich H. and Thieken A.H. (2009) Flood risk analyses – how detailed do we need to be?, *Natural Hazards*, **49**, 79–98.

BAFU (2010) Swiss Federal Office for the Environment (FOEN): EconoMe 2.0: Online-Berechnungsprogramm zur Bestimmung der Wirtschaftlichkeit von Schutzmassnahmen gegen Naturgefahren, Eidgenössisches Departement für Umwelt, Verkehr, Energie und Kommunikation UVEK, Bundesamt für Umwelt.

Bateman I., Carson R., Day B. *et al.* (2003) *Guidelines for the use of stated preference techniques for the valuation of preferences for non-market goods*, Cheltenham, UK: Edward Elgar Publishing.

Benson C. and Clay E. (1998) *The impact of drought on Sub-saharan African economies*, World Bank, Washington D.C.

Berrittella M., Hoekstra Y.A., Rehdanz K. *et al.* (2007) The economic impact of restricted water supply: A computable general equilibrium analysis, *Water Research*, **41**, 1799–1813.

Birol E., Karousakis K. and Koundouri P. (2006) Using economic valuation techniques to inform water resources management: A survey and critical appraisal of available techniques and an application, *Science of the Total Environment*, **365**, 105–122.

BMLFUW- Federal Ministry of Agriculture, Forestry, Environment and Water Management (2008a) *Richtlinien für die Wirtschaftlichkeitsuntersuchung und Priorisierung von Maßnahmen der Wildbach- und Lawinenverbauung gemäß § 3 Abs.* 2 Z 3 Wasserbautenförderungsgesetz 1985, Vienna.

BMLFUW- Federal Ministry of Agriculture, Forestry, Environment and Water Management (2008b) *Kosten-Nutzen-Untersuchungen im Schutzwasserbau Richtlinie*, KNU gemäß § 3 Abs. 2 Ziffer 3 Wasserbautenförderungsgesetz, Vienna.

Boarnet M.G. (1998) Business losses, transportation damages, and the Northridge earthquake, *Journal of Transportation and Statistics*, **1**, 49–63.

Booker J.F., Michelsen A.M. and Ward, F.A. (2005) Economic impact of alternative policy responses to prolonged and severe drought in the Rio Grande Basin, *Water Resources Research*, **41**, W02026, doi:10.1029/2004WR003486.

Booker J.F. (1995) Hydrologic and economic impacts of drought under alternative policy responses, *Water Resources Bulletin*, **31**, 889–906.

Bouwer L.M., Papyrakis E., Poussin J. *et al.* (2013) The costing of measures for natural hazard mitigation, *Natural Hazards Review*, 10.1061/(ASCE)NH.1527-6996.0000133.

Bouwer L.M., Poussin J., Papyrakis E. *et al.* (2011) *Methodology report on costs of mitigation*, CONHAZ Report, http://conhaz.org/CONHAZ%20REPORT%20WP04_2.pdf.

Boyd R. and Ibarrarán M.E (2009) Extreme climate events and adaptation: an exploratory analysis of drought in Mexico, *Environment and Development Economics*, **14**, 371–395.

Brouwer R. and Schaafsma M. (2009) The economics of flood disaster management in the Netherlands, in: *Natural Disasters: Do They Cost the Earth?* (eds P. Guha Sapir, I. Santos and A. Borde) Earthscan Publications.

Bubeck P. and Kreibich, H. (2011) *Natural Hazards: direct costs and losses due to the disruption of production processes*, CONHAZ Report, http://conhaz.org/CONHAZ%20REPORT%20WP01_2.pdf.

Büchele, B., Kreibich, H., Kron, A. *et al.* (2006) Flood-risk mapping: Contributions towards an enhanced assessment of extreme events and associated risks, *Natural Hazards and Earth System Sciences*, **6**, 485–503.

BUWAL- Bundesamt für Umwelt, Wald und Landschaft (1999) Risikoanalyse bei gravitativen Naturgefahren, Umweltmaterialien 107/I, Bern.

Cammerer H., Thieken A.H. and Lammel J. (2013) *Adaptability and transferability of flood loss functions in residential areas. Natural Hazards and Earth System Sciences (NHESS-D)* (http://www.nat-hazards-earth-syst-sci-discuss.net/1/3485/2013/nhessd-1-3485-2013.html)

Carroll N., Frijters P. and Shields M.A. (2009) Quantifying the costs of drought: new evidence from life satisfaction data, *Journal of Population Economics*, **22**, 445–461.

Cavailhes J., Foltete J-C., Joly D. and Tritz C. (2009) GIS based hedonic pricing of landscape, *Environmental and Resource Economics*, **44**(4), 571–590.

Cavallo E. and Noy I. (2009) *The economics of natural disasters - a survey*, Inter-American Development Bank Working Paper No 124.

Chao P.T., Floyd J.L. and Holliday W. (1998) *Empirical Studies of the Effect of Flood Risk on Housing Prices*, Institute for Water Resources - U.S. Army Corps of Engineers, IWR REPORT 98-PS-2.

COPA-COGECA (2003) Assessment of the impact of the heat wave and drought of the summer 2003 on agriculture and forestry, Committee of Agricultural Organisations in the European Union and the General Committee for Agricultural Cooperation in the European Union.

Corti T., Muccione V., Köllner-Heck P. *et al.* (2009) Simulating past droughts and associated building damages in France, *Hydrology and Earth System Sciences*, **13**, 1739–1747.

Daun C.M. and Clark, D. (2000) *Flood Risk and Contingent Valuation Willingness to Pay Studies: A Methodological Review and Applied Analysis, Institute for Urban Environmental Risk Management*, Marquette University, Milwaukee, WI.

DEFRA (2004) *The Appraisal of Human-Related Intangible Impacts of Flooding*, Environment Agency Flood and Coastal Defense, R&D Technical Report FD2005/TR.

DEFRA (2007) *The costs of the summer 2007 floods in England*, Project: SC070039/R1.

Department for Transport (2009) *Transport Analysis Guidance*, London, Department of Transport.

Easterling W. and Mendelsohn R. (2000) Estimating the economic impacts of drought on agriculture, in: Drought: A Global Assessment, (ed. D.A. Wilhite) vol. 1, pp. 256–268, London/New York: Routledge.

Elmer F., Thieken A.H., Pech I. and Kreibich H. (2010) Influence of flood frequency on residential building losses, *Natural Hazards and Earth System Sciences*, **10**, 2145-2159, doi: 10.5194/nhess-10-2145-2010.

EM-DAT (2013) *The OFDA/CRED International Disaster Database* – www.emdat.be – Université Catholique de Louvain Brussels – Belgium.

Emschergenossenschaft, Hydrotec (2004) *Hochwasser-Aktionsplan Emscher, Kapitel 1: Methodik der Schadensermittlung*, Report.

European Commission (EC) (2008) *Assessing the potential for a comprehensive community strategy for the prevention of natural and manmade disasters*, Final Report, http://ec.europa.eu/echo/civil_protection/civil/pdfdocs/stakeholders/potential_prevention_strategy.pdf, 2008.

FEMA (2011) *Hazus®-mh mr5*. Flood model, Federal Emergency Management Agency, Technical Manual.

Fink A., Brücher T., Krüger A. *et al.* (2004) The 2003 European summer heatwaves and drought – synoptic diagnosis and impacts, *Weather*, **59**, 209–216.

Fischer G., Shah M., Tubiello F.N. and van Velhuizen H. (2005) Socio-economic and climate change impacts on agriculture: an integrated assessment, 1990–2080, *Philosophical Transactions of the Royal Society of London, Series B Biological Sciences*, **360**, 2067–2083.

Fisher A., Fullerton D., Hatch N. and Reinelt P. (1995) Alternatives for managing drought: a comparative cost analysis, *Journal of Environmental Economics and Management*, **29**, 304–320.

Fuchs S., Heiss K. and Hübl J. (2007a) Towards an empirical vulnerability function for use in debris flow risk assessment, *Natural Hazards and Earth System Sciences*, **7**, 495–506.

Fuchs S., Thöni M., McAlpin, M.C. *et al.* (2007b) Avalanche hazard mitigation strategies assessed by cost effectiveness analysis and cost benefit analysis – Evidence from Davos, Switzerland, *Natural Hazards*, **41** (1), 113–129.

Glade T. (2003) Vulnerability assessment in landslide risk analysis, *Die Erde*, **134**, 123–146.

Grafton R.Q. and Ward M.B. (2008) Prices versus rationing: Marshallian surplus and mandatory water restrictions, *The Economic Record*, **84**, 57–65.

Green C., Viavattene C. and Thompson P. (2011) *Guidance for assessing flood losses*, CONHAZ Report, http://conhaz.org/CONHAZ%20REPORT%20WP06_1.pdf.

Grossmann M., Koch H., Lienhoop N., *et al.* (2011) *Economic risks associated with low flows in the Elbe River Basin (Germany)*: an integrated economic-hydrologic approach to assess vulnerability to climate change, GLOWA-Elbe working paper, http://www.glowa-elbe.de/pdf/schlussbericht_glowaii/Arbeitspapiere/grossmann.pdf.

H. John Heinz III Center for Science, Economics, and the Environment (2000) *The Hidden Costs of Coastal Hazards: Implications for Risk Assessment and Mitigation*, Washington, DC: Island Press, 252 pp.

Haimes Y., Horowitz B., Lambert J. *et al.* (2005) Inoperability input-output model for interdependent infrastructure sectors. I: Theory and methodology, *Journal of Infrastructure Systems* **11**, 67–79.

Hallegatte S. and Dumas P. (2008) Can natural disasters have positive consequences? *Investigating the role of embodied technical change, Ecological Economics*, **68**, 777–786.

Hallegatte S. and Ghil M. (2008) Natural Disasters Impacting a Macroeconomic Model with Endogenous Dynamics, *Ecological Economics*, **68**, 582–592.

Hallegatte, S. (2012) *A Cost Effective Solution to Reduce Disaster Losses in Developing Countries Hydro-Meteorological Services, Early Warning, and Evacuation, World Bank, Policy Research Working Paper 6058*, Washington DC.

Hallegatte S. (2008) An Adaptive Regional Input-Output Model and Its Application to the Assessment of the Economic Cost of Katrina, *Risk Analysis*, **28**, 779–799.

Hamilton M.J. (2007) Coastal landscape and the hedonic price of accommodation, *Ecological Economics*, **62**, 594–602.

Hanley N. and Spash C. (1993) *Cost Benefit Analysis and the Environment*, Cheltenham, UK: Edward Elgar Publishing.

Hanley N., Wright R.E. and Adamowicz W.L. (1998) Using choice experiments to value the environmental, *Environmental and Resource Economics*, **11**(3–4), 413–28.

Hartje V., Meyer I. and Meyerhoff J. (2001) Kosten einer möglichen Klimaänderung auf Sylt, in: *Sylt – Klimafolgen für Mensch und Küste*, (eds A. Daschkeit and P. Schottes)), Springer, Berlin, Germany.

Hensher D., Shore N. and Train K. (2006) Water supply security and willingness to pay to avoid drought restrictions, *The Economic Record*, **82**, 56–66.

Hochrainer S. (2009) *Assessing the Macroeconomic Impacts of Natural Disasters – Are There Any?*, World Bank Policy Research Working Paper 4968.

Holden S. and Shiferaw B. (2004) Land degradation, drought and food security in a less-favoured area in the Ethiopian highlands: a bio-economic model with market imperfections, *Agricultural Economics*, **30**, 31–49.

Holub M. and Fuchs S. (2009) Mitigating mountain hazards in Austria – Legislation, risk transfer, and awareness building, *Natural Hazards and Earth System Sciences*, **9**(2), 523–537.

Holub M., Suda J. and Fuchs S. (2012) Mountain hazards: reducing vulnerability by adapted building design, *Environmental Earth Sciences*, **66** (7), 1853–1870.

Horridge M., Madden J. and Wittwer G. (2005) The impact of the 2002–2003 drought on Australia, *Journal of Policy Modeling*, **27**, 285–308.

Huttenlau M. (2010) *Risk-based consequences of extreme natural hazard processes in mountain regions – Multi-risk analysis of extreme loss scenarios in Tyrol (Austria)*, Ph.D. thesis, University of Innsbruck, Austria.

ICPR (2001) *Atlas of flood danger and potential damage due to extreme floods of the Rhine*, Koblenz, International Commission for the Protection of the Rhine.

IPCC (2012) Summary for policymakers in: *Managing the Risks of Extreme Events and Disasters to Advance Climate Change Adaptation*, (eds C.B. Field, V. Barros, T.F. Stocker *et al.*). A Special Report of Working Groups I and II of the Intergovernmental Panel on Climate Change, Cambridge, UK, and New York, NY, USA: Cambridge University Press, pp. 3–21.

IWK- Institut für Wasserwirtschaft und Kulturtechnik, Universität Karlsruhe (1999) Auswertung der HOWAS-Datenbank.

Jaramillo C.R.H. (2009) *Do Natural Disasters Have Long-Term Effects on Growth?*, Universidad de los Andes, mimeo.

Karunasena G. and Rameezdeen R. (2010) Post-disaster housing reconstruction: Comparative study of donor vs owner-driven approaches, *International Journal of Disaster Resilience in the Built Environment*, **1**(2), 173–191.

Keiler M., Sailer R., Jorg P. *et al.* (2006) Avalanche risk assessment - a multi-temporal approach, results from Galtür, *Austria, Natural Hazards and Earth System Sciences* **6**, 637–651.

Keiler M., Zischg A., Fuchs S. *et al.* (2005) Avalanche related damage potential – Changes of persons and mobile values since the mid-twentieth century, case study Galtür, *Natural Hazards and Earth System Sciences*, **5**(1), 49–58.

Kind J.M. (2014) Economically efficient flood protection standards for the Netherlands, *Journal of Flood Risk Management*, **7**(2), 103–117.

Klaus J., Pflügner W., Schmidtke R.F. *et al.* (1994) Models for flood hazard assessment and nanagement, in: *Floods Across Europe, Hazard Assessment, Modelling and Management*, (eds E.C. Penning-Rowsell and M. Fordham), Middlesex University Press, London, UK, 67–106.

Kleindorfer P. and Kunreuther H. (1999) Challenges facing the insurance industry in managing catastrophic risks, in *The Financing of Catastrophe Risk*, (ed. K.A. Froot), University of Chicago Press, http://www.nber.org/books/froo99-1.

Kok M., Huizinga H.J., Vrouwenfelder A.C.W.M. and Barendregt A. (2004) Standard Method 2004. *Damage and Casualties caused by Flooding*. Client: Highway and Hydraulic Engineering Department.

Kreibich H., Christenberger S. and Schwarze R. (2011) Economic motivation of households to undertake private precautionary measures against floods, *Natural Hazards and Earth System Sciences*, **11**, 309–321.

Kreibich H., Seifert I., Merz B. and Thieken A.H. (2010) Development of FLEMOcs – A new model for the estimation of flood losses in the commercial sector, *Hydrological Sciences Journal*, **55**, 1302–1314.

Kreibich H., Thieken A.H., Petrow T.H. *et al.* (2005) Flood loss reduction of private households due to building retrofitting – Lessons learned from the Elbe floods in August 2002. *Natural Hazards and Earth System Sciences,* **5**, 117–126, SRef-ID: 1684-9981/nhess/2005-5-117.

Kroll C.A., Landis J.D., Shen Q. and Stryker S. (1991) *Economic Impacts of the Loma Prieta Earthquake: A Focus on Small Business, Studies on the Loma Prieta Earthquake*, University of California, Transportation Center.

Kulshreshtha S.N. and Klein K.K. (1989) Agricultural drought impact evaluation model: a systems approach, *Agricultural Systems*, **30**, 81–96.

Laframboise N. and Loko B. (2012) *Natural Disasters: Mitigating Impact, Managing Risks*, IMF Working Paper WP/12/245.

Langdon D. (2007) *Life cycle costing (LCC) as a contribution to sustainable construction: A common methodology*, Literature Review, Davis Langdon Management Consulting.

Laternser M. (2000) Der Lawinenwinter von 1999. *Fallstudie Goms (Konton Wallis)*. Versorgunsglage, Bewältigung der Krisensituation und wirtschaftliche Auswirkungen, Davos, Eidgenössisches Institut für Schnee und Lawinenforschung (SLF).

Leiter, A. and Pruckner G. (2007) *Dying in an Avalanche: current risks and Valuation, working paper*, University of Innsbruck, Austria.

Lequeux Q. and Ciavola P. (2011) *Methods for Estimating the Costs of Coastal Hazards, CONHAZ Report*, http://conhaz.org/CONHAZ%20REPORT%20WP07_1_FINAL.pdf.

Leschine T.M., Wellman K. and Green T.H. (1997) *The economic value of wetlands role in flood protection in western Washington*. Washington State Department of Ecology, Ecology publication No. 97–100.

Loayza N., Olaberria E., Rigolini J. and Christiansen L. (2009) *Natural Disasters and Growth-Going beyond the Averages*, World Bank Policy Research Working Paper 4980.

Logar I. and van den Bergh J.C.J.M. (2011) Methods for Assessment of the Costs of Droughts, CONHAZ Report, http://conhaz.org/CONHAZ%20REPORT%20WP05_1_FINAL.pdf.

Logar I. and van den Bergh J.C.J.M. (2013) Methods to assess costs of drought damages and policies for drought mitigation and adaptation: review and recommendations, *Water Resources Management*, **27**, 1707–1720.

MAFF- Ministry of Agriculture Fisheries and Food (1999) *Flood and Coastal Defence Project Appraisal Guidance*. Economic Appraisal, London.

Maiwald H. and Schwarz J. (2010) Von der Schadensaufnahme zur Verletzbarkeitsfunktion – ein Ansatz aus den Ingenieurwissenschaften, in: Thieken, A.H., Seifert, I., and Merz, B. (Eds.): Hochwasserschäden – Erfassung, Abschätzung und Vermeidung, oekom, Munich, Germany.

Markantonis V., Meyer V. and Schwarze R. (2011) *The intangible effects of Natural Hazards, CONHAZ Report*, http://conhaz.org/CONHAZ%20REPORT%20WP03_2.pdf.

Markantonis V., Meyer V. and Schwarze R. (2012) Valuating the intangible effects of natural hazards, Review and analysis of the costing methods, *Natural Hazards and Earth System Sciences*, **12**, 1633–1640.

Martin-Ortega J., González-Eguino M. and Markandya A. (2012) The costs of drought: the 2007-2008 case of Barcelona, *Water Policy*, **14**, 539–560.

McCarty C. and Smith S.K. (2005) *Florida's 2004 Hurricane Season: Local Effects*, Florida Focus, University of Florida, http://www.bebr.ufl.edu/sites/default/files/FloridaFocus1_3_2005_0.pdf.

Mechler R., Linnerooth-Bayer J. and Peppiatt D. (2006) *Microinsurance for Natural Disasters in Developing Countries: Benefits, Limitations and Viability*, ProVention Consortium, Geneva, http://www.proventionconsortium.org/themes/default/pdfs/Microinsurance_study_July06.pdf.

Meyer, V. and Mai, S., (2003) *Verfahren zur Berechnung der Schäden nach Deichbruch an der deutschen Nordseeküste. in Neue Ergebnisse der Küsten- und Meeresforschung. Tagungsband der 21. Jahrestagung des Arbeitskreises "Geographie der Meere und Küsten" (AMK) 2003 in Essen*. Essener Geographische Arbeiten 35. S. 169–178.

Meyer V., Becker N., Markantonis V. *et al.* (2013) Assessing the Costs of Natural Hazards - State-of-the-art and Knowledge Gaps, *Natural Hazards and Earth System Sciences*, **13**, 1351–1373.

Michelsen M.A. and Young R.A. (1993) Optioning agricultural water rights for urban water supplies during drought, *American Journal of Agricultural Economics*, **75**, 1010–1020.

Morton J., Barton D., Collinson C. and Heath B. (2005) *Comparing Drought Mitigation Interventions in the Pastoral Livestock Sector*, Natural Resources Institute report, Greenwich, NRI.

Munich Re (2006) Topics – Annual review: natural catastrophes 2005. München.

Munich Re (2012) *Topics Geo: Natural catastrophes 2011, Analyses, Assessments*, positions. Munich, Germany.

Munich Re (2013) Topics Geo Natural catastrophes 2012. München.

MURL (2000) Potentielle Hochwasserschäden am Rhein in Nordrhein-Westfalen, Ministerium für Umwelt, Raumordnung und Landwirtschaft des Landes Nordrhein-Westfalen.

Nadal N., Zapata R., Pagán I. *et al.* (2010) Building Damage due to Riverine and Coastal Floods, *Journal of Water Resources Planning and Management*, **136**(3), 327–336, doi: 10.1061/(ASCE)WR.1943-5452.0000036.

Nicholas J., Holt G.D. and Proverbs D. (2001) Towards standardizing the assessment of flood damaged properties in the UK, *Structural Survey*, **19**(4), 163–172.

Nöthiger C. (2003) *Naturgefahren und Tourismus in den Alpen - untersucht am Lawinenwinter 1999 in der Schweiz, Eidgenössisches Institut für Schnee und Lawinenforschung (SLF)*, Davos.

Noy I. and Vu T. (2009) *The Economics of Natural Disasters in Vietnam*. University of Hawaii Working paper: 09-03.

Noy I. and Nualsri A. (2007) *What do exogenous shocks tell us about growth theories? Working Paper 07-28*, University of Hawaii.

Noy I. (2009) The macroeconomic consequences of disasters, *Journal of Development Economics*, **88**, 221–231.

NR&M (2002) *Guidance on the assessment of tangible flood damages, Department of Natural Resources and Mines*, Queensland.

NRE (2000) *Rapid appraisal method (ram) for floodplain management, Victorian Department of Natural Resources and Environment*, Victoria, Melbourne.

Okuyama Y., Hewings G. and Sonis M. (2004) Measuring the economic impacts of disasters: Interregional input-output analysis using the sequential interindustry model, in: *Modeling Spatial and Economic Impacts of Disasters*, (eds Y. Okuyama and S. Chang) Heidelberg, Germany: Springer.

Olschewski R., Bebi P., Teich M. *et al.* (2011) *Avalanche protection by forests - A choice experiment in the Swiss Alps, Forest Policy and Economics, Available online 12 November 2011*, ISSN 1389-9341, 10.1016/j.forpol.2011.10.002.

Parker D., Green C. and Thompson C.S. (1987) *Urban Flood Protection Benefits: A Project Appraisal Guide*, Aldershot, UK: Gower Technical Press.

Pattanayak S.K. and Kramer R.A. (2001a) Pricing ecological services: Willingness to pay for drought mitigation from watershed protection in eastern Indonesia, *Water Resources Research*, **37**, 771–778.

Pattanayak S.K. and Kramer R.A. (2001b) Worth of watersheds: a producer surplus approach for valuing drought mitigation in Eastern Indonesia, *Environment and Development Economics*, **6**, 123–146.

Paudel Y., Botzen W.J.W. and Aerts J.C.J.H. (2013) Estimation of insurance premiums for coverage against natural disaster risk: an application of Bayesian Inference, *Natural Hazards and Earth System Sciences*, **13**, 737–754, doi:10.5194/nhess-13-737-2013.

Pauw K., Thurlow J. and van Seventer D. (2010) *Droughts and floods in Malawi: assessing the economy wide effects*, IFPRI Discussion paper 00962, Washington, DC, International Food Policy Research Institute, http://www.preventionweb.net/files/13792_ifpridp009621.pdf.

Pearce D.W. and Turner K. (1990) *Economics of Natural Resources and the Environment*, Harvester Wheatsheaf, Hemel Hempstead, Hertfordshire, UK: Johns Hopkins University Press, 378 pp.

Penning-Rowsell E., Johnson C., Tunstall S. *et al.* (2003) *The benefits of flood and coastal defence: techniques and data for 2003*. Enfield, Flood Hazard Research Centre.

Penning-Rowsell E.C. and Chatterton J.B. (1977) *The Benefits of Flood Alleviation: A Manual of Assessment Techniques*, Aldershot, UK: Gower Technical Press.

Pfurtscheller C., and Schwarze R. (2010) Kosten des katastrophenschutzes, in: *Hochwasserschäden - Erfassung, Abschätzung und Vermeidung*, (eds A.H. Thieken, I. Seifert and B. Merz, B.) oekem, München, Germany, 253–262.

Pfurtscheller C. and Thieken A.H. (2013) The price of safety: costs for mitigating and coping with Alpine hazards, Natural Hazards and Earth System Sciences, **13**, 2619–2637.

Pfurtscheller P., Lochner B. and Thieken A.H. (2011) Costs of Alpine Hazards, CONHAZ Report, http://conhaz.org/CONHAZ%20REPORT%20WP08_1_FINAL.pdf.

Przyluski V. and Hallegatte S. (2011) Indirect Costs of Natural Hazards, CONHAZ Report, http://conhaz.org/CONHAZ%20REPORT%20WP02_2_FINAL.pdf.

Raddatz C. (2009) *The Wrath of God: Macroeconomic Costs of Natural Disasters*, World Bank Policy Research Working Paper 5039, The World Bank, Washington D.C.

Reese S., Markau H.-J. and Sterr H. (2003) *MERK – Mikroskalige Evaluation der Risiken in überflutungsgefährdeten Küstenniederungen*, Abschlussbericht, Kiel.

Rijkswaterstaat (2004) *Droogtestudie Nederland: Samenvattend rapport fase 2a*. Inhoudelijke analyse, Resultaten droogtestudie Nederland.

Rogers D. and Tsirkunov V. (2010) *Costs and benefits of early warning systems. Global Assessment Report on Disaster Risk Reduction (GAR 2011)*, UN-ISDR and the World Bank, http://preventionweb.net/english/hyogo/gar/2011/en/bgdocs/Rogers_&_Tsirkunov_2011.pdf.

Rose A. and Liao S.-Y. (2005) Modeling regional economic resilience to disasters: A computable general equilibrium analysis of water service disruptions, *Journal of Regional Science*, **45**, 75–112.

Rose A. and Miernyk W. (1989) Input-output analysis: The first fifty years, *Economic System Research*, **1**, 229–271.

Rose A., Oladosu G. and Liao S.Y. (2007) Business Interruption Impacts of a Terrorist Attack on the Electric Power System of Los Angeles: Customer Resilience to a Total Blackout, *Risk Analysis*, **27**, 513–531.

Rosenberg N.J. (1993) A methodology called 'mink' for study of climate change impacts and responses on the regional scale: an introductory editorial, *Climatic Change*, **24**, 1–6.

Sächsische Staatskanzlei (2003) *Schadensausgleich und Wiederaufbau im Freistaat Sachsen*, Dresden, Germany.

Safecoast (2008) Coastal flood risk and trends for the future in the North Sea region. Results and recommendations of Project Safecoast, Synthesis report, 136 pp.

Scawthorn C., Flores P., Blais N. *et al.* (2006) HAZUS-MH Flood Loss Estimation Methodology. II. Damage and Loss Assessment, *Natural Hazards Review*, **7**(2), 72–81.

Schwarze R., Schwindt M., Weck-Hannemann H. *et al.* (2011) Natural hazard insurance in Europe: Tailored responses to climate change are needed, *Enviromental Policy and Governance*, **21**, 14–30.

Seifert I., Kreibich H., Merz B. and Thieken A.H. (2010) Application and validation of FLEMOcs - a flood loss estimation model for the commercial sector. *Hydrological Sciences Journal*, **55**(8), 1315–1324.

Sholwater P.S. and Lu Y. (2013) *Geospatial Techniques in Urban Hazard and Disaster Analysis.* Springer

Skidmore M. and Toya H. (2002) Do natural disasters promote long-run growth?, *Economic Inquiry*, **40**, 664–687.

SLF- Eidgenössisches Institut für Schnee und Lawinenforschung (2000) Der Lawinenwinter 1999. *Ereignisanalyse, Eidgenössisches Institut für Schnee- und Lawinenforschung (SLF)*, Davos, 588.

Smith, K. and Ward, R. (1998) *Floods: Physical Processes and Human Impacts*, John Wiley & Sons, Chichester, UK.

Smolka A. (2006) Natural disasters and the challenge of extreme events: risk management from an insurance perspective, *Philosophical Transactions of the Royal Society of London, Series A* **364** (1845), 2147–2165, doi:10.1098/rsta.2006.1818.

Stoffel L. (2005) Künstliche Lawinenauslösung, *Tec 21*, **131**(9), 4–7.

Strobl E. (2010) *The Economic Growth Impact of Hurricanes: Evidence from U.S. Coastal Counties*, Review of Economics and Statistics, **93**, 575-589, doi: 10.1162/REST_a_00082.

The World Bank and The United Nations (2010) *Natural hazards, unnatural disasters: the economics of effective prevention*, Washington D.C.

Thieken A.H., Müller M., Kreibich H. and Merz B. (2005): Flood damage and influencing factors: New insights from the August 2002 flood in Germany. *Water Resources Research*, **41**(12), W12430, doi: 101029/2005WR004177.

Thieken A.H., Olschewski A., Kreibich H. *et al.* (2008) Development and evaluation of FLEMOps – a new flood loss estimation model for the private sector, in: *Flood Recovery, Innovation and Response*, (eds D. Proverbs, C.A. Brebbia and E. Penning-Rowsell) WIT Press, pp. 315–324.

Thieken A.H., Seifert I. and Merz B., eds. (2010) *Hochwasserschäden. Erfassung, Abschätzung und Vermeidung*, München: oekom

Tierney K. (1997) Business Impacts of the Northridge Earthquake, *Journal of Contingencies and Crisis Management*, **5**, 87–97.

Totschnig R., Sedlacek W. and Fuchs S. (2011) A quantitative vulnerability function for fluvial sediment transport, *Natural Hazards*, **58**(2), 681–703.

Tsuchiya S., Tatano H. and Okada N. (2007) Economic loss assessment due to railroad and highway disruptions, *Economic Systems Research*, **19**, 147–162.

Turner K.R., Doktor P. and Adger N. (1993) *Key issues in the economics of sea level rise*, CSERGE Working Paper 93-04.

Van der Heide M.C., van den Bergh J.C.J.M., van Ierland E.C., Nunes P.A.L.D. (2008) Economic valuation of habitat defragmentation: A study of the Veluwe, the Netherlands, *Ecological Economics*, **67**(2), 205–216.

Viavattene C., and Faulkner H. (2012) An uncertainty index to measure the feasibility of whole life cycle costing approach in flood risk management, *Journal of Flood Risk Management*, **5**(3), 215–225.

Von Ungern-Sternberg T. (2004) *Efficient Monopolies - The Limits of Competition in the European Property Insurance Market*, Oxford, UK: Oxford University Press.

Wahlstrom M. (2013) Foreword, in: *The Economic Impacts of Natural Disasters*, (eds D. Guha-Sapir and I. Santos), USA: Oxford University Press, 368 pp.

Ward F.A., Booker J.F. and Michelsen A.M. (2006) Integrated economic, hydrologic, and institutional analysis of policy responses to mitigate drought impacts in Rio Grande Basin. *Journal of Water Resources Planning and Management*, **132**, 488–502.

Wegmann M., Merz H. and Meierhans Steiner K. (2007) Jährliche Aufwendungen für den Schutz vor Naturgefahren in der Schweiz - Strategie Naturgefahren Schweiz, Umsetzung des Aktionsplans PLANAT 2005 – 2008, Bern.

Wilhite D.A., ed. (2000) *Drought: A Global Assessment*, Volume I and II, London/New York, Routledge.

Wilhite D.A., Svoboda M.D. and Hayes M.J. (2007) Understanding the complex impacts of drought: a key to enhancing drought mitigation and preparedness, *Water Resources Management*, **21**, 763–74.

Wittwer G. and Griffith M. (2010) Closing the factory doors until better times: CGE modeling of drought using a theory of excess capacity, Paper presented at the GTAP 13th annual conference, Penang, Malaysia, https://www.gtap.agecon.purdue.edu/resources/download/5019.pdf.

Woo C. (1994) Managing water supply shortage: interruption versus pricing, *Journal of Public Economics*, **54**, 145–160.

Woo G. (2002a) Natural Catastrophe Probable Maximum Loss, Risk Management Solutions, London, UK.

Woo G. (2002b) Natural catastrophe probable maximum loss, *British Actuarial Journal*, **8** (5), 943–959, doi: 10.1017/S1357321700004037, published online June 2011.

Wünsch A., Herrmann U., Kreibich H. and Thieken A.H. (2009) The role of disaggregation of asset values in flood loss estimation: A comparison of different modeling approaches at the Mulde River, Germany, *Environmental Management*, **44**(3), 524–541.

Zhai G. and Ikeda S. (2006) Flood risk acceptability and economic value of evacuation, *Risk Analysis*, **26**, 683–694.

Zhai G., Fukuzono T. and Ikeda S. (2005) Modeling flood damage: case of Tokai Flood 2000, *Journal of the American Water Resources Association*, **41**(1), 77–92.

Zhongmin X., Guodong C., Zhiqiang Z. *et al.* (2003) Applying contingent valuation in China to measure the total economic value of restoring ecosystem services in Ejina region, *Ecological Economics*, **44**, 345–358.

4.2 Resilience and Adaptation to Hydrometeorological Hazards

Hugh Deeming[1], Maureen Fordham[1] and Åsa Gerger Swartling[2]

[1]*Engineering & Environment, Northumbria University, Newcastle upon Tyne, UK*
[2]*Stockholm Environment Institute, Stockholm, Sweden and Stockholm Resilience Centre, Stockholm University, Stockholm, Sweden*

4.2.1 Introduction

In this chapter we will discuss the related concepts of resilience and adaptation. This discussion emanates from the on-going EU FP7 emBRACE project[1] which used five case studies across Europe to investigate the role, structure and processes of 'community resilience' in the face of flooding, alpine hazards (e.g. avalanche and flash flood), heatwave and earthquake. As emBRACE case studies were (at the time of writing) in the early stages of implementation, the material for this chapter will rely largely on the theoretical analysis undertaken so far (see emBRACE, 2012a) with the exception of reference to some early pilot studies in northern England. Overall, the focus of the project is to identify any common components or drivers of resilience within the diverse and complex communities studied and to find ways to describe, conceptualise and measure that resilience in ways that will be useful for practitioners in civil protection, communities at risk and academics generally. Our primary focus in the emBRACE project is the community level but we employ the concept in its widest and relational sense; seeing communities inextricably linked to individuals, social groups, organisations and higher geographical and administrative levels (e.g. national, regional, global). In order to argue for a place for resilience and adaptation in discussions focused on the management of hydrometeorological hazards, it is important to first understand what the two terms have been proposed to mean and then, building on this, to elaborate our particular approach.

This chapter will first review the two main focal concepts under discussion, that is resilience and adaptation. Our approach to resilience requires a critical analysis of understandings of the concept of community – a concept often employed uncritically.

[1] 'Building Resilience Amongst Communities in Europe (emBRACE) is a 4-year FP7 project running from 10/2011 – 09/2015 (www.embrace-eu.org) Information given in this chapter reflects the authors' views only. The Community is not liable for any use that may be made of the information contained therein.

Hydrometeorological Hazards: Interfacing Science and Policy, First Edition. Edited by Philippe Quevauviller.

Finally, we employ a sustainable livelihoods approach (SLA) as the mechanism by which we elucidate the complex mix of components and processes of relevance to community resilience to hydrometeorological hazards.

4.2.2 Resilience

In his paper on the etymology of the word resilience, Alexander (2013), describes trends in its use, which he traces from ancient Greece, through the reign of the English king, Henry VIII, to its modern multi-disciplinary application, which includes the disciplines of materials science, systems theory (e.g. ecological and social-ecological), psychology and disaster risk reduction, to name just a few. The underpinning meaning that tracks through much of the early usage is the concept of 'bouncing back', that is resilience has been long associated with the concept of equilibrium, or rather, the return to equilibrium after perturbation (Manyena, 2006). From a flood risk management (FRM) perspective this type of 'engineering resilience' could be understood as transferring directly through into the development of technical measures that allow buildings to be restored to functionality rapidly after inundation, whereas those measures constructed to prevent any water ingress would be termed *resistance* measures (Defra, 2008). As Alexander (2013) and others (emBRACE, 2012a) suggest, however, the divergence of use across the disciplines, which has occurred in the last 50 years, bears some investigation. We discuss here a selection of such uses, concentrating on three of the most influential and relevant to the emBRACE project: psychological approaches; social-ecological systems (SES); and organisations and infrastructure.

4.2.2.1 Psychological approaches to resilience

The use of the resilience concept in psychology dates back to the 1940s/1950s, when researchers used individual history and prognosis to differentiate the developmental competencies of children with mental disorders; an approach that regarded resilience as a personality trait (de Bruijne, Boin and Van Eeten, 2010). Further work, however, investigated resilience as being derived from external conditions and moved towards the hypothesis that resiliency was more a *process* than a trait, which led to greater focus on examining personal resilience as a function of social interaction. From a disaster risk reduction (DRR) perspective, this shift was useful because, without social context, it was difficult to scale individuals' resilience to adversity to higher social levels (Alexander, 2013). This is not to say, however, that individual psychology does not offer value to the understanding of human resilience to hazards. Useful understandings have been developed in relation to two interrelated domains (i) resilience in terms of mental health outcomes (the psychological domain) and (ii) psychological factors at the individual and community levels, which are related to disaster preparedness/mitigation (individual, the social, economic, and physical domain) and, *ergo*, related to resilience in terms of capacity to respond (i.e. preparedness, protective actions and mitigation behaviours are assumed to lead to resiliency). Findings in relation to mental health have included some variation in empirical results related to the degree of psycho-pathological distress. While significant numbers of people do not experience psycho-pathological distress, even after such apparently traumatic events as marital bereavement (Bonanno *et al.*, 2002), Galea, Nandi and Vlahov's (2005) review of the prevalence of post-traumatic stress disorder (PTSD) after disasters, finds as much as 30–40% (across a range of studies) prevalence among direct victims of disaster. In the studies reviewed, PTSD risk factors included gender (women show a higher prevalence than men), pre-existing psychiatric morbidity, previous experience of traumatic or stressful events and lack of social support. Thus, morbidity is related to both individual psychological traits and a range of social factors; social support or embeddedness are widely found to be negatively related to mental health problems (Wind, Fordham and Komproe, 2011; Norris, Baker and Murphy, 2005).

In terms of disaster preparedness/mitigation, models have been developed to explain people's perceptions of their self- and community's efficacy in risk reduction and response (Grothmann and Patt, 2005, Lindell and Perry, 1992). From this perspective a particularly powerful determinant of an individual's perception of hazard risk has been found to be their personal experience of those hazards (Slovic, Kunreuther and White, 2001); a factor that is referred to as *availability* in the risk perception literature (Tversky and Kahneman, 1982). Whilst these phenomena have been identified in hazards studies, in terms of understanding longer-term suppression of psycho-social resilience following flooding in particular, another factor has been found to come into play. Research has indicated that these longer-term effects may not actually be related to any trauma suffered due to the hazard itself, but that they may, rather, result from the individual's poor recovery experiences (e.g. slow or inept restoration of housing) (Whittle *et al.*, 2010). The "retraumatising" effect (*ibid.*) of poor recovery management is where women's resilience can be severely tested (Enarson, 2001; Fordham, 1999; Peacock, Morrow and Gladwin, 1997) and where women and girls can experience a 'double disaster' (Bradshaw and Fordham, 2013) in the form of, *inter alia*, 'violence, psychosocial impact, deterioration in sexual and reproductive health, early and forced marriages, trafficking, migration, insecure employment and increased poverty, as well as changes in networks and familial support and changes in self-perception' (*ibid.* p. 15).

Given this evidence, it is clear that an understanding of the psychological domain is highly relevant to our understanding of impact, vulnerability and resilience. However, in terms of understanding the resilience of communities to natural hazards it is now important to shift the lens to focus on the other, interconnected, factors that influence the resilience of society (at various levels).

4.2.2.2 Social-ecological systems (SES)

While a high degree of interconnectedness between social and ecological systems is undisputed today, theorists from different backgrounds understand resilience in different ways (Bahadur, Ibrahim and Tanner 2010). When Holling (1973) independently incorporated resilience into the discipline of ecological systems science he did so in a way that also associated the notion of change with the concept. He did this by adopting the idea of the 'adaptive cycle' (of growth or exploitation; conservation; collapse or release; and reorganisation), which described the way ecosystems developed over time. He did acknowledge that resilience could be an indicator of a return to equilibrium (i.e. engineering resilience). However, using the adaptive cycle his principal suggestion was that resilience was better understood in terms of dynamic fluctuation, as systems demonstrated resilience though their capacity to cross thresholds between 'stability regimes' (i.e. alternate conditions). This introduction of thresholds was important, because it accommodated the understanding that a resilient system could be pushed dynamically from one regime into another, whilst still maintaining "essentially the same function" (Walker *et al.*, 2004). However, Holling also pointed out that these shifts or flips could occur suddenly as "surprises" (Holling, 1986), and that they could result in less favourable system conditions (Renaud *et al.*, 2010).

Berkes and Folke (1998) introduced the term social-ecological system (SES) to emphasise the integrated concept of humans in nature, stressing that the separation between social and ecological systems is artificial and arbitrary and pointing to the need to give the social and ecological dimensions equal weight in the analysis. The SES approach to resilience has since gained widespread scientific recognition, which is reflected in the extensive literature in the field (e.g. Folke *et al.*, 2005; Walker *et al.*, 2006), as well as the establishment of the Resilience Alliance, (www.resalliance.org) and a series of recent international resilience conferences (Stockholm, Sweden 2008; Arizona, U.S.A. in 2011; Montpellier, France 2014).

A growing field of research on the social dimension of ecosystems management has been introduced under the concept of adaptive governance of SES which investigates social sources of renewal and reorganisation during periods of abrupt change (crisis) (Folke *et al.*, 2005). Another related methodological approach to governance of SES is introduced under the concept of adaptive co-management (Armitage *et al.*, 2008, 2009; Plummer, Armitage and De Loë, 2013; Cundill and Fabricius, 2009).

Scale issues are dealt with in SES through the concept of panarchy (Gunderson and Holling, 2002), which describes the interrelationships between the rapid cycling processes experienced at the local (micro-) scale, to the slower cycling at the meso- and macro-scales. Panarchy is perhaps best thought of in terms of understanding the relative feedback processes operating across the scales of tree, stand, forest and beyond. Whilst compelling from an ecosystems perspective, panarchy's applicability to the social realm has been challenged because there remains no unified definition of social (or community) resilience (Alexander, 2013).

Whilst environmental changes (whether natural or anthropogenic) could be regarded as the principal drivers of non-human ecosystem persistence or regime shifts, human-social systems have a powerful capacity for learning and technological progress. Therefore, as with socio/psychological approaches, adaptability (i.e. the collective capacity of the human actors in the system to manage resilience: Gallopin, 2006; see also Walker *et al.*, 2004) has been proposed as a crucial social factor in the regulation of SES, even though adaptive behaviour and organisation within the system may be intentional or unintentional in relation to its effect on the system (note, social actions that push the system closer to, or over, a threshold could be considered under the category of maladaptations: Barnett and O'Neill, 2010). In addition to the closely related concepts of resilience and adaptation, a third concept has also been suggested as relevant to SES: transformability (Walker *et al.*, 2004, Olsson *et al.*, 2006). Transformation can be defined as the capacity to create a fundamentally new system when ecological, economic, or social (including political) conditions make the existing system untenable (Gallopin, 2006). Pelling (2010) discusses transformation in relation to social contracts and human security and suggests that it may be triggered by any economic, political or environmental shock, which leads to perceptions of manifest failure of a social contract.

In terms of DRR this is an interesting escalation of the threshold change idea within SES, because what emerges from the other side of a transformation will not exhibit the same appearance (i.e. the same conditions) as what went in. In effect, transformation is epitomised by 'grass-roots' change – rather than a shift – during which social systems can be subjected to sometimes violent, autocratically enforced control measures, or to more progressive regime changes (*ibid.*). In order to reduce the risk of transformatory shifts resulting in undesirable effects, Walker *et al.*, suggest that systems should undertake scenario planning as a means to organise information and to develop contingencies in a comprehensible (even if daunting) way "before it is too late" (Walker *et al.*, 2004: p. 6). Through its incorporation of social and emissions scenarios into its review work, it is clear that the Intergovernmental Panel on Climate Change (IPCC) is a leading proponent of such thinking in terms of climate-change associated hazards (IPCC, 2012, Vuuren and O'Neill, 2006).

In recent years the concept of 'social resilience' has gained increasing recognition in the literature (Adger *et al.*, 2002, Gunderson, 2003, Cinner, Fuentes and Randriamahazo, 2009). Adger (2000: p. 347) defines it as "the ability of groups or communities to cope with external stresses and disturbances as a result of social, political and environmental change". It refers more systematically and directly to the economic, spatial and social dimensions (ibid.) and differs from ecological resilience by embedding the added capacity of humans to anticipate and plan for the future (Moberg and Galaz, 2005).

4.2.2.3 Organisations and infrastructure

In his seminal publication "Searching for Safety", Wildavsky distinguishes between anticipation and resilience. He states that whilst anticipation focuses on avoiding and mitigating known and specific hazards, resilience is inhered as an organisation's "*capacity to cope with unanticipated dangers after they have become manifest, learning to bounce back*" (Wildavsky, 1988: p. 77). In organisational theory, it has been suggested that the majority of organisations exhibit a tendency toward anticipation, wherein the focus of attention rests on attempting to prepare for hazard effects that can be estimated beforehand (Kuhlicke, Kunath and Dressler, 2012). The problem with such an approach obviously lies in the fact that nobody has complete knowledge regarding future hazard effects, let alone knowledge of the consequences if such effects become linked or compounded (Government Office for Science, 2011).[2] In his analysis of the response to the disastrous 1998 ice storm in north-eastern North America, Murphy (2009) suggested that, given the incumbent risk management regime, the magnitude of that event was simply "unforeseeable" (p. 331) by those who had been tasked with developing contingency plans *ex ante*. This is not to say that the response the event provoked was deficient, but simply that an 'anticipation' strategy had proven ineffective in generating sufficient warning of the impacts to come. Given Murphy's admiration for the effectiveness of the improvisation that led the response to the ice storm, it is worth using this as a suitable hydrometeorological hazard-related opportunity to consider Weick and Sutcliffe's (2007) set of key principles for resilient organisations. These were

[2] Linked risks can arise from separate effects due to a common causal event, while compounding risks can occur where separate hazard events combine to increase the overall impact on a single area.

condensed from the authors' research into high-reliability organisations (HRO), with, for them, the mark of an HRO (a resilient organisation in their eyes) lying in their:

- preoccupation with failure;
- reluctance to simplify;
- sensitivity to operations;
- commitment to resilience[3]; and
- deference to expertise.

HROs, they suggest, operate so effectively because, by adhering to these principles, they exhibit a greater capacity for Wildavsky's resilience than anticipation. Furthermore, Weick and Sutcliffe also emphasise the importance of another attribute that is exhibited by such organisations and which links the principles in action. This is the concept of mindfulness.

> *"Mindfulness conveys a mental orientation toward continually refining and differentiating categories, an ongoing willingness and capability to invent new categories that carve events into more meaningful sequences, and a more nuanced appreciation of context and way to deal with it. In contrast, a tendency toward mindlessness is characterized by a style of mental functioning in which people follow recipes, impose old categories to classify what they see, act with some rigidity, operate on automatic pilot, and mislabel unfamiliar new context as familiar old one" (Weick and Sutcliffe, 2007: p. 88).*

What Weick and Sutcliffe's approach to HROs tells us is that even the most 'high-tech' systems (e.g. nuclear power stations and aircraft carriers) are still reliant on the 'mindful' use of key human analytical, organisational and operational skills. From this perspective it is clear that these and other such systems should not simply be considered as technological systems, but because of these technical, organisational, social and also economic, dimensions they should be regarded as *socio-technical* systems (de Bruijne, Boin and Van Eeten, 2010). Here, one of Weick and Sutcliffe's principles takes on a new light. In their eyes 'deference to expertise' describes an HRO's ability to sidestep a work system's *authority* ladder in favour of tapping into a *knowledge* hierarchy, which is that sometimes the greatest expertise is found in those with the greasiest hands, not the sharpest suits. Understanding this enables us to perhaps reflect on the true importance of building participatory socio-technical hazard warning systems, wherein locally-based and/or indigenous expertise is incorporated as an integral component of that system (Glantz, 2009; Parker and Handmer, 1998).

Taking this further, it is clear that society is becoming ever more reliant on the resistance and resilience capacities of its socio-technical systems (Comfort, Boin and Demcheck, 2010; Boin and McConnell, 2007; Murphy, 2009), with some of these systems becoming so important that they are referred to as critical infrastructure (CI). Traditionally, energy, water and transport infrastructure have been regarded as CI, with the acknowledgement of their importance increasingly recognised within governance

[3] Although it appears rather circular to cite a 'commitment to resilience' as a principle for resilient organisations, Weick and Sutcliffe define the aspired resilience in a way that is highly specific to the organisational context they are discussing. For them resilience is "the intrinsic ability of an organisation (system) to maintain or regain a dynamically stable state, which allows it to continue operations after a major mishap and/or in the presence of a continuous stress" (Weick and Sutcliffe, 2007: p. 14).

institutions. The Council of the European Union (2008) has even ratified a directive on the importance of critical infrastructure protection (CIP), which although principally focused on terrorist threats, does state its relevance across 'all hazards'. Taking a more general approach, Bruneau *et al.* (2003) suggest that resilient systems such as these exhibit three main traits:

- Reduced failure probabilities;
- Reduced consequences from failures (in terms of lives lost, damage, and negative economic and social consequences); and
- Reduced time to recovery.

In an example from the UK, guidance has been issued in relation to the development of CIP, which is based on the four principles of Resistance, Reliability, Redundancy, and Response and Recovery (Cabinet Office, 2011a). In relation to flooding, this guidance applies these principles against a benchmark that was set out by the Pitt Review into the 2007 UK floods (Pitt, 2008), which suggested that "as a minimum, essential services provided by critical national infrastructure (CNI) in the UK should not be disrupted by a flood event with an annual likelihood of 1 in 200 (0.5%)" (Cabinet Office, 2011a: p. 28). CI operators are, accordingly, suggested to consider the cost/benefits of such protection when considering which projects to fund. Such a deterministically-defined benchmark may prove useful in reducing many flood impacts, however, its conservative nature is revealed in the same document, which explains that the UK's private-sector electricity industry already has set protection standards of 1 in 1000 (0.1%) protection (ibid.). Taking this private-sector target and given the recent spate of low-probability/high-consequence flood events that have occurred across the UK in recent years (Doe, 2004; Eden and Burt, 2010) – which included the inundation of a major water-pumping station that temporarily disrupted the supply of potable water to 500 000 people (Pitt, 2008) – it is hard to say that this guidance equates to anything other than a very limited anticipatory stance.

This standard of protection issue does, however, raise another interesting point in a society that is recognising its exposure to an increasing number of risks related to natural hazards (IPCC, 2012). This relates to the definition of CI itself. For whilst, generally, CI is understood to include the energy and other infrastructures as already mentioned above, it has been suggested that the hazard-protection infrastructure (e.g. the flood defence assets themselves) should now be regarded as bearing critical status (e.g. IAoE, 2009). It is not clear, however, if this suggestion has found traction in the wider, more politicised, CI discourse. Whether this obscurity relates to the potentially high economic ramifications that such categorisation would bring is unclear (i.e. would CIP status necessitate additional assessment and maintenance costs and, if so, for whom.

4.2.3 Discussion

Overall the scope of resilience thinking has over the years been scaled up and extended back in time and into the future warning of large scale regime shifts (Rockström *et al.*, 2009). The widened scientific application has led to a stretching of its original meaning and definition to accommodate various phenomena over and beyond temporal and geographical scales. This poses some challenges for resilience because, if it is not defined rigorously, its real existence becomes harder to establish (Portes, 2010). Resilience is applied to such numerous phenomena (Folke, 2006) and is applied both analytically and normatively (Brand and Jax, 2007) that it risks losing any distinct meaning. To regain its heuristic value, several conceptual refinements and redefinitions have recently been introduced and debated; such as the distinction between different types of adaptive capacity (adaptability and transformability), as well as a distinction between types of resilience (general and specific resilience) (Folke *et al.*, 2010).

The most recent polysemous, usage (Reghezza-Zitt *et al.*, 2012) of the resilience concept, has, Alexander (2013) argues, resulted in a situation where no straightforward understanding of the concept exists and where conflicting interpretations (e.g. between stability and dynamism, or equilibrium vs. evolution) obtain. The emBRACE project has, in corroboration of this perspective, analysed the use of the concept in the literature and accounted for 81 diverse variables that have been proposed by respective authors to constitute resilience in particular contexts (emBRACE, 2012b). Such findings reinforce the contention that the development of an overarching paradigm for resilience has become unlikely (see also: Manyena, 2006), even if "*the concept shows promise in that it encourages the researcher to bridge the 'shear zone' between (dynamic) adaptation and (static) resistance*" (Alexander, 2013: p. 1273). Fordham *et al.* (2013), similarly conclude that the resilience concept is not yet fully formed to the extent enjoyed by the hazards or social-vulnerability paradigms (ibid. p. 21).

Given Alexander's and others' rather sceptical position, why should we still be considering resilience as a suitable aspiration in the development of methods of hydrometeorological-disaster risk reduction? It could be said that it is the very prevalence of debate about the concept, along with the widespread ambition to further develop the social dimension of resilience, that makes it most attractive. Brand and Jax (2007) critically assess this recent explosion in usage and agree that the concept has become too malleable. Their prescription is that it needs to be either applied with explicitly-defined characteristics (i.e. if attempting to operationalize the concept in, e.g. ecological science), or with a clear understanding that its deployment is as a 'boundary object' "adaptable to different viewpoints and robust enough to maintain identity across them" (Star and Griesemer, 1989: p. 387), which permits trans-disciplinary negotiation (Wenger, 2000).

Somewhat in accordance with this perspective, Kuhlicke (2013) suggests that the concept could now coalesce in certain circumstances as a 'myth' rather than as something solidly grounded an observable 'truth'. His insight comes from research following the devastating 2002 flooding in Germany. Here, actors looked retrospectively at their peer group's collective interpretations of how the municipality dealt with the severe and unanticipated impacts. They configured responses into a socially-divisive framework of being either the 'right'/resilient actions, which were those adopted, or the 'wrong' actions, which were those that had been considered but rejected. For Kuhlicke, this raised an important reflection in relation to the investigation of resilience 'post-event'; that it is important for any researcher to clearly question "who is benefiting from the 'myth of resilience' and to what end?" (Kuhlicke, 2013: p. 74). This is an important consideration, because it foregrounds the fact that many current frameworks for understanding resilience in a social setting fail to integrate considerations of 'power', that is, who chooses who and what should be resilient and against what? (Cannon and Müller-Mahn, 2010). Deeming (2013), explains that, in contrast to most non-human ecological systems, human social systems enact their resilience with due regard to broader *political* considerations and that this can mean that some aspects of social system resilience should be fought, rather than sought. For example, autocratic regimes of government tend to be extremely resilient, regardless of the popular antipathy they attract.

Power imbalance and inequity are not the only considerations in relation to understanding the complex interplay that affects the development of social resilience. Institutional inertia, which increases with scale, can result in path dependence. Therefore, it is always possible that innovative adaptive options for resilience building may be discounted in favour of more conservative measures; a process that has been described as 'muddling through' (Lindblom, 1964). It could be suggested that due to the complex nature of social interaction and exploitation of the floodplain, which has evolved over centuries or even millennia, most flood risk management policy could in fact be regarded as muddling through. This is because, despite huge improvements in technological expertise in structural engineering, losses from flooding (and repeated flooding) continue to mount (IPCC, 2012), and yet there are still relatively few examples of social retreat, even from proven highly-hazardous areas (note, FEMA, 2007).

The acknowledgement of the potential for surprise shifts and transformatory risks also compounds the 'dark side of resilience' issue (Berkhout, 2008). Berkhout points out that resilience-focused discussions are always socially contingent, in respect to which (or rather, whose) perspective is being taken. Jasanoff (2008) provides a useful example of this framing issue in her discussion of the technological disaster at Bhopal, which although not hydrometeorologically related, does illustrate the conflicts between critical infrastructure and community resilience framings of this event:

> *"The factory explosion drove home the non-resilience of material technological systems underpinning the green revolution. What proved resilient were more local capacities for social invention, legal innovation, interim relief, solidarity and spirit." (Jasanoff, 2008: p. 13)*

As inhered above, resilience is unequally distributed, with the greater resilience of one group likely affecting the resilience of another, in both positive and negative ways. Whilst rapid industrialisation of the global south would serve as a macro-illustration of this effect (i.e. economic growth (positive) versus air pollution (negative), a better hydrometeorological hazards example would be that of the (hopefully hypothetical) implementation of a solely internet-based flood warning system, which would raise the response capacity of those in possession of the required IT, but would effectively ignore the needs of those without (Glantz, 2009).

This realisation of the potential for unequal (and inequitable) access to the 'beneficial' aspects of resilience also reminds us about the issue of scale. In his assessment of natural hazards in the US, Mileti (1999) proposed that resilience should be regarded as "*the ability of a locale to withstand an extreme event, without suffering devastating losses ... and without a large amount of assistance from outside the community*" (p. 33). This definition is an illustration of the fact that, particularly in relation to hazards, the *place community*[4] has become the standard focus of attention (Twigg,

[4] See further discussion of definitions of community below.

2009). Yet, this idea that 'locales' can be regarded as separable from their wider social context, especially in terms of how they negotiate the disaster cycle, is problematic. It is, of course, an aspiration of emergency managers everywhere that the populations for whom they are responsible could be developed into effective and autonomous local communities, capable of enduring extreme hazards without outside assistance (King, 2000). It is also interesting that this is the exact interpretation of community resilience that has been adopted by the UK Government in its 'Strategic National Framework for Community Resilience' (Cabinet Office, 2011b).[5] Unfortunately, however, this aspiration is effectively founded on resilience being operationalised purely in terms of a sub-population's response capacity. It fails to acknowledge the fact that locales should no longer be considered as discrete from wider society. A flood may isolate a town from assistance for a time; however, this very isolation will itself draw assistance from outside the locale, and once the waters subside it is more than likely that some formal assistance from 'outside' will be essential.

Renaud *et al.* (2010) discuss this in relation to the recovery of coastal communities in Sri Lanka after the 2004 tsunami and in Myanmar following cyclone Nargis. They point out that due to the saline contamination of groundwater, parts of coastal Sri Lanka's autonomous recovery would have been essentially impossible. The population were only able to 'bounce back' from this devastating event due to the considerable logistical operation that was put into place by the Government and NGOs. The situation in Myanmar, related much more to the 'locale' concept, because, due to the Government's refusal to sanction a substantive international aid effort, large communities were left almost totally dependent upon their own solidarity, which underpinned their coping capacity.

By contrast, and in a European context, the importance of locale has recently been emphasised in terms of UK flood risk management (FRM). For many years the Environment Agency funded FRM schemes on a national priority-scoring system, which meant, simplistically, that every proposed scheme was compared for its cost-benefit against others, with the most beneficial ratios being rewarded. However, this approach has recently been adjusted in order that schemes that can offer partnership funding (sourced from the 'community') are now more likely to receive central funds than previously (Defra, 2011). Several schemes have already been implemented in this way, but criticism has been levelled at the fact that much of the 'community' contribution has actually come from already-stretched local authority budgets and that smaller, often rural, schemes still could not generate sufficient private funding to become viable (Defra, 2013).

Given recent interest in building the resilience of communities to natural hazards (e.g. Cabinet Office, 2011b; emBRACE, 2011) we are faced with a challenge. With the concept of resilience already so contested (Alexander, 2013; Brand and Jax, 2007), the uncritical introduction of the equally argued-over concept of 'community' (Gilchrist, 2003; Pitchford, 2008; Stacey, 1969) could appear to make the task unmanageably complex. However, we argue for the necessity of its deconstruction and accordingly, we apply a similar level of scrutiny and review to this concept, from DRR and hydrometeorological extremes perspectives, in the next section.

[5] The working definition adopted by the UK Cabinet Office is: "Communities and individuals harnessing local resources and expertise to help themselves in an emergency, in a way that complements the response of the emergency services."

4.2.3.1 Community

'Community' is an apparently everyday, common sense term which in daily usage tends to go unexamined and unchallenged; and yet, it has been the subject of scholarly debate for some time. When Raymond Williams included 'community' in his critical analysis of 'Keywords' in 1976 he concluded that "Community [...] seems never to be used unfavourably" (Williams, 1976: p. 76). However, from the standpoint of the twentyfirst century and much subsequent scholarship, we can see that 'community', like 'resilience' has a contested meaning incorporating a darker side. Jennifer Davis, in the Keywords Project (undated), acknowledges its embodiment of processes of both inclusion and exclusion. Thus, while the word may be "warmly persuasive" (Williams 1976: p. 76), it also serves to label and exclude through a process of 'othering' (de Beauvoir, 1949). Common across many definitions and usages are three components: place or locale; institutions or networks; and a sense of belonging (Bell and Newby, 1972: p. 16–17); (Blackshaw, 2010: p. 5).

In his discussion of the concept, Delanty (2003) proposes that the experience of community in modern society is becoming an increasingly individualised affair, with place bearing less relevance in people's perceptions of their associational (or community) activities. For example, IT and social media are now connecting *communities of interest* around the globe with unprecedented ease (e.g. Twitter, Facebook). He argues, therefore, that community should no longer be defined by place (if it ever could be: Stacey, 1969), but in terms of capacity to communicate. In respect of DRR, however, whilst the capacity to communicate is fundamental to the development of useful contingencies (e.g. networks through which warnings can be transmitted and confirmed: Fitzpatrick and Mileti, 1994), geographical location remains a fundamental predictor of hazard exposure and, therefore, a vital factor in understanding how any population anticipates, prepares for, responds to and recovers from hazard events (e.g. think of a population resident on a floodplain). This being said, taking Delanty's lead, this application of the term 'community of place' should be regarded as naïve, because the aggregation of people into a singular cohort simply to produce a useful social metric, ignores the fact that society is not composed of any sort of set of ordered groups but can in fact be considered as a much more chaotic social mosaic (Marsh and Buckle, 2001). Such aggregation also risks dangerous consequences (particularly for some), because by defining communities in this way it becomes far harder to identify where within such an artificial construction reside those who lack the multiple resources needed to fully engage in DRR activities. Therefore, whilst place communities could be said to bear principal relevance in this discussion, the incorporation of other types of community (e.g. interest, circumstance, identity and practice), into our considerations should be regarded as essential (emBRACE, 2012a). Such an understanding also re-emphasises the importance of understanding that any 'place' resilience will likely be a function of more than just the sets of resources available *in that locale*.

In emBRACE Working Paper 1.1, we adopted a typology of five community types: geographical communities; communities of interest; communities of circumstance; communities of practice; and communities of identity (emBRACE, 2012a: pp. 35–38 and see Table 4.2.1).

Table 4.2.1 Typology of community types.

Community types	Typical characteristics (not all necessarily apply)
Geographical communities	• Identifiable geographical or administrative boundaries • Physical proximity (e.g. a street or an apartment block)
Communities of interest	• Groups of people who have an association • Interaction through a shared interest (e.g. sports clubs, parent groups, faith groups, online communities) • Interaction through work (including business groups) • People typically engaged together voluntarily to achieve a shared outcome • May or may not share a world view
Communities of circumstance	• Formed through the sharing of experience (e.g. when people are affected by the same incident or circumstances) • May or may not share the same interests or geographical location but develop a community subsequent to the shared experience
Communities of practice	• Shared or mutual engagement in a practice of interest (e.g. in this context, from organisations (both statutory and voluntary) providing disaster-related services and support (e.g. disaster managers, police officers, fire-fighters, Red Cross) • Based on participation • Self-organising and flexible in structure • Not determined or bound by organisational affiliations and can span institutional boundaries and structures (Wenger 1998, p. 2–3) • May also share a geographical location and may be affected in the same way as the communities they support/serve
Communities of identity	• Highly varied and fluid • May represent personal characteristics (e.g. gender, sexuality, dis/ability) • Nationality or ethnicity • Culture • Less to do with geography but more to do with 'imagined communities' (Anderson 1991) • Representative of a particular sub-group interest, often highlighting gaps in service provision or recognition (e.g. identifying needs, or marginalization, of non-heterosexuals in disasters (Gaillard in Fordham 2012: p. 400)

To illustrate the respective influence of these community types, it may be useful to consider the following hypothetical example. A hazard-exposed community (geographical) may contain people with direct experience of that hazard (circumstance). These people may work jointly and/or may engage with others outside the area to develop effective risk mitigation measures (practice) (e.g. a Flood Action Group working with local authorities). Such practice communities may attract membership from individuals or from existing groups (interest) (e.g. Rotary International). However, they may fail to represent minority views, interests or vulnerabilities (identity) if effort is not expended in terms of generating inclusive engagement.

What this example and the preceding discussion seeks to illustrate is that although the idea of resilient communities could be said to be largely grounded in an optimistic aspiration of social integration, a nuanced understanding of the complexity of the concept must be considered if it is to be applied in a useful and usable way (i.e. in terms of informing effective civil protection practice).

We turn now to the other part of our discussion; adaptation.

4.2.3.2 Adaptation

The past decade has witnessed the emergence of the global adaptation research and policy agenda in response to climate variability and change. The IPPC defines adaptation as "*any adjustment in natural or human systems in response to actual or expected climatic stimuli or their effects which moderates harm or exploits beneficial opportunities*" (IPCC, 2001 Annex B, p. 365). Projected temperature rises of 2 °C or more above the pre-industrial levels are likely to cause major disruptions that would challenge our ability to adapt at affordable economic, environmental and social cost. Thus adaptive capacity is an important element in societal resilience. It is defined by IPCC (2012) as: "The combination of the strengths, attributes, and resources available to an individual, community, society or organization that can be used to prepare for and undertake actions to reduce adverse impacts, moderate harm, or exploit beneficial opportunities" (p. 556).

The concepts of adaptation, adaptive capacity, vulnerability, resilience, exposure and sensitivity are interrelated and have wide application to global change science (Smit and Wandel, 2006). Oftentimes, however, silos of knowledge on vulnerability, adaptation and resilience and their applications have been produced, failing to help make communities and systems more robust to extremes and to change (Vogel *et al.*, 2007). In the research on social-ecological systems, adaptive capacity represents a core feature of resilience (e.g. Gallopin 2006; Berkes, Colding and Folke, 2003; Folke *et al.*; 2002). A similar term, adaptability, is also used to describe the required capacity to adjust responses to changing external drivers and internal processes and thereby allow for development along the current trajectory (stability domain) (Folke *et al.*, 2010). In contrast, the social science dominated research tradition on adaptation in relation to environmental change takes an actor-centered perspective; focusing on the processes of negotiation, decision-making and action, as well as the agency of social actors to respond to specific environmental stimuli and emphasising the reduction of vulnerabilities. The systems-based analysis is viewed as a complement to the actor-based analytical approach by examining the implications of these processes for the rest of the system. (Nelson, Adger and Brown, 2007).

Climate change adaptation (CCA) and disaster risk management (especially disaster risk reduction, DRR) are critical elements of long-term sustainability for economies, societies, and environments across scales (Wilbanks and Kates, 2010). Together with environmental management and poverty alleviation, these are engaged in reducing socio-economic vulnerability to natural hazards. However, the Central European floods of 2002 and the European heatwave of 2003 have shown that successful preparation and adaptation cannot be taken for granted, even in the most developed societies. Similarly, the effects of two windstorms (Gudrun in 2005 and Per in 2007) in southern Sweden illustrate that the ability to deal with extreme weather events is also lacking in Northern Europe. Preparing for climate change is important to ensure a continued high quality of life, profitability of economic sectors, and security from floods, storms and other climate risks.

In recent years there have been calls for better collaboration and learning across, in particular, the research and policy communities of DRR and CCA (Thomalla *et al.*, 2006). The authors suggest that international development policy, supporting all forms of risk reduction through integrating DRR and CCA strategies (ibid.), as well as attention to the root causes of hazard vulnerability (Blaikie *et al.*, 1994; Wisner *et al.*, 2004), are necessary precursors for resilience. The 2012 IPCC report on *Managing the risks of extreme events and disasters to advance climate change adaptation* (the so called SREX report) is a recent example of the growing commitment at global scale to establish such collaborations and cross-fertilization between the DRR and CCA communities. The report focuses on strategies for reducing exposure and vulnerability and increasing resilience to the potential adverse impacts of climate extremes, even though risks cannot be fully eliminated. (IPCC, 2012).

Today many adaptation measures exist that could reduce vulnerability to extreme events including hydrometeorological hazards. Examples include: physical measures such as dams to provide flood protection and water supply, dykes and coastal restoration for protection against coastal surges; place-based services such as improved construction standards, land-use planning to reduce exposure, improved warning systems and evacuation procedures; and economic and financial measures such as broader availability of insurance and emergency relief (IPCC, 2007, Chapter 18). Some countries have been identified as frontrunners in the development of national and regional adaptation strategies, such as the UK and the Netherlands (PEER, 2009). However, there is far less attention paid to those adaptation measures more distant from the hazard. For example, interest has been growing in community-based disaster management organising (related to the social and political domain) but this has been slow to develop in a European context. Least well served in a resilience and adaptation context are those measures related to human resources (education, health, skills, etc.), which have largely relied upon somewhat top-down, hazard awareness raising and disaster preparedness activities. Additionally, post-disaster health interventions have traditionally focused more on physical than mental health (WHO, 2001; Ahern *et al.*, 2005) and have been more reactive (aiding coping after the event) than anticipatory.

One research domain that has received considerable scholarly interest recently is the role of learning in facilitating climate change adaptation (Larsen *et al.*, 2012; Nilsson *et al.*, 2012), as well as for building adaptive capacity (Diduck, 2010; Pelling and High, 2008; Rudberg, Wallgren and Gerger Swartling, 2012) and social-ecological resilience (Krasny, Lundholm and Plummer, 2011; Fazey *et al.*, 2007). In particular, learning approaches appear to attract a number of research disciplines, including natural resource management (Pahl-Wostl *et al.*, 2007; Ison, Roeling and Watson, 2007; Diduck, 2004), education sciences (Wals, 2008; Lundholm and Plummer, 2010) and environmental governance (Plummer, Armitage and De Loë, 2013, Armitage and Plummer, 2010; Diduck, 2010). The ongoing emBRACE project from which this chapter draws early empirical results, is undertaking case studies on social learning in the context of DRR in the UK: the community response and recovery after heatwave in London and flood risk management in the Morpeth area, Northumberland. These empirical results are not discussed in this chapter, but will be presented, along with others, in a forthcoming volume of this Wiley series.

4.2.3.3 Resilience to hydrometeorological extremes

When considering community resilience in the context of hydrological extremes it is tempting to focus on the hazard/disaster and narrowly constrained emergency response, preparedness and mitigation matters (i.e. in the context of the disaster management cycle). However, if one could perfect a warning system, or flood control measure, or response capability, would that make people resilient? In this chapter we have introduced a number of lenses through which it can be seen that resilient people, communities and systems require a much wider and more complex frame in which to understand how resilience is produced, reproduced, maintained or lost, than can be provided within even quite advanced socio-technical risk-management systems (Comfort, Boin and Demchak, 2010). Social, collective or 'community' resilience is not something that can be created simply with altruistic intent, or through political intervention (e.g. Granatt and Macintosh, 2001; O'Brien and Read, 2005). To be resilient to hazards such as flooding, a community requires entitlement and access to a range of capital resources, along with the capability to deploy them in ways that allow them to "anticipate, absorb, accommodate or recover from the effects of a hazardous event in a timely and efficient manner" (Resilience definition: UNISDR, 2009).

From this perspective, the theory of resilience and its (broader) conceptualisations also underscore the fact that crises can be an important trigger for change and reorganisation (Sutcliffe and Vogus, 2003), which clearly differentiates resilience from other concepts (e.g. vulnerability), which by contrast view crises and disaster in largely negative terms.

4.2.3.4 The sustainable livelihoods approach (SLA)

The complexity of factors that have been discussed here necessarily pushes us towards understanding community resilience as a dynamic process that links available community resources and adaptive capacities to a "positive trajectory after disturbance", rather than thinking about it as an outcome or as a system property. This understanding is based largely on the work of Norris *et al.* (2008), but is given further structure by our proposal that resilience should be viewed using an SLA (Chambers and Conway, 1991). Considering resources and capacities from an SLA perspective, involves categorising them, typically, in terms of human, social, natural, technological/physical and financial/economic capital. However, we also follow Tobin's (1999) suggestion and Kuhlicke's (2013) warning, that to understand resilience across any scale of society, there is an imperative to also explore the undeniably social concept of the 'Political' (and the 'political').

> *"Political capital is the power and capacity to influence political decision-making through formal and informal participation and/or access to political processes. It therefore includes the ability to represent oneself or others, the freedom and capacity to become collectively organised to claim rights and to negotiate access to resources and services. It also extends to the right to hold government and service providers accountable for quality and access"*
>
> (Poverty-Wellbeing Platform, 2007: 3.3.2 - p. 2)

Acknowledging the multi-scaled influences of political capital is vital in this context, because rather than just assuming that 'resilience' to hazards can be achieved simply – if only we could just do the right thing with the resources we have at hand we would be resilient to disasters – the inclusion of political capital into our framework necessitates, as already implied, an appreciation of distributional effects and the potential for social in/equity. This inclusion parallels the interpretation of SLA adopted by the UK Department for International Development (DFID, 2011) in whose new resilience approach, sensitivity and adaptive capacity are determined by the pool of assets and resources that can be mobilised in the face of shocks and stresses. Assets and resources can be 'hard' and tangible or 'soft' and intangible (Sanderson, 2012: p. 659). The former include such things as physical buildings or infrastructure while the latter refer, for example, to knowledge, skills or health. The resource sets included are familiar (social, human, technological, physical, economic, financial, environmental, natural and political) and whether a system or a process is resilient is a function of its sensitivity and adaptive capacity. The outcomes vary from 'collapse' to 'bounce back better' through processes of surviving, coping, recovering, learning and/or transforming (DFID, 2011).

Out of the range of assets, capitals or resources discussed in the disasters literature, much research has focused on social capital and its influence on vulnerability reduction or building community resilience (e.g. in terms of the bonding, bridging and linking within and between communities and its potential for generating negative as well as positive benefits). A key contribution of the emBRACE Project is to go beyond social capital in disasters to understand the contributions of other capitals and processes in building community resilience (emBRACE, 2012a: p. 56). In our discussion we use the term 'resources' in preference to 'assets' or 'capitals' because the latter concepts are narrower in scope and have an unwanted suggestion of monetisation.[6] Table 4.2.2 presents

Table 4.2.2 Resource sets for resilience.

Human Resources and Capacities	Health (physical and mental), work, knowledge, skills, education, self-esteem and wellbeing. These are fundamental resources for anybody and without which it is difficult to make use of the other resource sets.
Social Resources	Family, friends and informal networks; more formal membership of groups; trust relationships that assist in collective action and knowledge-sharing.
Political Resources	Obviously associated with social resources, but worthy of emphasis, political resources are manifest in the power and capacity to influence political decision-making (through formal and informal participation in and/or access to political processes); hazard management legislation and standards.
Financial Resources	Earned income, pensions, savings, credit facilities, benefits, access to insurance.
Natural/ Place-based (Wilding, 2011)	Protecting and developing ecosystem services (especially in this context those that offer degrees of flood protection such as an operable floodplain, appropriate flood defences); land, water, forests and fisheries (for direct exploitation as well as more indirectly for personal wellbeing etc.); cultural/heritage resources; local public services, amenities, and access to jobs and markets (the availability of access rather than having employment which is covered by human resources).
Physical Resources	Adequate housing, roads, water and sanitation systems, transport, communications and other infrastructure; tools/equipment/premises for 'work' or livelihood; structural hazard mitigation.

[6] See also Fine (2001; 2002) who argues that Social Capital is a form of colonisation of social science by economists.

our initial resource sets and their characteristics, as applied in a pilot study undertaken by two of the present authors in Cumbria, UK in 2012: Deeming and Fordham, in preparation). We have refined the 'natural' resource set through the addition of 'place-based' resources (see Wilding, 2011) to better reflect the services available in an economically developed context.

In May 2013, two of the present authors undertook a small-scale pilot study in Cockermouth, Cumbria with the help of a team of postgraduate students from the University of Northumbria's MSc Disaster Management and Sustainable Development (Deeming and Fordham, in preparation). This brief research study aimed to investigate the utility of the SLA approach in this community resilience context. We provide a brief summary of this research below.

4.2.3.5 Pilot study location and event: Cockermouth, Cumbria, UK

In November 2009, significant rain fell on already saturated ground in the county of Cumbria, northwest England. Within a 24 hour period, 314 mm of rain was recorded (Cumbria County Council, 2011, p. 8), as a consequence of which many parts of Cumbria suffered extreme flooding (estimated as a one in 550 year event) (Environment Agency, 2011).

The 2009 floods resulted in severe travel disruption on both roads and railways and several bridges collapsed or had to be closed. Across Cumbria 2239 properties were flooded: 80% were residential properties, 20% were shops and businesses, and many schools were forced to close (Riding, 2011, p. 15; Cumbria Intelligence Observatory, 2010, pp. 25–26). Power supplies and telecommunications were interrupted (including contact with the emergency services in some areas). Cockermouth was the worst affected town where the depths of floodwaters reached 2.5 metres and affected 80% of businesses (Cumbria Intelligence Observatory, 2010, p. 25) and over 800 properties (Environment Agency, 2009, p. 6).

Building on previous networking and discussion with stakeholders in Cockermouth, a small team (9 students and 2 staff) visited Cockermouth between 2nd and 3rd May 2012 and carried out a number of interviews and focus group discussions with individuals, businesses and service providers. A common research interview schedule was used by all with the understanding that a flexible approach would be adopted in order to draw out the most from our participants concerning the validity of our proposed SLA approach. Responses confirmed the utility of the approach with all resource sets used to a greater or lesser degree in participants' accounts of their vulnerability, resilience, or adaptation. In Section 4.2.3.6 the experiences of two local small business owners are compared to briefly illustrate the complexity of factors that combine to produce a disaster resilience outcome.

4.2.3.6 Comparison of two small businesses in Cockermouth, Cumbria

For confidentiality reasons the two small businesses interviewed are represented as Small Business 'a' (SBa) and Small Business 'b' (SBb) in Table 4.2.3.

Table 4.2.3 Resource availability and capacity to exploit by two small businesses in Cockermouth, Cumbria, UK.

Resource sets	SBa (small business 'a')	SBb (small business 'b')
Human Resources	Skills in technology, financial accounting; knowledge of possibility of opportunities; effective social skills for business; self-image as shy but capable	Older owner; health issues; limited knowledge and skills related to technology; limited social skills for business; self-image as embattled
Social and Political Resources	Business as family concern; well-connected and networked; active and productive participation in trade bodies and community at large;	Isolated, without visible family support for business; poor experience with trade bodies (led to cancelled membership);
Financial Resources	Availability of savings and credit; effective insurance; ownership of premises; financially productive use of post-flood services (e.g. temporary trading space);	Low capital resources; ineffective insurance (unresolved and unresolvable claims); rented premises; financially unproductive use of post-flood services (e.g. temporary trading space);
Natural/ Place-based Resources	Flood defences overtopped in extreme event; business has firm attachment to place; capitalises on river and cultural-historic location; effective use of post-flood services (e.g. temporary trading space);	Flood defences overtopped in extreme event; limited business attachment to place; ineffective use of post-flood services (e.g. temporary trading space);
Physical Resources	Historic building close to river limits resilience measures but many building alterations made to improve resilience; lives outside the town but access soon reinstated;	Rejected original flooded premises (on safety grounds) and moved to alternative location; safer (on higher ground) but less good for footfall; lives outside the town but access soon reinstated;

These are both local small business owners, sharing the availability of place-based resources (e.g. flood defences (overtopped in this event), disaster response, municipal commitment to town recovery and regeneration (including provision of services such as skips, etc.), charity and volunteer aid to the town). Both were hit badly by the floods and both businesses continued to operate during and after the flood. However, they had a different trajectory of recovery and differing resilience outcomes which preliminary analysis identifies as dependent upon a complex mix of factors touching upon a range of resource sets and capacities. SBa generally recovered well with an expanded business, incorporation of many resilience measures (bounce forward) and a recognised place within the community. SBb recovered less well, with a business of similar size (bounce back) and a reduced sense of wellbeing and community embeddedness.

The differences between the recovery (and resilience status) of the two businesses are related strongly to the differences between the two business owners in terms of availability of economic capital, business acumen, social networks and individual psychology. In relation to one of our key components of resilience, social learning, these two small business owners had both benefitted (albeit to differing degrees) from the social learning which had taken place in the town, however, other factors served to limit its effectiveness or application. For example, although they were both beneficiaries of community level information and knowledge regarding the hazard and appropriate adaptation measures, they were limited in their adaptive capacity, not by a failure to learn but by matters such as the externally imposed constraints on physical alteration of historic buildings; or the personal limitations imposed on preferred action through limited financial means.

4.2.4 Conclusions

As health, according to the WHO (1946) definition, is 'a state of complete physical, mental and social well-being and not merely the absence of disease or infirmity', so resilience might be seen as 'a state of complete physical, mental and social well-being and not merely the absence of vulnerability to a hazard, shock or stress'. The emBRACE project uses a working approach to resilience as something bigger, more active and more connected than often conveyed in common usage of the term. To be resilient to disasters is more than to be in receipt of effective emergency management services – although this is not to underestimate their importance; it is also more than to be part of a well connected community. We expect to show over the course of the project that resilience is better described as a dynamic process than a state. Given this, there is an opportunity to effect change and help build greater resilience amongst communities if first we better understand what resilience and adaptation mean.

Acknowledgements

The authors are indebted to the work of the emBRACE partner teams with whom we have worked collaboratively on the ideas expressed here. Gratitude is also expressed to the EU FP7 programme, under which emBRACE was financed (contract 283201). The authors remain responsible for any errors that may have occurred.

References

Adger N., Kelly N., Winkels A. *et al.* (2002) Migration, Remittances, Livelihood Trajectories, and Social Resilience. *AMBIO* **31**(4),358–366. DOI: 10.1579/0044-7447-31.4.358.

Adger N. (2000) Social and ecological resilience: are they related? *Progress in Human Geography,* **24**, 347.

Ahern M., Kovats R.S., Wilkinson P. *et al.* (2005) Global health impacts of floods: Epidemiologic evidence, *Epidemiologic Reviews* Vol. **27**, 2005 DOI: 10.1093/epirev/mxi004.

Alexander D. (2013) Resilience and disaster risk reduction: an etymological journey. *Natural Hazards and Earth System Sciences, Discussion*, 1257–1284.

Armitage D. and Plummer R., eds. (2010) *Adaptive Capacity and Environmental Governance.* Springer Series on Environmental Management, New York: Springer.

Armitage D., Plummer R., Berkes F. *et al.* (2009) Adaptive co-management for social–ecological complexity. *Frontiers in Ecology and the Environment* **7**, 95–10.

Armitage D., Marschke M. and Plummer R. (2008) Adaptive co-management and the paradox of learning. *Global Environmental Change* **18**, 86–98.

Bahadur A.V., Ibrahim M. and Tanner T. (2010) *The resilience renaissance? Unpacking of resilience for tackling climate change and disasters*. Strengthening Climate Resilience Discussion Paper 1. Brighton : Institute of Development Studies, September 2010.

Barnett J. and O'Neill S. (2010) Maladaptation. *Global Environmental Change,* **20**, 211–213.

Bell C. and Newby H. (1972) *Community Studies,* London, George Allen and Unwin.

Berkes F. and Folke C., eds. (1998) *Linking Social and Ecological Systems: Management Practices and Social Mechanisms for Building Resilience*. Cambridge: Cambridge University Press.

Berkes F., Colding J. and Folke C. (2003) *Navigating Social-ecological Systems: Building Resilience for Complexity and Change*. Cambridge: Cambridge University Press.

Berkhout F. (2008) Order in socio-technical systems: The dark side of resilience. in *Re-framing Resilience* (ed. M. Leach) Sussex University Press, UK, pp. 11–13.

Blackshaw T. (2010) *Key Concepts in Community Studies,* London, Sage.

Blaikie P., Cannon T., Davis I. and Wisner B. (1994) *At Risk: Natural Hazards, People's Vulnerability, and Disasters*. London: Routledge.

Boin A. and Mcconnell A. (2007) Preparing for critical infrastructure breakdowns: The limits of crisis management and the need for resilience. *Journal of Contingencies and Crisis Management,* **15**, 50–59.

Bonanno G.A., Wortman C.B., Lehman D.R. *et al.* (2002) Resilience to loss and chronic grief: A prospective study From preloss to 18-months postloss. *Journal of Personality and Social Psychology,* **83**, 1150–1164.

Bradshaw S. and Fordham M. (2013) Women, Girls and Disasters: A review for DFID. London: Department for International Development (DFID).

Brand F.S. and Jax K. (2007) Focusing the meaning(s) of resilience: Resilience as a descriptive concept and a boundary object. *Ecology and Society,* **12**, 23.

Bruneau M., Chang S.E., Eguchi R.T. *et al.* (2003) A framework to quantitatively assess and enhance the seismic resilience of communities. *Earthquake Spectra,* **19**, 733–752.

Cabinet Office (2011a) *Keeping the Country Running: Natural Hazards and Infrastructure*. London: Cabinet Office.

Cabinet Office (2011b) *Strategic National Framework on Community Resilience*. London: HMG.

Cannon T. and Müller-Mahn D. (2010) Vulnerability, resilience and development discourses in context of climate change. *Natural Hazards,* **55**, 621–635.

Chambers R. and Conway G. (1991) Sustainable Rural Livelihoods: Practical Concepts for the 21st Century. Brighton, UK: Institute of Development Studies (IDS).

Cinner J.M., Fuentes M.M.P.B. and Randriamahazo H. (2009) Exploring social resilience in Madagascar's marine protected areas. *Ecology and Society* **14**(1), 41.

Comfort L.K., Boin A. and Demchak C.C. (2010) *Designing Resilience: Preparing for Extreme Events*, University of Pittsburgh Press.

Council of the European Union (2008) *Council Directive on the identification and designation of European critical infrastructures and the assessment of the need to improve their protection.* Brussels: European Commission.

Cumbria Intelligence Observatory (2010) Cumbria Floods November 2009: An Impact Assessment. Online at: www.cumbria.gov.uk/floods(accessed 21 November 2012).

Cundill G. and Fabricius C. (2009) Monitoring in adaptive co-management: Toward a learning based approach. *Journal of Environmental Management*, **90**(11) 3205–3211.

Cumbria County Council (2011). *Cumbria Floods November 2009 Learning From Experience: Recovery Phase Debrief Report*. Cumbria Resilience Forum, Cumbria County Council, Carlisle, UK.

De Beauvoir S. (1949) *The Second Sex,* Harmondsworth, UK., Penguin.

De Bruijne M., Boin A. and Van Eeten M. (2010) Resilience: Exploring the concept and its meanings. in *Designing Resilience: Preparing for Extreme Events*. (eds L.K. Comfort, A. Boin and C.C. Demchak) University of Pittsburgh Press, pp. 13–32.

Deeming H. (2013) Clarifying resilience: an invited comment. *Natural Hazards Observer*, XXXVII, **1**, 14–16.

Deeming H. and Fordham M., *Resilient Livelihoods: a Case-Study Report from a Flooded Community*, EU FP7 emBRACE Project Report, in preparation.

Defra (2008) *Developing the evidence base for flood resistance and resilience: summary report*. London: Defra.

Defra (2011) *Flood and Coastal Resilience Partnership Funding* (Online). London: Defra. Available: http://www.defra.gov.uk/environment/flooding/funding-outcomes-insurance/funding/.

Defra (2013) *Managing Flood Risk: Third Report of Session 2013–14, Volume 1*. London: House of Commons Environment, Food and Rural Affairs Committee.

Delanty G. (2003) *Community,* London, Routledge.

Diduck A. (2010) The learning dimension of adaptive capacity: Untangling the multi-level connections. in *Adaptive Capacity and Environmental Governance* (eds D. Armitage and R. Plummer) Springer Series on Environmental Management, Springer, pp. 199–221.

Diduck A. (2004) Incorporating participatory approaches and social learning. in *Resource and Environmental Management in Canada*, (ed. B. Mitchell) third edn. Oxford University Press, Don Mills, Ontario, pp. 497–527.

DFID (2011) *Defining Disaster Resilience: A DFID Approach Paper*. London: Department for International Development (DFID).

Doe R.K. (2004) Extreme precipitation and run-off induced flash flooding at Boscastle, Cornwall, UK – 16 August 2004. *Journal of Meteorology* **29**, 319–333.

Eden P. and Burt S. (2010) Extreme monthly rainfall: November 2009. *Weather,* **65**, 82–83.

emBRACE (2011) *EU FP7 Project emBRACE: Building Resilience Amongst Communities in Europe* (online). emBRACE Consortium. Available: www.embrace-eu.org (accessed 13 August 2013).

emBRACE (2012a) Deliverable 1.1: Work Package 1 – Early Discussion and Gap Analysis on Resilience. EU FP7 emBRACE Consortium.

emBRACE (2012b) Working Paper on the Systematization of Different Concepts, Quality Criteria, and Indicators. emBRACE consortium.

Enarson E. (2001) What women do: gendered labor in the Red River Valley flood. *Environmental Hazards* **3**, 1–18.

Environment Agency (2011) 'River Maintenance in Cumbria', in Our Approach to River Maintenance. Online at: http://www.environment-agency.gov.uk/homeandleisure/floods/131663.aspx (accessed 8 October 2012).

Fazey I., Fazey J.A., Fischer J. *et al.* (2007) Adaptive capacity and learning to learn as leverage for social–ecological resilience. *Frontiers in Ecology and the Environment* **5**, 375–380.

FEMA (2007) Mitigation in Wisconsin: Soldiers Grove Reflects on Moving. *FEMA: Best Practices*, Fall 2007.

Fine B. (2002) *The World of Consumption. The Material and Cultural Revisited*. London: Routledge.

Fine B. (2001) *Social Capital versus Social Theory: Political Economy and Social Science at the Turn of the Millennium*. London: Routledge.

Fitzpatrick C. and Mileti D.S. (1994) Public risk communication in: *Disasters, Collective Behaviour, and Social Organisation*. (eds R.R. Dynes and K.J. Tierney) London: Associated University Press, pp. 71–84.

Folke C. (2006) Resilience: The emergence of a perspective for social-ecological systems analyses, *Global Environmental Change*, **16**, 253–267.

Folke C., Carpenter S.R.,Walker Br. *et al.* (2010) Resilience thinking: Integrating resilience, adaptability and transformability. *Ecology and Society,* **15**(4), 20.

Folke C., Carpenter S., lmqvist T. *et al.* (2002) Resilience and sustainable development: Building adaptive capacity in a world of transformations. *AMBIO*, **31**(5), 437–440.

Folke C. Hahn T., Olsson P. and Norberg J. (2005) Adaptive governance of social–ecological systems. *Annual Review of Environment and Resources* **30**, 441–473.

Fordham M. (2012) Gender sexuality and disaster in *Routledge Handbook of Hazards and Disaster Risk Reduction*, (eds B. Wisner, J.C. Gaillard and I. Kelman, London: Routledge, pp. 395–406.

Fordham M. (1999) The intersection of gender and social class in disaster: Balancing resilience and vulnerability. *International Journal of Mass Emergencies and Disasters,* **17**, 15–36.

Fordham M., Lovekamp W.E., Thomas D.S.K. and Phillips B.D. (2013) Understanding social vulnerability. in *Social Vulnerability to Disasters* 2nd Edn (eds D.S.K. Thomas, B. D. Phillips, W.E. Lovekamp and A. Fothergill) CRC Press, pp. 1–29.

Galea S., Nandi A. and Vlahov D. (2005) The epidemiology of post-traumatic stress disorder after disasters. *Epidemiologic Reviews,* **27**, 78–91.

Gallopin G.C. (2006) Linkages between vulnerability, resilience, and adaptive capacity. *Global Environmental Change,* **16**, 293–303.

Gilchrist A. (2003) *The Well-Connected Community: A Networking Approach to Community Development,* Bristol, UK, The Policy Press.

Glantz M.H., ed. (2009) *Heads Up! Early Warning Systems for Climate-, Water- and Weather-Related Hazards,* New York: United Nations University Press.

Governement Office for Science (2011) *Blackett Review of High Impact Low Probability Risks*. London: Government Office for Science.

Granatt M.S.D. and Macintosh J.P. (2001) *Building Resilience: a New Approach to Crisis [Classified Document]*. London: Cabinet Office.

Grothmann T. and Patt A. (2005) Adaptive capacity and human cognition: The process of individual adaptation to climate change. *Global Environmental Change,* **16**, 199–213.

Gunderson L. (2003) Adaptive dancing: Interactions between social resilience and ecological crises. in BERKES, F., COLDING, J., FOLKE, C. (eds.) Navigating Social-ecological Systems: Building Resilience for Complexity and Change (eds F. Berkes, J. Colding and C. Folke) Cambridge: Cambridge University Press, pp. 33–52.

Gunderson L.S. and Holling C.S. (2002) *Panarchy: Understanding Transformations in Human and Natural Systems,* Washington, D.C., Island Press.

Holling C.S. (1973) Resilience and stability of ecological systems. *Annual Reviews of Ecological Systems,* **4**, 1–23.

Holling C.S. (1986) The resilience of terrestrial ecosystems: local surprise and global change. in *Sustainable Development of the Biosphere*. (eds W.C. Clark and R.E. Munn) Cambridge University Press, London, pp. 292–320.

IAoE (2009) *Ireland at Risk: Adaptation for Climate Change*. Dublin: The Irish Academy of Engineering (IAoE).

IPCC (2001) *Climate Change 2001: Impacts, Adaptation, and Vulnerability*. Contribution of Working Group II to the IPCC Third Assessment Report 2001. Cambridge University Press.

IPCC (2007) *Climate Change 2007: Impacts, Adaptation, and Vulnerability*. Contribution of Working Group II to the IPCC Third Assessment Report 2001. Cambridge University Press.

IPCC (2012) *Managing the Risks of Extreme Events and Disasters to Advance Climate Change Adaptation*. A Special Report of Working Groups I and II of the Intergovernmental Panel on Climate Change (eds C.B. Field, V. Barros, T.F. Stocker *et al.*) Cambridge, UK, and New York, NY, USA: Cambridge University Press.

Ison R., Roeling N. and Watson D. (2007) Challenges to science and society in the sustainable management and use of water: investigating the role of social learning. *Environmental Science and Policy* **10** (2007), 499–511.

Jasanoff S. (2008) Survival of the fittest. in *Re-framing Resilience* (ed. M. Leach) Sussex University, UK, p. 13.

KEYWORDS Project. Undated. *Keyword Community* [Online]. Available: http://keywords.pitt.edu/keywords_defined/community.html (accessed 26 August 2013).

King D. (2000) You're on your own: Community vulnerability and the need for awareness and education for predictable natural disasters. *Journal of Contingencies and Crisis Management,* 8 December, 223.

Krasny M.E., Lundholm C. and Plummer R., eds. (2011) *Resilience in Social-Ecological Systems: the Roles of Learning and Education*. London: Routledge.

Kuhlicke C. (2013) Resilience: a capacity and a myth: findings from an in-depth case study, *Disaster Management Research*, **67**, 1.

Kuhlicke C., Kunath A. and Dressler G. (2012) Resilience and Organisations. in *Working Paper: Work Package 1 - Early Discussion and Gap Analysis on Resilience*. (ed. The emBrace Consortium) EU FP7 Project emBRACE - 283201.

Larsen R.K., Gerger Swartling Å., Powell N. *et al.* (2012) A framework for facilitating dialogue between policy planners and local climate change adaptation professionals: Cases from Sweden, Canada and Indonesia. *Environmental Science and Policy,* **23**, 12–23.

Lindblom C.E. (1964) The science of "muddling through". in *The Making of Decisions: A Reader in Administrative Behaviour* (eds W. J. Gore and J.W. Dyson) New York: The Free Press, pp. 79–88.

Lindell M.K. and Perry R.W. (1992) *Behavioral Foundations of community Emergency Planning,* Washington, DC, Hemisphere Press.

Lundholm C. and Plummer R. (2010) Resilience and learning: A conspectus for environmental education. Special issue Resilience in Social-Ecological Systems: the Role of Learning and Education. *Environmental Education Research* **16**(5-6) 645–663.

Manyena S.B. (2006) The concept of resilience revisited. *Disasters,* **30**, 434–450.

Marsh G., Buckle P. (2001) Community: the concept of community in the risk and emergency management context. *Australian Journal of Emergency Management,* **16**(1), 5–7.

Mileti D.S. (1999) *Disasters by Design: A Reassessment of Natural Hazards in the United States,* Washington DC, Joesph Henry Press.

Moberg F. and Galaz V. (2005) *Resilience: Going from Conventional to Adaptive Freshwater Management for Human and Ecosystem Compatibility*, Stockholm: *Swedish Water House Policy Brief* 3, SIWI.

Murphy R. (2009) *Leadership in Disaster: Learning for a Future with Global Climate Change,* London, McGill-Queens University Press.

Nelson R.N., Adger, W. N. and Brown, K. (2007). Adaptation to Environmental Change: Contributions of a Resilience Framework. *Annual Review of Environment and Resources*, **32**, 395–419.

Nilsson A.E., Gerger Swartling Å. and Eckerberg K. (2012) Knowledge for local climate change adaptation in Sweden: challenges of multilevel governance, *Local Environment: The International Journal of Justice and Sustainability*, **17**(6), 751–767.

Norris F., Stevens S., Pfefferbaum B. *et al.* (2008) Community resilience as a metaphor, theory, set of capacities and strategy for disaster. *American Journal of Community Psychology,* **41**, 127–150.

Norris F.H., Baker C.K. and Murphy A.D. (2005) Social support mobilization and deterioration after Mexico's 1999 flood: effects of context, gender, and time. *American Journal of Community Psychology,* **36**, 15–28.

O'Brien G. and Read P. (2005) Future UK emergency management: new wine, old skin? *Disaster Prevention and Management,* **14**, 353.

Olsson P., Gunderson L.H., Carpenter S.R. *et al.* (2006) Shooting the rapids: navigating transitions to adaptive governance of social-ecological systems. *Ecology and Society*, **11**(1), 18.

Pahl-Wostl C., Sendzimir J., Jeffrey P. *et al.* (2007) Managing change toward adaptive water management through social learning. *Ecology and Society*, **12**(2), 30.

Parker D.J. and Handmer J.W. (1998) The role of unofficial flood warning systems. *Journal of Contingencies and Crisis Management*, **6**, 45–57.

Peacock W.G., Morrow B.H. and Gladwin H. (1997) *Hurricane Andrew: Ethnicity, Gender and the Sociology of Disasters* London, UK, Routledge.

PEER - Partnership for European Environmental Research (2009). Europe adapts to climate change: Comparing National Adaptation Strategies,. PEER Report No 1. Helsinki: Partnership for European Environmental Research.

Pelling M. and High C. (2008) Shadow spaces for social learning: a relational understanding of adaptive capacity to climate change within organisations. *Environment and Planning A,* **40**, 867–884.

Pelling M. (2010) *Adaptation to Climate Change: from Resilience to Transformation,* London, Routledge.

Pitchford M. (2008) *Making Spaces for Community Development,* Bristol, UK, Policy Press.

Pitt M. (2008) *Learning lessons from the 2007 floods: An independent review by Sir Michael Pitt.* The final report. London: Cabinet Office.

Plummer R., Armitage D.R. and De Loë R.C. (2013) Adaptive co-management and its relationship to environmental governance. *Ecology and Society,* **18**(1), 21.

Portes A. (2010) Migration and social change: some conceptual reflections. *Journal of Ethnic and Migration Studies* **36**(10), 1537–1563.

Poverty-Wellbeing Platform (2007) *Working with a Sustainable Livelihoods Approach, Version 1.1.* (online). Available: http://www.poverty-wellbeing.net/media/sla/docs/3-3.htm.

Reghezza-Zitt M., Rufat S., Djament-Tran G. *et al.* (2012) What resilience is not: uses and abuses. *CyberGeo,* **621**, 1–23.

Renaud F., Birkmann J., Damm M. and Gallopin G. (2010) Understanding multiple thresholds of coupled social–ecological systems exposed to natural hazards as external shocks. *Natural Hazards,* **55**, 749–763.

Riding K. (2012) *The role of the third sector in helping communities in Cumbria recover from the November 2009 floods*, Cumbria Voluntary Services, Penrith, UK.

Rockström J., Steffen W., Noone K. *et al.* (2009) A safe operating space for humanity. *Nature,* **461**, 472–475.

Rudberg P., Wallgren O. and Gerger Swartling Å. (2012) Beyond generic adaptive capacity: exploring the adaptation space of the water supply and wastewater sector of the Stockholm region, Sweden. *Climatic Change*, **114**(3-4), 707–721.

Sanderson D. (2012) Livelihood protection and support for disaster in Wisner, B., Gaillard, JC and Kelman, I. *The Routledge Handbook of Hazards and Disaster Risk Reduction* (eds B. Wisner, J.C. Gaillard and I. Kelman) London: Routledge, pp. 697–710.

Slovic P., Kunreuther H. and White G. (2001) Decision processes, rationality and adjustment to natural hazards. in *The Perception of Risk*. (ed. P. Slovic) London: Earthscan, pp 1–31.

Smit B. and Wandel J. (2006) Adaptation, adaptive capacity and vulnerability. *Global Environmental Change,* **16**, 282–292.

Stacey M. (1969) *Methods of Social Research,* Oxford, UK, Permagon Press.

Star S.L. and Griesemer J.R. (1989) Institutional ecology, 'translations' and boundary objects: Amateurs and professionals in Berkeley's Museum of Vertebrate Zoology 1907–39. *Social Studies of Science,* **19**.

Sutcliffe K.M. and Vogus T.J. (2003) Organizing for resilience. in *Positive Organizational Scholarship* (eds K. Cameron, J,E, Dutton and R.E. Quinn San Francisco: Berrett-Koehler, pp. 94–110.

Thomalla F., Downing T., Spanger-Siegfried E., Han G. and Rockström J. (2006) Reducing hazard vulnerability: towards a common approach between disaster risk reduction and climate adaptation. *Disasters*, **30**(1), 39–48.

Tobin G.A. (1999) Sustainability and community resilience: the holy grail of hazards planning? *Global Environmental Change Part B: Environmental Hazards,* **1**, 13–25.

Tversky A. and Kahneman D. (1982) Judgment under uncertainty: heuristics and biases. *Science*, **185**, 1124–1131.

Twigg J. (2009) *Characteristics of a Disaster Resilient Community: A Guidance Note*. London: Aon Benfield UCL Hazard Research Centre.

UNISDR (2009) *Terminology: Basic terms of disaster risk reduction* (Online). UN International Strategy for Disaster Reduction. Available: http://www.unisdr.org/eng/library/lib-terminology-eng%20home.htm (accessed 3 July 2009).

Vogel C., Moser S., Kasperson R. and Dabelko G.D. (2007) Linking vulnerability, adaptation, and resilience science to practice: Pathways, players, and partnerships. *Global Environmental Change,* **17**, 369–374.

Vuuren D. and O'Neill B. (2006) The consistency of IPCC's SRES Scenarios to 1990–2000 Trends and Recent Projections. *Climatic Change,* **75**, 9–46.

Walker B., Holling C.S., Carpenter S.R. and Kinzig A. (2004) Resilience, adaptability and transformability in social-ecological systems. *Ecology and Society,* **9**, art. 5.

Walker B., Anderies J.M., Kinzig A.P. and Ryan P. (2006) Exploring resilience in social-ecological systems through comparative studies and theory development: introduction to the special issue. *Ecology and Society* **11**(1), 12.

Wals A.E.J. (ed.) Social Learning Towards a Sustainable World. Wageningen: Wageningen Academic Publishers, pp. 537.

Weick K.E. andSutcliffe K.M. (2007) *Managing the Unexpected: Resilient Performance in an Age of Uncertainty,* San Francisco, Jossey Bass.

Wenger E. (2000) Communities of practice and social learning systems. *Organization,* **7**, 225–246.

Whittle R., Medd W., Deeming H. *et al.* (2010) *After the Rain – learning the lessons from flood recovery in Hull, final project report for "Flood, Vulnerability and Urban Resilience: a real-time study of local recovery following the floods of June 2007 in Hull" Lancaster*, UK: Lancaster University.

WHO (2001) *The World health report* 2001. *Mental health: new understanding, new hope*. Geneva, Switzerland: WHO.

Wilbanks T. and Kates R. (2010) Beyond Adapting to Climate Change: Embedding Adaptation in Responses to Multiple Threats and Stresses. *Annals of the Association of American Geographers,* **100**(4), 719–728.

Wildavsky A. (1988) *Searching for Safety*, New Jersey: Rutgers University.

Wilding N. (2011) *Exploring Community Resilience In Times of Rapid Change*. Dunfermline: Fiery Spirits Community of Practice supported by Carnegie UK Trust.

Williams R. (1976) *Keywords: A Vocabulary of Culture and Society,* London, Fontana.

Wind T.R., Fordham M. and Komproe I.H. (2011) Social capital and post-disaster mental health. *Global Health Action*, **4**.

Wisner B., Blaikie P., Cannon T. and Davis I. (2004) *At Risk : Natural Hazards, People's Vulnerability and Disasters*, 2nd edn, London: Routledge.

Part Five
Conclusions

5

Conclusions, Outlook

Philippe Quevauviller[1]
Department of Hydrology and Hydrological Engineering, Vrije Universiteit Brussels, Brussels, Belgium

5.1 Contextual developments

The increasing rate of hydrometerological extreme events over the last two decades has led to various policy initiatives, the development of which are described in Part Two of this book. In Europe, the adoption of the Water Framework Directive (and related directives) and of the Flood Directive has led to a reframing of the way water-related risks (floods and droughts) are being tackled (Chapter 2.3). It is not yet clear to what extent Europe will have a harmonising effect in the way for example flood risks are managed, but the EU policy has set – for the first time – a coordinated framework for risk evaluation and mapping, monitoring and design of programmes of measures (that will, for flood risks, be implemented alongside the 2nd river basin management planning of the Water Framework Directive, that is from 2015 to 2021). Besides this, UN initiatives are in full development (in close coordination with the EU) through the so-called Hyogo Framework for Action which encompasses strategies for reducing disasters of all types, including hydrometeors (Chapter 2.1). Finally, policy developments take into consideration recommendations expressed by the IPCC, in particular the 5th Assessment Report (Chapter 2.2).

Policies relevant to hydrometeorological extreme events are evolving from being merely "reactive" to disasters to a more comprehensive and long-term risk management approach which covers the full crisis management cycle (from prevention/preparedness to response/recovery). This change of paradigm relies on the scientific knowledge base about climate change impacts and other features related to different hydrometeorological events (Chapters 1.2 and 1.3) as well as the awareness about how resilience has evolved in history and how vulnerability should be perceived, not only in relation to climate change but also to urban (mis)development (Chapter 1.1). To complement this, recognition of the need for a structured science–policy dialogue (already in place within IPCC) is now well anchored in the way research and policies are evolving (Chapter 1.4).

[1] The views expressed in this chapter are purely those of the author and may not in any circumstances be regarded as stating a formal position of any EU institution

Hydrometeorological Hazards: Interfacing Science and Policy, First Edition. Edited by Philippe Quevauviller.

5.2 Scientific developments

The community of scientists involved in studies of hydrometeorological extreme events has given significant effort to common methodological problems that are most relevant to risk management operations, examples of which are the role of hydroinformatics in flood management (Chapter 3.1), better preparedness for drought hazards (Chapter 3.2) and enhanced prediction of storm impacts (Chapter 3.4). Regional studies are also manifold, an example of which deals with drought patterns in the Mediterranean (Chapter 3.3). Other scientific developments concern social and economic considerations, in particular cost assessments of natural hazards (Chapter 4.1) and the way society may adapt to hydrometeorological hazards (Chapter 4.2).

Examples given in this book naturally only reflect a snapshot of the numerous studies carried out worldwide. The overall Book Series on Hydrometeorological Extreme Events will actually provide many more examples of research developed in this area through comprehensive volumes dealing with floods, droughts, (coastal) storms and resilience aspects.

5.3 Outlook

Without attempting to make a comprehensive 'SWOT' analysis (strengths, weaknesses, threats, opportunities) of the hydrometeorological features described in this book, a 'SWOT' format may be a convenient way to discuss the current outlook and associated challenges.

5.3.1 Strengths

- Policy backing by a range of legally binding instruments, including EU directives (e.g. the Flood Directive) and international policy developments (e.g. the current and post-2015 Hyogo Framework for Action);
- Technological advances related to disaster risk management have opened up new possibilities;
- Strong scientific investment in the methodological basis for all aspects of crisis management, including hydrometeorological extreme events, in the past decade (in particular through projects funded under the 7th EU Framework Programme on Research and Development);
- Range of modelling tools and relevant techniques for improved assessments of hydrometeorological risks broadened significantly.

5.3.2 Weaknesses

- Despite the established policy framework at EU and international level, risk management of hydrometeorological extreme events is dependent upon national/regional authorities. Coordination and exchange of good practices are on-going but there is much room for improvement in this regard;

- While technological developments are on-going, communication gaps and insufficient transfer of knowledge often hamper a proper implementation of new technologies;
- The same can be said about scientific developments, that is research outputs are rarely communicated in a way which can be taken up by different stakeholders, and efforts are needed to improve links between science, policy and industry end user's communities to ensure that scientific progress is made known in a timely and efficient fashion.

5.3.3 Threats

- Gaps between different decision-making levels (EU, national, regional) may have consequences for the way hydrometeorological extreme events are managed, either due to insufficient knowledge of regulated practices developed at international level (top-down) or the lack of consideration or awareness of local knowledge by decision-makers (bottom-up), this may result in losing efficiency in various operations linked to the disaster management cycle;
- Research is sometimes disconnected from real-case scenarios and may lead to recommendations that cannot be taken onboard by decision-makers if nothing is done to demonstrate the applicability and usefulness of research outputs;
- Linked to the above, the wide range of actors involved in managing hydrometeorological events (first responders, firemen, police, civil protection units etc.) enhances the risks of insufficient coordination of operations on the ground.

5.3.4 Opportunities

- On-going intensification of risk assessments of hydrometeorological extreme events at international level provides opportunities to squeeze as much information as possible out of the available data, proving their value, and data providers can build their case on the assessment results.
- Viewing data series (including historical series) of verified quality, that is requiring validation prior to their use in modelling, is an essential component of risk assessment and mapping, scenario building and design of appropriate prevention strategies.
- Linked to this, research focusing on modelling tools, methods and techniques related to early warning, prevention and preparedness, as well as tools for improving response and recovery to disasters, represents an invaluable opportunity for strengthening the knowledge-base and helping operational units to improve their practices. This has, however, to be done in a concerted way, that is ensuring that end users are working closely with scientists from the start to the end of research projects.
- The demonstration of the usefulness and applicability of new techniques/methods is a crucial element for raising awareness about research outputs at the level of implementers and local actors. This essential continuation of research developments,

directed toward pilot case studies (field testing), training and capacity-building is increasingly supported at international and national levels, an example of which is the way demonstration activities are highlighted as a key component of the new EU Research Framework Programme called Horizon 2020 that will run from 2014 until 2021.

Index

Hydrometeorological Hazards: Interfacing Science and Policy, First Edition. Edited by Philippe Quevauviller.